Advanced Trends in Wireless Communications

Advanced Trends in Wireless Communications

Editor

Dan Hallmen

scitus
academics

Advanced Trends in Wireless Communications
Edited by **Dan Hallmen**

ISBN: 978-1-68117-278-1
Library of Congress Control Number: 2016937450

SCITUS Academics LLC,
www.scitusacademics.com
Box No. 4766, 616 Corporate Way,
Suite 2, Valley Cottage,
NY 10989

Notice

Printed in United States of America on Acid Free Paper

Preface

Wireless communication is one of the most important technological contributions to mankind. Wireless communication means information transmission over a distance without the help of wires, cables or other electrical conductors. The transmitted distance can be between a few meters (e.g. TV remote control) and several thousand kilometers (e.g. Communication by radio). It covers different types of fixed, mobile and portable applications, including two-way radios, mobile phones, personal digital assistants and wireless networking. Other radio wireless applications include GPS units, garage door openers, computer mouse, keyboards and headsets, headphones, radio receivers, satellite TV, and cordless telephones. Wireless operations allow long-range communications services that cannot be implemented using cables. The term is commonly used in the telecommunications industry to refer to telecommunications systems that use certain forms of energy—radio waves, acoustic power, etc.—for the transmission without cables. This transmits information both over short and long distances.

This book provides an overview of the main themes in mobile and wireless communications, cell architecture and technology. The book chapters deal with a wide range of wireless communication topics.

Table of Contents

Chapter 1 Building programmable wireless networks: *an architectural survey* 1

Chapter 2 Enabling 5G mobile wireless technologies 63

Chapter 3 Comparison of Indoor Localization Systems Based on Wireless Communications 97

Chapter 4 A New Adaptive EDCA Approach to QoS of Wireless Communications 139

Chapter 5 A Novel Outage Capacity Objective Function for Optimal Performance Monitoring and Predictive Fault Detection in Hybrid Free-Space Optical and RF Wireless Networks 159

Chapter 6 METIS research advances towards the 5G mobile and wireless system definition 189

Chapter 7 Survey on Wireless Sensor Network Technologies for Industrial Automation: The Security and Quality of Service Perspectives 223

Chapter 8 Reconfigurable Microstrip Double-Dipole Antennas for Personal Wireless Communications 261

Chapter 9 Critical Infrastructure Surveillance Using Secure Wireless Sensor Networks 281

Index 325

1

CHAPTER

BUILDING PROGRAMMABLE WIRELESS NETWORKS: *AN ARCHITECTURAL SURVEY*

Junaid Qadir[1*], Nadeem Ahmed[1] and Nauman Ahad[1,2]

ABSTRACT

In recent times, there is increasing consensus that the traditional Internet architecture needs to be evolved for it to sustain unstinted growth and innovation. A major reason for the perceived architectural ossification is the lack of the ability to program the network as a system. This situation has resulted partly from historical decisions in the original Internet design which emphasized decentralized network operations through colocated data and control planes on each network device. The situation for wireless networks is no different resulting in a lot of complexity and a plethora of largely incompatible wireless technologies. With traditional architectures providing limited support for programmability, there is a broad realization in the wireless community that future programmable wireless networks would require significant architectural innovations. In this paper, we will present an unified overview of the programmability solutions that have been proposed at the device and the network level. In particular, we will discuss software-defined radio (SDR), cognitive radio (CR), programmable MAC processor, and programmable routers as device-level programmability solutions, and software-defined networking (SDN), cognitive wireless networking (CWN), virtualizable wireless networking (VWN) and cloud-based wireless networking (CbWN) as network-level programmability solutions. We provide both a self-contained exposition of these topics as well as a broad survey of the application of these trends in modern wireless networks.

1 INTRODUCTION

Wireless networks have become increasingly popular due to the inherent convenience of untethered communication. They are deployed ubiquitously in myriad of networking environments ranging from cellular mobile networking, regional or city-wide networking (e.g., through worldwide interoperability for microwave access (WiMAX) technology), and local area or even personal networking environments (e.g., through Wi-Fi and Bluetooth technology, respectively)[1]. With the usage of wireless networks promising to increase in the future, both in demand and application diversity[2], the issue of devising and implementing flexible architectural support becomes all the more important.

While newer wireless technologies have been emerging at a prolific rate, the architecture of wireless networking has largely been static and difficult to evolve. The malaise of architectural 'ossification' is not unique to wireless networking though but applies more generally to networking. Before we can describe the reasons of this ossification, we operationally define the *data plane* to be responsible for forwarding packets at line speed, and the *control plane* for figuring out, and instantiating, the forwarding state that the data plane needs. Various reasons have been offered to explain the Internet's architectural ossification such as: *i*) vertical integration and coupling of the data plane and the control plane at node level; *ii*) lack of abstractions and modularization of the control plane, and finally, resulting from the preceding two reasons; and *iii*) lack of programmability of the network as a whole. These reasons, subtly related to each other, have collectively discouraged networking growth and innovations[3].

To manage the complexity of computer systems, computer scientists have long recognized the potency of the concept of abstraction[4]. It has been argued that the most formidable challenge to the networking industry is posed by the paucity of useful abstractions[5]. With a lack of foundational abstractions, networking reduces to a 'plethora of protocols and tools' without any underlying architectural base[5, 6]. There are three main benefits of using *abstractions*: *i*) modularity, which allows managing complex problems scalably through reuse of modules offering common functionality; *ii*) separation of concerns through loose inter-module coupling which is implementation agnostic; and *iii*) innovation, since new developments can focus on the module that needs fixing without 'reinventing the entire wheel'[7]. The Open System Interconnection (OSI) layered model - which is composed of layers representing modular subcomponents that interact through well-defined abstract interfaces - is often considered as a major success story of computer networking. The OSI layering model, however, relates mostly to the Internet's data plane which has evolved to offer various useful abstractions[6]. The Internet's control plane, on the other hand, has developed mostly in an ad hoc fashion and has lacked well-developed abstractions until quite recently[8].

Broadly speaking, *programmable networks* denote networks that can tune itself, or reconfigure itself, through a software-based adaptation - typically managed through an application programmer interface (API). Incorporation of functionality in software allows networks to innovate at the rate of software development cycle which is a lot more agile than the sluggish rate of hardware development[9]. Applications of programmability include rapid provisioning of services[10], flexible resource management[11], efficient resource sharing[12], and support for new architectures such as cloud computing, Internet of things (IoT)[13], etc.

Traditionally, networking devices have offered some ability to configure the system/network but with limited programmability. In particular, the devices mostly offered a limited set of *configuration* options which may be envisioned as 'knobs' that can be tuned to suit the operator. The vision of a programmable device is to allow the operator to *program* in any way desired; i.e., the operator should be free to define new custom tunable knobs as desired to support niche applications or services and not be limited to the configuration options made available by the device's vendor. Programmable devices thus offer far greater flexibility than configurable devices.

Interestingly, programmable networking is not entirely a recent concept. The lack of programmability of networks has long been recognized, and various approaches have been proposed to address this deficiency[14, 15]. In a remarkably prescient paper, published by Campbell et al. in 1999[14], the impending impact of numerous programming trends is anticipated using surprisingly modern terminology. In particular, it was predicted that higher levels of network programmability will result from separation of hardware from software, availability of open network interfaces, virtualization of networking infrastructure, and rapid creation and deployment of network services. These predictions have come to fruition exactly as forecasted in the forms of software-defined networking (SDN), standardized APIs, network virtualization, and cloud computing. Similarly, another insightful paper[16], the early version of which dates to 1996, talks about applying programming language perspective to networks and their protocols and talks about their aim of creating the 'Smalltalk of networking'. These ideas are having a renaissance in the modern era in the context of SDN programming languages[17][a].

To develop programmable wireless networks, it is imperative that we emphasize the development of both *programmable wireless data planes* and *programmable wireless control planes*. In this paper, we will provide a unified holistic overview of programmable wireless networking and will highlight overarching themes and insights. We will see that there are four prominent technological trends that underlie most of the current research focusing on enabling future programmable wireless networking. These technological trends are: *i*) software-defined wireless networks (SWNs), *ii*) cognitive wireless networks (CWNs), *iii*) virtualizable wireless networks (VWNs), and *iv*) cloud-based wireless networks (CbWNs). We will provide a background on these trends and will provide a broad survey of the applications of these trends in the literature.

The *main contribution of this paper* is that we provide a unified overview of the emerging field of programmable wireless networks. We demonstrate that the seemingly disparate fields of active networking, software-defined radios, cognitive radios, software-defined networking, and wireless virtualization, are in fact kindred disciplines. We develop this idea and propose new directions of future programmable wireless networks. Our paper is different from other survey papers[11, 18, 19] in its focus on *wireless* programmable networks while the previous papers had focused mainly on generic (wired) programmable networks.

The organization of the remaining paper is as follows. In the next section, we describe the basic *device-level* building blocks for building future programmable wireless networks. Thereafter, we discuss the various architectural approaches of building *generic programmable networks* in Section 3. We will thereafter discussing four dominant categories of programmable *wireless* networks, i.e., SWNs, CWNs, VWNs, and CbWNs, and highlight works belonging to each category in Sections 4.1, 4.2, 4.3, and 4.4, respectively. In Section 5, we discuss various open research issues and future directions of research. We conclude the paper in Section 6.

2 BUILDING BLOCKS FOR PROGRAMMABLE WIRELESS NETWORKING

Programmable devices are envisioned to be a key component of future programmable networks. In this section, we discuss various techniques and architectures that have been proposed to realize the benefits associated with programmable wireless networks. In particular, we elaborate upon the trends of software-defined radio, cognitive radio, MAC programmable wireless devices, programmable wireless testbeds, and programmable radios in this particular order.

2.1 Software-defined radios (SDR)

The defining characteristic of a SDR is that it implements most of the basic building blocks of physical (PHY) layer radio communication in software. With the hardware stripped down to the elements essential to all radio communication, custom blocks traditionally implemented in hardware - e.g., filters, amplifiers, modulators, demodulators, mixers, etc. - are now implemented in software (Figure1). This implies that appropriate programming of the generic radio hardware can in principle allow it to support arbitrary technologies. The SDR technology was a significant paradigm shift ushering in a new era of programmable wireless devices. Thus, using an SDR, an operator could program a wireless device to support any of the myriad of wireless technologies[20]. This opened up an unprecedented opportunity for creating a programmable *wireless* device for the first time[21–23].

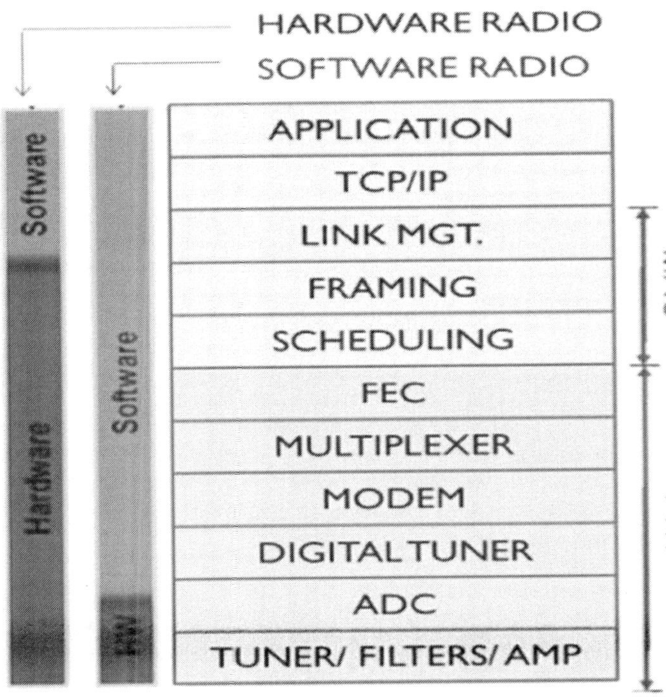

Figure 1 The hardware/software division of radio tasks in software-defined radio vs. a traditional hardware radio.

The precise definition of SDRs is debated, with no clear consensus on how reconfigurable must a radio be to be deemed an SDR. Clearly, it is a bit of a stretch to call every radio with a digital signal processor (DSP) as an SDR. A working definition provided in[21] is that an SDR is 'a radio that is substantially defined in software and whose physical layer behavior can be significantly altered through changes to its software'. The SDR forum defines an 'ultimate software radio' (USR) as 'a radio that accepts fully programmable traffic and control information and supports a broad range of frequencies, air-interfaces, and applications software'.[21]. In[24], two extremes SDR platforms are discussed: The first type is an SDR that is composed of programmable components, such as field programmable gate arrays (FPGAs), DSPs, etc., which are *programmed* directly; the other extreme is a highly *configurable* chipset-based SDR which is 'programmed' by setting configuration registers in the chip to determine the choice of frequency, coding, and PHY- and MAC-level protocol details. Most practical SDRs lie between these two extremes[24].

While traditionally SDRs have mainly been used in military settings due to excessive cost, the technology has now matured to a stage where its form and cost is amenable to non-military markets[24]. While the SDR of 1990s was the size of a small refrigerator and could easily cost more than $100,000, today the size of an SDR is akin to the size of a computer battery and it can cost less than $500, extrapolating the trend, it is reasonable to assume that future pricing and form factor of SDRs will match that of a typical consumer electronic device[24]. The democratization of SDR technology will conceivably revolutionize wireless and mobile networking, e.g., a consumer will not be limited to any single wireless protocol with a wireless device. This will lead to unprecedented flexibility as technologies (such as Wi-Fi and Bluetooth) will no longer be 'baked' into the hardware, but will be software applications, or applets, that any SDR could support. Due to their versatile nature, SDRs are radio chameleons potentially running a telephony protocol (such as CDMA) at a given time and switching to a completely different data communication protocol (such as Wi-Fi or WiMAX) next moment[24].

A prominent and popular example of SDR platform is to use the universal software radio peripheral (USRP) hardware kit[25] along with the open-source GNU Radio software toolkit[26] that implements in software various necessary signal processing blocks. The USRP hardware digitizes the received analog signal and imports it into a computer so that it may be processed by GNU Radio software (or similar toolkits such as the OSSIE framework based on the 'joint tactical radio system' (JTRS) software communications architecture[27]). Such an arrangement allows building a *custom radio* that can be programmed to support an arbitrary wireless technology through appropriate signal processing blocks that operate on the signal received or to be transmitted. Some example of SDR platforms include Implementing Radio in Software (IRIS)[28], USRP[29], Wireless Open-Access Research Platform (WARP)[30], Airblue[31], SORA[32], OpenRadio[33], and Ziria[34, 35]. We provide a brief description of these platforms in Table1 and refer to the original papers for details.

Table 1. Representative summary of important programmable networking components and platforms

Component category	Project and reference	Year	Brief summary
Software-defined radio (SDR) platforms			
	IRIS [28]	2004	Implementing Radio in Software (IRIS) project developed at Trinity College, Dublin
	USRP [29]	2005	Flexible SDR development platform, often used with GNUradio, manufactured by Ettus/ NI
	WARP [30]	2008	Wireless Open-Access Research Platform (WARP) developed by Rice University
	Airblue	2010	An FPGA based SDR platform that can run at high speeds compatible to commodity hardware
	SORA [32]	2011	Programmable SDR platform, developed by Microsoft, for commodity multi-core PCs
	OpenRadio [33]	2012	Programmable wireless data plane that can programmed across the wireless stack
	Ziria [34,35]	2013	SDR framework developed at Microsoft, comprising a programming language and an optimizing compiler, that can synthesize efficient SDR code from high-level PHY description
Cognitive radio (CR) platforms			
	BEE2 [36]	2005	Reconfigurable hardware platform developed at University of California, Berkeley
	KNOWS [37]	2007	CR hardware platform, for operation in TV white spaces, developed by Microsoft
	WinC2R [38]	2008	CR hardware platform developed by the WINLAB at Rutgers University
Programmable network components			
	Virtual Switches [39]	1996	Proposed virtualizing ATM switches as part of the xbind [39] project (OpenSig framework)
	Switchlets [40]	1998	Proposed dynamically loadable code on a (partition of) ATM switches as part of the Tempest [40] project subscribing to OpenSig framework
	Virtual base stations [41]	1998	Proposed as part of the Mobiware [41] project subscribing to the OpenSig framework
	Routelets [42]	1999	proposed in the Genesis [42] project subscribing to the OpenSig framework
	Click [43]	1999	Software architecture for building flexible and configurable routers
	XORP [44]	2003	An open programmable router platform for research experimentation
	SoftMAC [45]	2005	Proposed a software system for developing new MAC protocols on commodity 802.11 NICs
	MultiMAC [46]	2005	Dynamically reconfigures MAC and PHY properties to select appropriate MAC component on per node/ flow basis
	NetFPGA [47,48]	2007	Programmable and extensible router with embedded FPGA
	RouteBricks [49]	2009	Software router architecture (Click based) that parallelizes router functionality
	SwitchBlade [50]	2010	FPGA-based platform for deploying custom protocols with programmability and performance
	Ansari et al. [51]	2010	Programmable decomposable MAC framework
	TRUMP [52]	2011	Programmable component-based MAC framework
	Wireless MAC processor [53]	2012	Composition of custom MAC protocols by programming with basic MAC commands
	MAClets approach [54]	2012	Programmable framework that allows installing MAC stacks as 'applets'

To summarize, SDRs are envisioned as an essential component of future programmable wireless devices. Their importance can be gauged from the fact that almost all advanced wireless programmability techniques (such as cognitive radio and programmable wireless processors, etc.) are SDR-based.

2.2 Cognitive radios (CR)

CRs have evolved from the concept of SDRs[55], and it represents the next major milestone after SDR in the drive towards powerful programmable wireless *devices*. Joseph Mitola coined the term 'cognitive radio' in 1999 when he envisioned a broadening of the SDR concept. In particular, Mitola anticipated that incorporation of substantial artificial intelligence (AI) in the form of machine learning, knowledge reasoning, and natural language processing will transform SDRs into cognitive radios that will optimize network performance by sensing, learning, and reacting to environmental stimuli[56]. In other words, CR allows an SDR to reprogram itself autonomously based on network conditions. In a modern setting, this is achieved by incorporation of a cognitive engine that employs various AI-based techniques to build a knowledge base, based on which reasoning is performed to make 'optimal' decisions[57].

CRs are viewed as an essential component of next-generation wireless networks[58, 59] and have a wide range of applications including intelligent transport systems, public safety systems, femtocells, cooperative networks, dynamic spectrum access, and smart grid communications[57, 58]. CR can dramatically improve spectrum access, capacity, and link performance while also incorporating the needs and the context of the user[57]. Although cognitive behavior of CRs can enable diverse applications, perhaps the most cited application of cognitive radio networks (CRNs), which are networks where nodes are equipped with CRs, is dynamic spectrum access (DSA)[60]. DSA is proposed as a solution to the problem of 'artificial spectrum scarcity' that results from static allocation of available wireless spectrum using the command-and-control licensing approach[60]. Under this approach, licensed applications represented by primary users (PUs) are allocated exclusive access to portions of the available wireless spectrum prohibiting other users from access even when the spectrum is idle. With most of the radio spectrum already being licensed in this fashion, innovation in wireless technology is constrained. The problem is compounded by the observation, replicated in numerous measurement-based studies world over, that the licensed spectrum is grossly underutilized[58, 60]. The DSA paradigm proposes to allow secondary users (SUs) access to the licensed spectrum subject to the condition that SUs do not interfere with the operations of the primary network of incumbents.

While programmable wireless devices (such as SDRs and CRs) do serve as the building block for programmable wireless networking infrastructures, it is pertinent to note here that the task of building programmable wireless *networks* is much more nuanced. Various vexing problems (such as routing[61], security[62], etc.) need to be solved while taking into account the network-wide behavior[63]. Historically, most of the CR research has focused on optimizing at a device level, with network-level programmability being a recent afterthought[64]. In the subsequent sections, we will see how trends of software-defined networks (Section 4.1) and cognitive networks (Section 4.2) allow us to extend the programmability concepts to network proportions.

2.3 MAC programmable wireless devices

In the past few years, numerous new wireless technologies, with distinct MAC protocols, have been proposed to serve a variety of niche wireless applications. Since there is no universal and one-size-fits-all MAC protocol that will work equally well for all such scenarios, there is a lot of interest in creating programmable wireless devices which will implement, what may be effectively called, *software-defined MAC*. A majority of current wireless devices do not support SDRs, or even software-defined MAC, and effectively can support only a single technology. Although SDRs offer great flexibility in altering its PHY later characteristics, supporting programmatic MAC on SDRs also entails significant research challenges[32, 65].

In recent times, there has been work in supporting programmable, or software-defined MAC, on *commodity wireless devices*. In 2005, Neufeld et al. proposed SoftMAC[45] as a software system that allows development of new MAC protocols on commodity 802.11 network interface cards (NICs) to build a SDR with predefined PHY characteristics but with a flexible MAC layer. SoftMAC was extended by the MultiMAC framework[46], proposed by Doerr et al., to provide support for DSA applications in CRNs. In particular, MultiMAC supports multiple concurrent MAC layers that can be switched with minimal impact. MultiMAC aimed at dynamically reconfiguring MAC and PHY layer properties to select appropriate MAC component on per node/flow basis thereby acting as a stepping stone to the concept of 'cognitive MAC' which supports intelligent reconfiguration of MAC and PHY layers.

Other ways have also been proposed for building MAC programmable wireless devices. One way of doing this is by creating an abstraction of a *wireless MAC processor* with an instruction set representing common MAC actions, events, and conditions which can be programmed through an API to compose any custom MAC protocol[53]. Another approach, known as the MAClet approach, is to conceive the entire MAC protocol stack akin to a Java applet which can be loaded onto a MAC processor and run[54]. While these approaches could be conceivably implemented on FPGA-based SDR platforms, such as WARP[30] or USRP[25] in a straightforward manner, the main contribution of the works[53, 54] has been to implement these approaches on a commodity Broadcom wireless NIC. In[66], a new service-oriented architecture for programmable wireless interfaces is proposed which replaces the traditional PHY and MAC layers with a *i*) *a platform layer*, which exposes static primitives for managing hardware events and frame transmissions and *ii*) *three layers of functionalities* - state machines, functions, and services - that expose a programmable interface to upper layers. The proposed approach differs from SDR solutions since the adaptation and customization is accomplished through programmable interfaces exposed at a layer higher than the PHY layer. Besides these aforementioned works[53, 54, 66], there have been other *component oriented design*[67] efforts for composing customizable MAC protocols from a set of basic functional components[51, 52]. For a detailed survey of dynamically adaptable protocol stacks in general, the interested reader is referred to[68].

2.4 Programmable routers

Programmable routers have been developed that incorporate programmable datapath processing capabilities in commodity ×86 architectures to perform custom protocol operations and arbitrary payload processing more flexibly and at a lower cost. These programmable routers are not specific to wireless technologies but we discuss this technology in this section because these routers can potentially be very useful in the context of programmable wireless networking. The Click programmable router[43] is an early influential software router which snaps together various modular 'elements' to assemble the router logic. Although Click offers the capability of rapid prototyping and deployment and decent performance for a software router running on a PC, any purely software-based approach will be hard pressed to satisfy the demanding performance requirements of modern networks. However, with the recent advances in the processing and input/output (I/O) performance of commodity servers, software routers running on ×86 architecture offers a disruptive lower cost and more flexible proposition.

More recently, programmable routers with reprogrammable hardware such as FPGAs have been proposed to simultaneously address the needs for flexibility, extensibility, and performance for the forwarding plane. Prominent projects in this category include the NetFPGA project[47], the RouteBricks project[49], and the SwitchBlade project[50]. Extensible open-source *control plane software* also exists with the Extensible Open Router Platform (XORP) open-source software suite[44] being a prominent example; XORP defines a fully extensible platform, suitable for both research and deployment, which builds upon the extensible Click framework in its forwarding plane.

To wrap this section, we note that we have provided a representative summary of various architectural components of programmable networking, including a summary of programmable MAC devices, is provided in Table1. The characteristics of SDR, CR Programmable MAC, and programmable routers are also juxtaposed in Table2 for a ready comparison.

Table 2. Comparison of SDR, programmable MAC, and CR paradigms

	Software-defined radio (SDR)	Cognitive radio	Programmable MAC	Programmable routers
Raison d'être	Software defined ability to adapt/program PHY and MAC layer characteristics.	Using 'cognition' to drive the capability of adapting (typically providing by SDRs) to optimize the network performance.	Supporting custom creation of MAC protocols, rather than hard-wired MAC through appropriate vendor-independent abstractions.	Supporting custom data plane processing to facilitate customized processing of the packets in the data line.
Applications	Reconfigurability, building block of cognitive radio, interoperability, more degrees of freedom.	Dynamic spectrum access (DSA); interoperability and improved handovers; link optimization (modulation, power, topology, etc.); better resource utilization; increased capacity, reliability, and security; technology neutral coexistence.	Defines primitives for composing custom MAC protocol logic which can program the whole radio protocol stack independently of the platform (analogously to Java Applet).	Useful for developing software-defined routers with customized data planes that allows custom protocol operations and/or any arbitrary payload processing at the network-layer.
Strengths	Lower life cycle cost, increased interoperability with multiple waveforms, field upgradable, demonstrates hardware flexibility.	They inherit the strengths of SDR (being SDR-based typically). They can learn about the environment and self-optimize by modifying radio parameters accordingly to ensure certain QoS. Hardware and policy flexibility.	Provides some capabilities of a full SDR on commodity WLAN hardware allowing support for arbitrary MAC protocols. This results in a cheaper solution that is also simpler. The main strength of this approach is programmability.	Flexibility in programming arbitrary processing in the forwarding plane. This can allow routers to diversifying beyond a forwarding only regime into the territory of *middleboxes* to define customized data plane processing of packets.
Weaknesses	Basic hardware of SDR is typically more expensive that single-mode hardware radio.	Numerous technical hurdles must still be overcome for CR to be ready for implementation in a real world scenario. A commercial fully functional CRN is yet to emerge.	More work needs to be done to support more sophisticated abstractions widely on commodity devices.	Not deployed widely. May not match the processing ability of fixed routers with predefined data plane processing.

3 REVIEW OF PROGRAMMABLE NETWORKING ARCHITECTURES

3.1 Traditional programmable networking architectures

With the lack of programmability complicating networking innovations, it was the early 1990s when work on creating programmable network started in earnest[19]. At the time there were two major, slightly differing schools, that advocated programmable networks: the first group proposed the 'OpenSig' approach[19] while the second group furthered the 'Active Networking' approach[69]. The general consensus that emerged was that the programmability solution lies in separating the control software from the hardware and in having open interfaces for management and control. The building blocks for creating programmable networks started appearing thereafter in the form of various programmable networking components (such as the Click modular router[43], etc.). More recently, with the emergence of datacenters, virtualization, and cloud computing technology, the requirement of programmability has become mainstream. In the remainder of this subsection, we will outline these developments in more detail.

3.1.1 The OpenSig approach

In the mid 1990s, the *OpenSig approach*[70] advocated both the separation of the data plane and the control plane for *ATM networks* and the usage of *open interfaces* for signalling between these two planes. The main idea was that with separated control and data planes, and an open standard interface, the ATM switches would become remotely programmable and thus more manageable. The OpenSig community actively worked on standardizing such an open interface, and a number of experimental networks set up in various research institutes explored these proposals. The Tempest framework[40], based on the OpenSig philosophy, allowed multiple control planes to simultaneously control a single network of ATM switches. The main reason of OpenSig approach could not quite become mainstream was the static nature of the interfaces it defined[14].

3.1.2 Active networking

The *Active Networking* (AN) approach[16, 69] was popularized at the same time as OpenSig, i.e., in the mid 1990s, when the Internet was rapidly commercializing and experiencing the need of more flexible control. The AN approach aimed at creating programmable networks that can allow rapid network innovations. The AN research - driven mainly by the efforts of the Defense Advanced Research Projects Agency (DARPA) - was motivated by the need to rapidly commission new services and dynamically configure networks in run-time. It was perceived that the static nature of OpenSig networks could not support these needs.

The main idea of AN is to *actively* control network nodes so that the network nodes may be programmed to execute arbitrary mobile code as desired by the operator[16]. The value proposition of such an approach was that it would enable new innovative applications, that leverage computation with the network, and that it would increase the rate of innovations by decoupling services from the underlying infrastructure[71, 72]. The flexibility offered by such an approach, on the other hand, was also accompanied by concerns about its performance and security implications.

The AN approach consisted of two programming models: *i*) the *capsule model*, where the data packets contained not only the data to be communicated but also in-band instructions to execute; and the *ii*) *programmable switch model*, in which the out-of-band mechanisms were utilized to execute code at various nodes[18, 69]. While it is the capsule model - which was the more radical approach, significantly different from the traditional operational paradigm of networking - that is most closely associated with AN, it is fair to say that both these models have bequeathed valuable legacies inherited by modern programmable networking frameworks. In particular, the capsule approach attracted interest since it could provide a clean method of upgrading data plane processing along an entire network path[18, 72]. Using the capsule approach, numerous services such as active load balancing, multicasting, caching, etc. could be supported[69, 73].

The AN framework was vigorously pursued by the research community in the mid and late 1990s - helped by the interest and generous funding of DARPA. Various influential projects were initiated in this time frame with some prominent AN projects being the ActiveWare project[74] at MIT, the CANES[75] project at Georgia Tech, the SwitchWare project[76] at University of Pennsylvania, the ANTS project at University of Washington[77, 78], and the Tempest project[40] at Cambridge University. More details about these, and other important AN projects, can be seen in Table3 and in the survey paper[69].

Table 3. Representative summary of important programmable networking concepts

Project	Framework	Year	Summary
Active control of network nodes			
ANTS [77,78]	Active Networking	1997	Java-based active networking toolkit proposed as part of the MIT's ActiveWare [74] project.
SwitchWare [76]	Active Networking	1998	Active networking project at Uni. of Pennsylvania which focused on both security and performance issues.
CANEs [75]	Active Networking	1998/9	Composable Active Network Elements (CANEs) project at Georgia Tech.
Separation of the control Plane (CP) and the data plane (DP), and remote control of DP			
Tempest [40]	OpenSig	1996	Programmable framework for safe control of ATM switches. Allowed multiple control architectures to coexist on the same network, and a safe partitioned environment for third party, or dynamically loaded active code.
GSMP [79]	Pre-SDN	2002	General Switch Management Protocol (GSMP) proposed by an IETF working group to control a label switch.
ForCES [80]	Pre-SDN	2004	It defines a standardized interface between a network's control elements (CEs) and forwarding elements (FEs).
RCP [81]	Pre-SDN	2004	This work proposed separating routing from routers and outsourcing it to a separate router control platform (RCP).
SoftRouter [82]	Pre-SDN	2004	SoftRouter proposed separation of the control plane functions from the packet forwarding functions.
4D [83]	Pre-SDN	2005	4D proposed an architecture with *decision, dissemination, discovery,* and *data* - i.e. the 4D- planes, to separate decision logic from distributed systems issues.
Routing as a service [84]	Pre-SDN	2006	Proposed the provision of offering customized route computation as a service by third party providers.
PCE architecture [85]	Pre-SDN	2006	Path-computation-element (PCE)-based architecture (RFC 4655) where the PCE is an application located within a network node, or on an out-of-network server.
CogNet [63]	Pre-SDN	2005	CogNet proposed separated CP and DP with an extensible global CP controlling the separated DPs through an API.
IRSCP [86]	Pre-SDN	2006	Proposed intelligent Route Service Control Point (IRSCP) that allowed route selection to be performed outside the routers through external network intelligence.
SANE [87]	Pre-SDN	2006	SANE is a enterprise network security/ protection architecture implemented through a 'logically centralized' server.
Ethane [88]	Cusp of SDN-era	2007	Ethane proposed fine-grain control of simple flow-based Ethernet switches through a centralized controller.
OpenFlow [8]	SDN	2008	OpenFlow defines a southbound API/protocol standardized by ONF through which a separated dedicated controller can control multiple DPs remotely.
Open APIs			
xbind [39]	OpenSig	1996	Toolkit developed at Columbia Uni. for creating *broadband kernels* - that program broadband ATM nets like PCs.
Mobiware[41]	OpenSig	1998	Programming QoS-aware middleware for mobile multimedia networking developed at Columbia University.
NetScript	OpenSig	1999	Language for programmable processing of packet streams.
OpenFlow [8]	SDN	2008	Southbound API standardized by ONF.
Floodlight API [8]	SDN	2012	A RESTful northbound API between the controller platform and the SDN Applications.
Juniper APIs [89]	SDN	2012	JunOS SDK, XML API (NetConf), and Contrail REST API supported.
Cisco ONE [90]	SDN	2012	Network APIs (including southbound API) specified by Cisco; supports EEM (tcl), Python Scripting.
OpenStack APIs [91]	Cloud/ SDN	2012	OpenStack Neutron (formerly Quantum) is a OpenStack subsystem for managing networks in a cloud environment.
Network virtualization			
Virtual switches [39]	OpenSig	1996	Proposed virtualizing ATM switches as part of the xbind [39] project.
Switchlets [40]	OpenSig	1998	Proposed dynamically loadable code on a (partition of) ATM switches as part of the Tempest [40].
Virtual base stations [41]	OpenSig	1998	Proposed as part of the Mobiware [41] project subscribing to the OpenSig framework.
Routelets [42]	OpenSig	1999	Proposed in the Genesis [42] project.
PlanetLab [92,93]	Overlays Networks	2003	Proposed overlays, virtualized slicing, and programmability for accelerating innovations in the Internet community.
FlowVisor [94]	SDN	2009	Virtualizes OpenFlow-based SDNs by carving out virtualized '*slices*' out of production networks [95].
SecondNet [96]	SDN/ Datacenters	2012	Proposes a virtual datacenter (VDC) abstraction as the unit of resource allocation for multiple tenants in the cloud.
Network virtualization platform (NVP) [97]	SDN/ Datacenters	2014	Proposed a overlay-based network virtualization platform (NVP) for use in production multitenant datacenters.

The AN paradigm was the first in a series of *clean-slate Internet redesign*proposals[18]. Many of the programmable networking concepts that appear eminently modern - such as separation of control plane and data plane, remote control of data planes, virtualization, network APIs, etc. - have in fact germinated from the AN community. However, despite the ground breaking nature of the AN paradigm, the AN paradigm failed to really catch on. One reason for its failure to capitalize on the intense interest around was the lack of a compelling application use case in the AN approach that could work pragmatically within the existing framework. Secondly, AN emphasized the flexibility of providing network end users the chance to program the network which never became a popular use case.

3.2 Virtualization and cloud computing

Virtualization is a technique, fundamental to various disciplines of computer science, which allows *sharing* of resources while providing *abstractions* identical, for all practical purposes, to that of the original resource.

Virtualization has been especially influential in the modern era of large-scale datacenters. Prior to the popularization of virtualization technology, various concerns (such as security, isolation, performance) dictated that servers be dedicated for particular applications (e.g., dedicated web servers, database servers, etc.) and provisioned for peak load. This led to gross underutilization with 10% to 20% utilization of resources being commonplace. This has motivated the creation of a new 'virtual machine' (VM) abstraction using which multiple VM instances, completely isolated from each other, can be created on the same physical machine. These virtual machines provided an interface to end applications that was identical to that of the underlying physical server. With the programmability features of VM cloning and mobility, which allows taking VM snapshots and transporting to any physical server that is currently underutilized, physical resources can now be shared both efficiently and securely. Due to these desirable properties, virtualization has truly become an indispensable component of modern computing.

The popularity of compute virtualization in the datacenter environment has spawned two further trends: *i*) cloud computing and *ii*) network virtualization.

The main insight of *cloud computing* is to provide services in a virtualized datacenter, provisioned programmatically through APIs via the web, as a service in *utility computing*style[98]. Although utility computing was conceived as early as 1961 by John McCarthy[99], it is only recently that cloud computing has turned this vision into a reality. The cloud paradigm is differentiated from traditional datacenters mainly in the dynamism of service provisioning which has been made feasible by virtualization technology and advances in web APIs. The ability to *program* services has led to great innovations and has democratized computing largely by making computing resources available on per-use pricing. The 'holy grail' of the cloud computing paradigm is the vision of installing a generic 'network fabric'[100] which can be then programmed to provide any service without any need of manual configuration of network nodes. The implementation of such a fabric-based virtualized datacenter has proven itself elusive, due to the complexity of virtualizing networks, so much so that it is now a common sentiment in the networking industry that networking is the bottleneck in datacenter innovations[b]. With traditionally vertically integrated network devices, supporting cloud-era applications entails the undesirable burden of manually configuring various network switches through vendor-specific command-line-interfaces (CLIs) - a process that is cumbersome and error prone[101].

With the presence of 10s or 100s of VMs per machine, a software-based *hypervisor switch*, inside the physical server, takes care of inter-VM networking. A significant tipping point was recently witnessed when estimated number of physical ports were overtaken by virtualized ports - a significant inflection point in networking history indeed[102]. This has significant architectural implications. In particular, it has been highlighted that using an hypervisor overlay with a networking fabric constructed out of SDN technology (to be covered in Section 3.3) can become the functional equivalent of the traditionally influential *end-to-end principle*[103]. In addition, the virtualization/SDN hybrid architecture will also subsume the functionality of MPLS and middleboxes to offer a clean split between the core and the edge. In this new architecture, the SDN-based fabric will become the new core, while the hypervisor switches will be the new edge. We shall see later that these edge devices consist of hypervisor switches (e.g., Open vSwitch[104]) that are *software defined* and thus are *programmable* (using protocols such as OpenFlow[8]). This paradigm shift to software control fundamentally changes the pace of innovation and opens up a world of new possibilities.

While VMs have unshackled applications from being tied to particular physical servers, traditional *network virtualization* techniques (such as VLANs, VPNs, and overlay networks) do not offer an analogous*virtual network* (VN) abstraction that decouples the network from the physical infrastructure. This VN abstraction should, like the VM abstraction does for the server, ensure detachment of the virtual network from the physical infrastructure as well as isolation between multiple tenants sharing the same infrastructure, while providing the same interface as the original network. There was a notable early work on network virtualization in the OpenSig era: the Genesis project[42] proposed, in 1999, a *virtual network kernel* that was capable of *spawning* virtual network architectures on-demand. The term 'spawning' is used in Genesis as a metaphorical reference to the use of this term in the field of operating systems where it refers to the process of creating a new process that runs on the same hardware - analogously, spawning a network means creating a new network architecture on the same infrastructure. This concept, although important and novel, is distinct from the modern virtualization abstraction of a VN. Just like a VM is a software container - encapsulating logical CPU, memory, storage, networking, etc. - providing an interface identical to a physical machine to an application, a VN is a software container - encapsulating logical network components, such as routers, switches, firewalls, etc. - that presents an interface identical to a physical network to network applications. The VN abstraction for wireless networks is visually depicted in Figure2. This abstraction allows great flexibility to IT managers as the physical network can now be managed as a 'fabric' offering some transport capacity that can be used, programmed, and repurposed as needed.

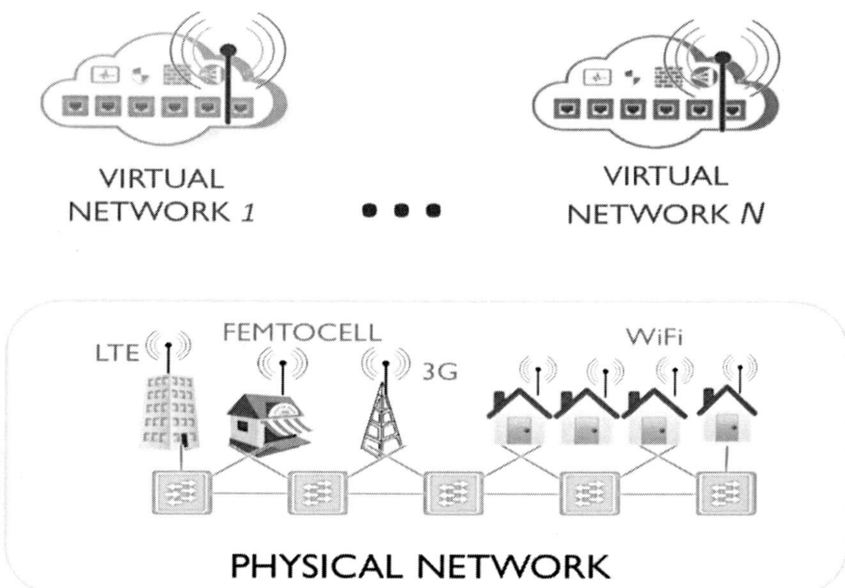

Figure 2. Using virtualization, multiple virtual networks can coexist on the underlying physical infrastructure substrate in a decoupled fashion.

Virtualization is also a popular method in the Internet community for introducing innovations in production networks with minimal intervention through the use of *overlay networks*[92, 93]. An overlay network is essentially a virtualized logical network built on top of a physical network with tunnels interconnecting edge devices. The overlay network is typically decoupled from the underlying physical network through dual address spaces representing a tunnel encapsulation with the virtual address space on the inside and the physical address space on the outside. The overlay network appears to the nodes connecting to it as a native network with the possibility of multiple overlays existing on the same underlying physical infrastructure (which allows support for multi-tenancy). Each overlay network is effectively a distinct logical network which can support service properties such as an arbitrary policy of L2, L3, and access control list (ACL) processing distinct from the physical network. This makes overlay networking a popular technique for supporting disruptive innovations in networks without requiring interventions in the core network[92, 93].

Interested readers are referred to comprehensive surveys on network virtualization[105, 106] and cloud computing[98] for more details.

3.3 Software-defined networking (SDN)

The major insight of 'software-defined networking' (SDNs) is to allow horizontally integrated systems by allowing the separation of the control plane and the data plane[11, 107] while providing increasingly sophisticated set of abstractions. SDN has revolutionized the networking industry by providing architectural support for 'programming the network'. SDN promises to be a major paradigm shift in networking landscape leading to improved and simplified networking management and operations. While conceived mainly in academia, SDN has been taken up by the industry by gusto with numerous success stories[108, 109]. SDN has also been seen recent successful industrial deployments[12]. Although SDN and active networking paradigms share a common motivation, i.e., of creating programmable networks, both of them are different in their focus; active networks strived more for data-plane programmability while SDN's focus has been on control-plane programmability[18].

Although the term SDN has only been coined in 2009, the idea of SDN has a rich intellectual history. In particular, it is the culmination of many varied ideas and proposals in the general field of programmable networks[11, 18] with many of the initial ideas of programmable networking (of 'open interfaces' and 'separation of control and data plane') espoused by the OpenSig and Active Networking community now maturing in the form of the SDN) architecture. While the SDN architecture is very similar to the AN architecture, it has become more popular due to technological advances, compelling use cases, and importantly, certain pragmatic design choices. In particular, SDN has become popular largely due to the need of virtualization in modern datacenters and cloud computing which require network virtualization support due to their dependence on automated provisioning, automation, and orchestration. While AN focused on developing radically new data plane abstractions, the SDN approach has focused more on newer control plane abstractions (which arguably addresses a bigger pain point). The SDN architecture is different from the AN architecture since the former has emphasized on the separation of the control plane and the data plane[18] which was not integral to the AN architecture.With the growing popularity of SDN, various industrial stakeholders have jumped on the SDN bandwagon to exploit its early success, and the term 'SDN' has seen a considerable broadening. To analyze and reason about SDN, it is, therefore, vital that we define it precisely. There are three key characteristics of SDN. Firstly, there is a separation of the data plane and the control plane. Secondly, a single control plane (or controller) may control multiple data planes (or the datapath of switches/router). Lastly, SDNs incorporate modularity in the control plane through which high-level abstractions can be used by network control programs. The distinction between traditional and SDN network architecture can be clearly observed in Figure3b. To summarize these views, SDN deals with abstractions and mechanisms for creating a general, horizontal networking platform.

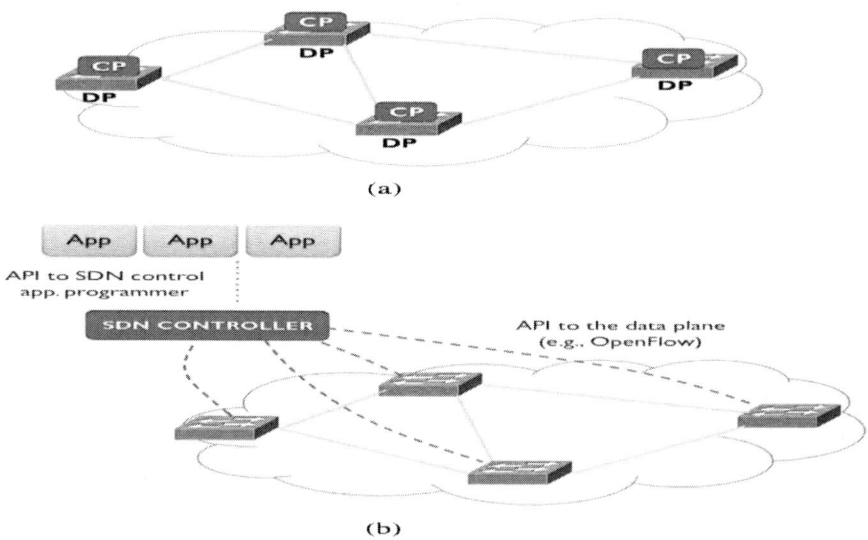

Figure 3. Comparison of traditional and SDN network architectures. **(a)** In traditional networking, the control planes (CP) and the data planes (DP) are colocated on devices to ensure decentralized network control. **(b)** In SDNs, the DPs and CPs are separated with a centralized controller controlling

multiple DPs while supporting a *southbound* API to the DPs and a *northbound* API to the SDN applications.

By providing abstraction layers, it is possible to program new applications on central controllers for a wide variety of purposes. These *applications* include setting up virtual networks, enforcing quality of service (QoS), explicit routing, etc. The most fertile application area of SDN has been datacenters and campus networks[8]; however, SDN has also been proposed in many other settings such as service providers, carrier networks, and wireless networks. Specifically, SDN has been applied to wireless networks in varied settings such as wireless sensor networks (WSNs) and wireless mesh networks (WMNs)[110].

Although networks have always been software defined[c], writing, managing, and updating the networking software could be done only by the developers employed by the vendors. This 'closed' network architecture stifled innovation as the network was inherently non-programmable, and all new feature requests had to be routed to the networking vendor for implementation. SDN changes this paradigm by opening up the network through the simple, yet powerful, concept of separation of the control plane and the data plane. This separation, coupled with newer control abstractions, forms the core of the new SDN architecture. The development of SDNs is supported by a burgeoning open-source community. With the separation of the control plane from the data plane, it is possible for third party/open-source developers to write program applications for the controller. This allows networks to employ programmable commodity hardware rather than closed vendor hardware, increasing flexibility and development while reducing costs.

The Open Network Foundation (ONF) is an organization that is working on developing and maintaining standards for SDNs. Broadly speaking, there are two main APIs in the SDN architecture: *i*) the *southbound API* defines an interface between a centralized network controller[d] and networking devices[8], while *ii*) the *northbound API* defines the interface exposed by the controller to the network applications. There also has been activity in the Internet Engineering Task Force (IETF) community on standardizing SDN (e.g., with the I2RS[111] and ALTO[112] projects). In addition, there is the OpenDaylight[e] project under the Linux Foundation which is a community-led and industry-supported framework which aims at fostering innovation in the SDN space through the best practices of open-source development best practices.

OpenFlow[8] represents perhaps the most readily recognized protocol associated with SDN[f]. OpenFlow is an example standard southbound API which has been standardized by the ONF. The standardization of OpenFlow has propelled it as the principal SDN control plane abstraction, enabling thereby numerous innovations[113]. With the control logic implemented in a separate controller and a standardized control API between the controller and the data planes, the vision of programming the network using a high-level control language can be achieved. While the current control API defined by OpenFlow is fairly primitive (and has been compared to Assembly language), it is a matter of time before higher level control languages are developed that offer more sophisticated abstractions. Indeed, work in this direction is already underway[17]. The seemingly innocuous refactoring of the functionality from individual devices to the centralized controller unleashes a powerful new paradigm offering abstractions which facilitate simplified, efficient, and scalable management of network operations and services.

The development of the SDN architecture has also led to the development of the 'network operating system' abstraction. The role of an operating system (OS) is to manage the complexity of various components, that a computer is composed of, and to present a simplified programming interface to the application programmer. In a similar vein, it is envisioned that the network OS (NOS) will manage the various tasks necessary to manage the network (such as exchange of distributed state and computation of routes, etc.) and present a simplified interface to network application programmers. The NOS is typically implemented at the SDN controller(s). An NOS is expected to implement a *state management layer*, managing distributed state in the network to provide a consistent network view, and provide an API to network applications to facilitate high-level programming. Various NOS have been implemented for SDN including the seminal work for NOX[114] and subsequent efforts for ONIX[115] and ONOS[116].

Since the proposal of the SDN architecture[8] in 2008, many research works have focused on the development of higher layer protocols and applications that can leverage and exploit the programmability offered by the SDN architecture. In particular, routing, transport layer, and management frameworks have been proposed that work with OpenFlow and SDN. The routing proposals include *i*) Quagflow[117], which partners the open-source routing software Quaqqa[118] with OpenFlow and *ii*) RouteFlow approach[119] which can be used to provide 'virtual routers as a service' in SDN environments[120]. The transport layer protocol proposals include the work on OpenTCP[121]. Finally, there has been work on supporting multimedia delivery with QoS with the OpenQoS[122].

The SDN architecture, as has been highlighted before, uses many of the programmability concepts of earlier projects. In particular, it builds upon earlier proposals to *i*) separate the control plane and the data plane (early proposals include SoftRouter[82], 4D[83], RCP[81], and work in the ForCES working group[80]), *ii)* control multiple data planes from a separate controller (like the Tempest framework[40]), and *iii*) utilize open interface for communications between the separate controller and the data planes (like the OpenSig framework[70]). A representative summary of various programming concepts, many of which SDN exploits, is shown in Table3.

Despite the fact that SDN utilizes many of the active networking projects, SDN has become more popular than its predecessors due to the various technology pushes (e.g., advances in computing and networking technology) and application pulls (e.g., datacenter and cloud services, network virtualization, etc.) and greater industrial acceptance due to certain pragmatic design choices[18]. The long-term success of SDN would require innovations in new abstractions for the control and data plane balanced with a pragmatic strategy for its deployment.

4 DOMINANT TRENDS IN PROGRAMMABLE WIRELESS NETWORKING

In this section, we focus on prominent trends in wireless networking that have the potential to play a major part in creating future programmable wireless networks. In particular, we will discuss SWNs, CWNs, and VWNs, and CbWNs (Table4). Our generalized treatment of wireless networking will subsume discussions on both technologies that have evolved from their telecom roots (such as 4G networks such as WiMAX and long-term evolution (LTE)) and also those that have predominantly data networking foundations (such as Wi-Fi).

4.1 Trend 1: software-defined wireless networks (SWNs)

With increasing deployment and diversification of wireless technology, managing wireless networks has become very challenging. SDN is a promising architecture that can be used for conveniently operating, controlling, and managing wireless networks. As discussed in Section 3.3, the defining characteristic of SDN is generally understood to be the separation of the control and data plane. The presence of programmable controllers enables us to call these networks 'software defined'.

Table 4. Representative summary of important trends in wireless networking

Trend 3: virtualizable wireless networks (VWNs)		
Applications		
Multi-tenancy support	MobileFlow [135]	Proposed a virtualized SDN-based framework suitable for multi tenant mobile networks
Multi-provider support (infrastructure sharing for MVNOs)	Virtualization of 4G/ 5G RAN [156], WiMAX BS [157], LTE [158]	Virtualization allows better support for multi-tenancy and multi-provider and infrastructure sharing, which is convenient both in terms of user experience and economics
Virtualized NIC abstraction	Commodity WLAN card [159]	TDM-based wireless virtualization to create a virtual WLAN using commodity hardware
Virtual-APs	Hamaguchi et al. [160]	Virtual AP that uses virtualization technology to optimize deployment of AP
Network settings		
WLAN-based VWNs	Commodity WLAN card VWN [159]	Virtualization of commodity WLAN technology
	Virtual Wi-Fi [161]	Virtual Wi-Fi to support fully functional wireless functionality inside VMs
	multi-purpose AP (MPAP) [162]	Proposed MPAP for virtualizing heterogeneous technologies on a SDR
SDN-based VWNs	LVAP (based on Odin) [132]	Proposed Odin, based on SDN, to allow orchestration of programmable WLANs
	OpenAPI [163]	Proposed virtualizing the access network via Open APIs
	Virtual router as a service [120]	Proposed virtual-routers-as-a-service based on the RouteFlow architecture [119]
	eNodeB virtualization [164]	Proposed using OpenFlow for eNodeB virtualization in 4G-LTE networks
Cellular mobile VWNs	Virtualization of RAN [156]	Proposed network virtualization substrate (NVS) to be used in LTE RANs
	WiMAX BS [157]	Proposed virtualizing resources in a cellular WiMAX base station to enable MVNOs
	LTE [158]; eNodeB virt. [164]	Proposed virtualization of LTE environments
CRN-based VWNs	Spectrum Virtualization Layer [165]	This work proposed a virtualized layer for supporting DSA in general wireless networks
Trend 4: cloud-based wireless networks (CbWNs)		
Applications		
Computation offloading	Yang et al. [166]; CloneCloud [167]	Utilized computation offloading through elastic execution between mobile device and cloud
Centralized (remote) management	RFC 5412 [168]; Dalvi et al. [169]	Proposed centralized cloud-based approaches for managing WLAN
Real-time reconfiguration	Misra et al. [170]	QoS-guaranteed bandwidth redistribution among gateways in mobile cloud computing
Mobile cloud computing	Survey [171]	A comprehensive survey of mobile cloud computing technology and applications
	Wireless network as a service [172]	Investigates the pragmatism of having wireless networking as a service
	Wireless network cloud [173]	Proposed wireless network cloud (WNC) to operate a wireless access network in cloud mode
	RAN as a service [174]	Proposed an architecture for offering cloud-based RAN-as-a-service
Cloud-based virtualized WNs	CloudMAC [175]	Proposed an OF-based architecture for cloud processing of 802.11 MAC
Network settings		
Cloud-based cognitive WNs	TV white space and clouds [176]	CWNs can perform increasingly complex tasks by offloading these computations to the cloud
Cloud-based cellular WNs	Cloud-based 5G RAN [177]	Proposed using cloud technologies for flexible 5G radio access networks
	Cloud-based LTE [178]	Investigated how cloud computing can be applied to LTE cellular systems. Also, evaluated OpenStack, Eucalyptus, and OpenNebula for this task

The development of *programmable wireless networks* requires changes not only in the control plane but also in the data plane. In particular, the wireless data plane needs to be redesigned to define new, more useful, abstractions. To put things into perspective, the current data plane abstraction offered by OpenFlow supporting switches is based on primitive *match-action* paradigm. To lend greater support to innovations in control plane functionality, the data plane functionality has to evolve to support more sophisticated, and useful, abstractions. Research on newer data plane abstractions is being vigorously pursued with the use of programmable hardware being popularly proposed[179]. The vision of programmable wireless networks thus requires synergy in multiple related domains and would require innovations in both the data plane and the control plane of wireless networks. Using SDN technology for wireless networking will extend the benefits of SDNs - simplification, flexibility, evolvability, and rapid innovation - to wireless environments[139].

In this section, we will initially discuss applications of SWNs in Section 4.1.1. We will then discuss architectures for providing software programmable wireless data planes (Section 4.1.2) as well as how SDN techniques have been applied in various settings of wireless networking (Section 4.1.3). While the finer details are application dependent, all SWNs seek to *i*) attempt to make management of networks a lot more easier, *ii*) allow the same hardware to support multiple wireless protocols, and *iii*) provide an abstraction layer to allow all, or some part, of the wireless architecture to be programmable. These aims are facilitated through the separation of the control and data planes which allows a separate controller to programmatically reconfigure network properties. The theme of providing abstractions for programmability thus pervades the SWN approaches we have discussed in this section.

4.1.1 Applications of SWNs

The biggest promise of SDN is to simplify and improve network management and operations[180]. We discuss some main applications of SDNs in this regard next.

- *Coexistence of diverse technologies*. SDN can be used to integrate diverse wireless technologies and facilitate optimized management and coexistence of diverse wireless technologies. For example, Yap et al.[128] have proposed supporting diverse wireless technologies using the OpenFlow protocol in the *OpenFlow Wireless* project. SDN can also be used to coexistence of diverse technology through the support of *mobile offloading*, or Wi-Fi roaming, using which traffic from a mobile cellular network is outsourced to a Wi-Fi data network to improve the quality-of-experience (QoE) of the end user.
- *Mobile offloading*. SDN can also facilitate handovers between heterogeneous technologies and across service provider. There has been work in using OpenFlow-based controller in managing mobile offloading applications in mobile networks such as 3G or 4G/LTE. There are currently various roaming solutions which facilitate homogeneous Wi-Fi handover (such as IEEE 802.11f, 802.11k, and 802.11r) as well as heterogeneous Wi-Fi handover such as IEEE 802.21. Yi et al.[131] have proposed utilizing SDN techniques to implement heterogeneous technology handover. In another recent work, Ding et al. have proposed SoftOffload[129] as a SDN-based programmable framework that enables mobile traffic offloading.
- *Efficient resource utilization*. We have already pointed out that OpenFlow Wireless[128] can be used to support diverse wireless technologies. In addition, OpenFlow Wireless, among various other SDN-based works, can facilitate more efficient resource utilization and better *infrastructural sharing*. The trend of VWNs, which we will

discuss later in Section 4.3, can complement SDN in facilitating better resource utilization.

- *Better QoS/QoE*. The separation of the control and data planes and the usage of a centralized controller can facilitate better provisioning of network-wide QoS/QoE policies. As a particular example, Sivaraman et al.[163] have proposed to virtualize the access infrastructure of internet service providers (ISPs) and home network using SDN APIs to enable better QoS/QoE through dynamic controlled sharing among user traffic streams.

- *Traffic engineering or adaptive routing*. SDN can be used to dramatically improve the network utilization through better traffic engineering (with Google's traffic engineering work B4[12] being a prime example on how traffic engineering can lead to vastly improved performance). The SDN architecture can facilitate efficient traffic engineering and/or adaptive routing in wireless and mobile networks[130].

- *Distributed mobility management (DMM)*. DMM is a an architectural framework for evolving mobile IP services from the currently deployed mobile core networks (which have serious scalability/reliability issues due to their hierarchical centralized architecture) to a more scalable model with distributed operations[127, 181]. Current IP mobility frameworks force a mobile network operator (MNO) to deploy central entities (like the home agent, local mobility anchor, packet gateway, etc.) in charge of coordinating the mobility management. The current architecture suffers from the problems of suboptimal routing, lack of scalability, and lack of reliability. SDN-based solutions can be used to implement a partially distributed model in which the control plane and data planes are managed separated. The split of control/data plane using SDN principle allows mobility anchors to optimally route traffic[181].

- *Programmatic network service orchestration*. There is also the growing trend of using SDN for performing dynamic programmatic service orchestration such as proposed in the recent work proposed by Schulz-Zander et al.[132].

- *Real-time analytics/reconfiguration*. There is renewed interest in the Internet community in deploying the centralized SDN architecture for optimized reconfiguration of network infrastructure based on real-time analytics performed at the SDN controller(s). In a recent work, Metsch et al.[124] have proposed using real-time analytics for service operation and management in mobile networks.

- *Security enhancement*. The clean separation of data and control planes in the SDN architecture and the consolidation of the control functionality at the centralized controller allows improved implementation of enterprise-level security policy. In this regard, Ding et al.[133] have proposed an SDN-based framework for security enhancement in wireless mobile networks. Some applications of SDN techniques in mobile networks for improving security include *i*) implementation of 'intrusion detection systems' (IDS), *ii*) prevent denial of service (DOS) attacks near the wireless edge, and *iii*) implement secure handoffs in mobile networks[133].

4.1.2 Software-defined programmable wireless data planes

In this section, we will discuss two prominent architectures of software-defined programmable wireless data planes named OpenRadio and OpenRoads, respectively.

4.1.2.0 OpenRadio

The *OpenRadio*system[33] defines a novel design of a wireless data plane that allows programming of the entire wireless stack through a modular and declarative programming interface. OpenRadio proposes to refactor the functionality of wireless protocols into two parts. The *processing plane* deals with programs and algorithms that process data using the underlying hardware. The *decision plane*, on the other hand, is responsible for making logical decisions on the data being processed by the processing planes. It should be observed that the concepts of the processing and decision planes are subtly analogous to that of the data and control planes in the SDN world, respectively. OpenRadio is themed in the mold of both SDRs and SDNs. OpenRadio uses an abstraction layer for managing wireless protocols on generic hardware configured through software like SDRs, while also allowing the separation of protocol from hardware similar to SDNs. OpenRadio can support different wireless protocols, like Wi-Fi, WiMAX, and LTE, etc. though a common hardware, thereby significantly reducing costs and making it easier to configure, optimize, and even define protocols. OpenRadio's major strength is its ability to detach protocol from hardware and to bind the former with software to allow increased flexibility. With newer wireless protocols regularly being rolled out, the ability to reprogram functionality centrally and programmatically is of great convenience. OpenRadio can also be used for cell-size based optimization in cellular networks and for management of frequency spectrum in the presence of multiple heterogeneous cell stations to avoid interference[33].

4.1.2.0 OpenRoads or 'OpenFlow Wireless'

A seminal development in the field of programmable SWNs has been the *OpenRoads*project[123] - known also as *OpenFlow Wireless*[128]. OpenRoads provides a complete platform that can be used to apply SDN principles in wireless environments and thereby create a programmable wireless data plane. One particularly appealing benefit of OpenRoads is that it allows efficient handover between diverse wireless technologies, by 'flattening' multiple vertically integrated wireless technologies, to allow seamless mobility for clients of mobile wireless networks. In[128], the feasibility of this SDN-based approach is explored for mobility management with vertical handovers between IEEE 802.11 and IEEE 802.16 networks. OpenRoads employs OpenFlow and the Simple Network Management Protocol (SNMP) on wireless routers. OpenFlow provides means to manage the forwarding plane while SNMP allows configuration of these wireless devices. FlowVisor and SNMP demultiplexer are used to divide and make the control more scalable. Each controlling flow is given a particular 'flow value' to ensure that different controlling flows are isolated from one another so that only those flows that are intended for particular devices would be installed. High-level control interfaces are used upon OpenFlow to provide communication between different devices, configuration of these wireless devices, and management of flow.

4.1.3 Network-specific categorization of SWNs

We will now detail different wireless networking projects that have incorporated SDN principles. These projects vary in the manner in which they employ SDN principles as well as in the nature of wireless networks (wireless local area networks (WLANs), cellular networks, sensor networks, and personal area networks).

4.1.3.0 WLAN-based SWNs

Odin[134] is a proposed SWN architecture that employs the principles of SDNs in WLANs. In its popular form, WLAN decisions are made by clients and not the WLAN infrastructure itself. For example, a client decides which access point it prefers to join rather than the infrastructure deciding it for the client. In WLANs, association of clients with specific access points keeps on changing with client mobility. This poses a significant challenge to any potential SDN-oriented WLAN architecture as it would be difficult for controller programmers to keep track of the ever changing association between access points and clients. The Odin architecture suggests the usage of light virtual access points (LVAPs). LVAPs virtualize access point-client association and decouples it from physical access points. Whenever a client connects to the WLAN network, it is allotted an identification number on its LVAP that remains fixed regardless of its associated physical access point. The complexities of the physical access point are thus hidden from central controller programmers. The Odin program offers many *advantages*. Odin provides seamless mobility between access points as the need to constantly establish new connections with physical access points changes. Additionally, flexible routing policies further allow load balancing. Furthermore, with an improved overview of the network, it is possible to reduce interference and eliminate issues such as hidden node problems.

4.1.3.0 Cellular mobile SWNs

Recently, there has been significant interest in improving the performance of cellular mobile networks through SDN principles. In particular, frameworks have been proposed that incorporate SDN principles into 3GPP evolved packet core (EPC) mobile carrier networks (the MobileFlow project[135]) and 4G LTE cellular networks (the SoftRAN project[136] and the SoftCell project[137]). The main *advantages* of the cellular mobile SWN approach include better management of radio resources, more flexible routing, real-time monitoring, better mobility support, and the ability to offload data to Wi-Fi networks[182, 183].

There are a few problems with the current LTE architecture: *i*) centralized data flow as all the data passes through the packet gateway (P-GW), *ii*) centralized monitoring and control is not scalable and is expensive, and *iii*) base stations and infrastructure are difficult to configure. The first two problems are related with the central control and monitoring of LTE networks. Thus, a possible solution would need to distribute some of the control and monitoring responsibilities leading to a hybridized control plane. This seems to be a departure from one of the fundamental principles of SDNs, i.e., centralized control. Solution to the third problem lies in adapting an SDN-based architecture so that remote applications may be used for the tasks. As discussed earlier, OpenRadio provides an ideal modular interface to configure base stations remotely and conveniently.

4.1.3.0 Wireless sensor network (WSN)-based SWNs

WSNs have been popular within the research community but have always being considered as an application-specific technology. Treating WSNs as application-specific technology leads to the problem of resource underutilization with potentially multiple application-specific WSNs being deployed over a shared area where a single WSN may have sufficed.

Incorporating SDN in sensor networks would provide solution to these problems[138]. Separation of control and data planes would provide abstraction, helping to manage and control the network. By employing sensor network-based SWN, network controllers could set policies and quality of services to support multiple potential applications. This would also allow usage of the same sensor nodes for multiple application/ purposes. This again increases resource utilization and optimization.

4.1.3.0 Low-rate personal area network (LR-PAN)-based SWNs

SDN attributes can also be used to great advantage in LR-PANs[139]. All LR-PANs essentially employ the same 802.15 data link layer[184, 185]. Differences in higher layers of their respective protocol stacks lead to different LR-PAN protocols such as ZigBee, Bluetooth, etc. This leads to incompatibilities in communication between different nodes. By using the same tools that are used in OpenRadio, we could separate hardware from protocol and use an abstraction layer to program and define different wireless protocols. This would allow us to run different wireless LR-PAN protocols on the same wireless device. In this way, it would be possible for nodes to be dynamically associated with many networks at a time, allowing us to use network resources more efficiently. The separation of data and control planes extends the usual SDN benefits of simpler management, flexible control, and efficient resource utilization to LR-PAN SWNs.

4.2 Trend 2: cognitive wireless networks (CWNs)

It has been highlighted earlier that the predominant focus of most of the existing CRN research has been on enabling intelligent device-level behavior, with a notable exception being some work on *cognitive networks*[63, 64, 140–142]. Cognitive networks, in contrast to cognitive radios, are characterized by their *network-level* intelligent and self-aware behavior. In this paper, we refer to such cognitive networks as 'cognitive wireless networks' (CWNs). CWNs employ a cognition loop (as can be seen in Figure4) to observe the environment, orient itself, and thereafter decide/plan to arrive at the best decision according to the networking/user and application context.

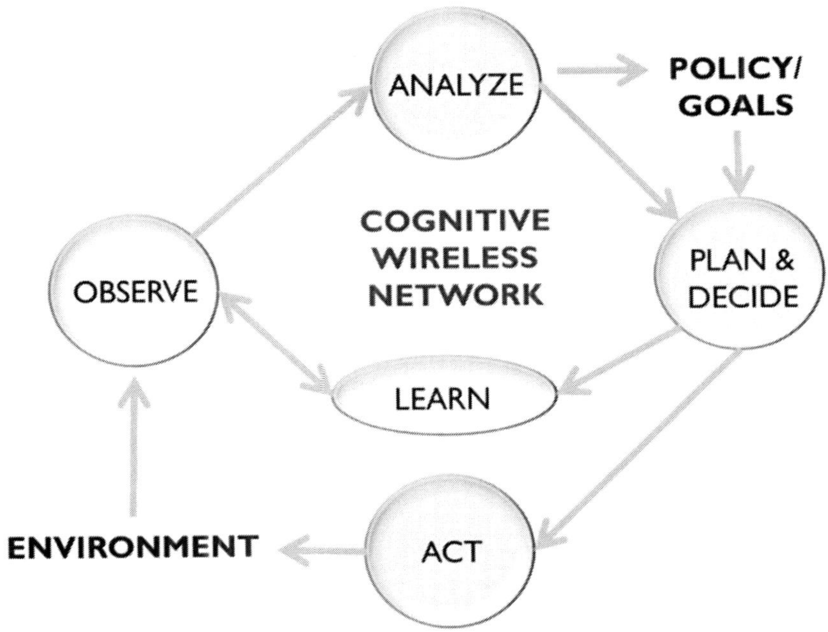

Figure 4. Cognitive wireless networks (CWNs) which include an embedded cognitive engine. Which can observe network conditions, orient itself with the context, learn from experience, and decide to act, all while taking into account end-to-end network goals.

While device-level reconfiguration capabilities (e.g., SDRs and CRs) and network-level reconfiguration capabilities (e.g., SDN) will undoubtedly be a big part of future programmable wireless networking, the resulting programmable wireless architecture will still not be fully automated unless AI techniques are incorporated into the core of the framework. In addition to programmable data plane and programmable control plane, both offering various useful abstractions, future wireless programmable networking also requires a *knowledge plane*[186]. Since CRNs inherently embody AI techniques with wireless communications, it seems natural to explore using CRs, along with the capabilities of SDN and SDRs, and to provide mechanisms for implementing the knowledge plane of future programmable wireless devices.

In future work, the hybrid use of SDN and CRN technology could plausibly lead to a more powerful programmable wireless networking paradigm. While the CogNet project[63] did propose an architectural model of separated control and data planes with an extensible global control plane controlling the separated data planes through an API - which is similar in spirit to the SDN architecture - no concrete proposal has followed this initial conception. This area is ripe for further exploration to exploit the best of CRN and SDN worlds.

We note here that CWNs are *autonomously self-programmed networks*, i.e., CWNs incorporate the ability of autonomously adapting, or *programming*, itself so that operation parameters are optimized to fulfill the desired goals of performance. This conception of programmability in CWNs is distinct from the traditional view of programmability (which also applied to SWNs) which implicitly assumes non-autonomous programming. Future programmable wireless networking will arguably employ both autonomous and non-autonomous programming to reap the benefits of both approaches.

In the following subsections, we will introduce applications of CWNs and will provide a network-specific categorization of CWNs.

4.2.1 Applications of CWNs

- *Dynamic spectrum access*[144]. An important adaptive feature of CWNs is DSA which allows reconfiguration of operating frequency of a SU to allow communication in licensed spectrum. This depends critically on spectrum sensing (which is performed to detect the presence of primary users, or PUs) which is used to ensure that incumbent licensed users, represented by the PUs, are not interfered with. In certain cases, spectrum sensing can be avoided and a database lookup specifying the activity pattern of the PU suffices[144]. Many IEEE standards (such as IEEE 802.11, 802.15, and 802.16) incorporate some basic cognitive radio functionality such as dynamic frequency selection (DFS) and power control (PC) which facilitate *coexisting networks* sharing the same frequency[187]. While DSA is the most popularly cited application of CRNs, developing network-level intelligence in CRNs will enable numerous other applications - including the ability to reprogram itself optimally according to the network conditions.
- *Cognitive networking*. Cognitive networking broadly encompasses models of cognition and learning that have been defined for CRs while emphasizing an end-to-end network-wide scope. Such cognitive networks can perceive current conditions to plan, decide, and act while catering to the *overall network*'s end-to-end goals[64, 142]. Figure4 serves to illustrate the vision of cognitive networks. To help CRNs become *cognitive networks* (CN), it is imperative that intelligence be integrated into the fabric of CRN architecture and protocols across the stack. The cognitive networking vision foresees an intelligent network

capable of setting itself up given high-level instructions and which can continually adapt and manage itself according to changing environmental conditions to optimize network-wide performance metrics.

- The grand vision of cognitive networking can enable various interesting applications. In particular, the applications of *network-optimized MAC protocols*[146, 147], *adaptive routing*[61, 143], and *improved security*[149] hold special promise and can enable significant advancement in the state of the art in wireless networking services.

- *Parameter optimization.* In typical wireless networks, it has been observed that PHY and MAC layers offer many *knobs* that can be tweaked to optimize performance which may be measured through some *meters*. In[60], many examples of knobs and meters at the PHY and MAC layers have been provided in the context of CWNs. Since wireless networks often operate in dynamic conditions, configuring the knobs optimally is not a trivial problem. Various AI-based techniques have been proposed in the literature to assist CWNs in their quest of performing autonomous parameter optimization adaptations in such settings[145]. Apart from artificial intelligence, CWNs also borrow techniques and tools from various other fields such as game theory, control theory, optimization theory, metaheuristics, etc.[59].

- *Enhanced reliability.* To be widely deployed, it is important that next-generation wireless networks provide enhanced reliability independently of the state of the network and the radio transmission medium. Unfortunately, the wireless transmission medium is particularly error prone. CRs have been proposed as a solution to this problem, and they can combat various failures through efficient failure recovery and prevention mechanisms. This can allow CWNs to provide consistent QoS under all circumstances. A detailed survey of various failure conditions and how CRs can be used to provide enhanced reliability in the face of such failures is provided in[148].

- *Adaptive optimization.* CWNs offer the ability to adaptive optimize against the varying wireless environmental conditions using AI-based techniques. CWN typically utilizes adaptive frameworks such as reinforcement learning, learning automata, game theory, etc., for adaptively optimizing the parameters of the network. There are numerous optimization-based applications of CWNs including dynamic spectrum access[144], parameter optimization[145, 188], optimized MAC[146] and routing[61, 189], enhanced reliability[148] and security[149, 189], QoS assurance and management[190, 191], channel assignment[192], etc

4.2.2 Network-specific categorization of CWNs

The use of CR technology has been proposed in various network settings. This allows the CWN trend to manifest itself in different network settings providing in each case the benefits of cognitive networking. There has been work in applying cognitive technology to *IEEE 802.11 networks*. Niyato et al. have proposed using CRs for opportunistic channel selection in IEEE 802.11-based WMNs. There also has been work in implementing white space networking with commodity IEEE 802.11 WLAN NICs[151]. CRs have also been used in the IEEE 802.22 wireless regional area network (WRAN) standard which is proposed to be used in the TV bands. Other white space networking projects include the work of Yuan et al.[37] which has focused on using white spaces in the TV band for DSA. There also has been working in embedding CR technology in WSNs and vehicular networks (VANETs) to help realize *cognitive sensor networks*[154] and *cognitive vehicular networks*, respectively.

4.3 Trend 3: virtualizable wireless networks (VWNs)

Virtualization has transformed both the operational efficiency and the economics of the compute industry and more recently, the datacenter environment. With the growing role of virtualization in networking, it is highly likely that future programmable wireless networks will be virtualization based.

There is also growing interest among major network service providers to decouple the functionality of telecom devices from dedicated devices to enable 'network functions virtualization' (NFV) (as can be seen in Figure5) which will enable implementation of network functions (such as mobile network node, etc.) on servers in datacenters[193]. The concept of NFV extends to any data plane packet processing and control plane function in mobile or fixed networks including, but not limited to, mobile network nodes and traditional switching devices such as routers, switches, home gateways, etc. NFV technology also allows the providers to make the data plane programmable which will facilitate in orchestrating middlebox functionality efficiently.

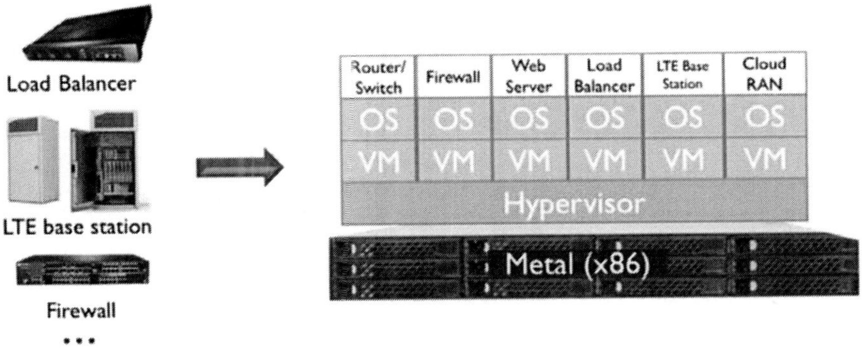

Figure 5.NFV is used to convert fixed-function hardware appliances into virtualized cloud software instances. That run on commodity infrastructure hosted in cloud datacenters.

Traditionally, it is not uncommon for a single packet to undergo processing in the data plane through multiple middleboxes that are used to augment data plane processing by L2 and L3 switches. This is in contrast to the traditional mainframe like model of building middleboxes based on monolithic platforms housed in fixed location in the network (typically at the edge of carrier's core network). This traditional architecture suffered from being rigid, static, and resistant to attempts at automation and orchestration. The functionality of network infrastructure orchestration can be implemented in the NFV framework, on the other hand, by dynamic 'network service chaining' (NSC)[194] through virtual network functions (such as ACLs, QoS, load balancing, etc.) running as virtualized cloud instances in datacenters.

There has been a lot of work done in exploiting SDN principles to enable network virtualization solutions. An initial SDN use case, espoused in[8], was allowing researchers to run experimental protocols in virtualized 'slices' of the production network. While the concept of slicing network through virtualization technology predates SDN - e.g., it has been used in VINI[195], PlanetLab[92], Emulab[196], and more recently in the NSF funded Global Environment for Network Innovations (GENI) project[197] - Casado et al. have extended this idea further by proposing a *network hypervisor* to virtualize the network's forwarding plane[198]. The concept of a network hypervisor is analogous to the traditional hypervisor concept. The network hypervisor implements a network-wide software layer through which it is aimed that multiple virtualized networks, that are decoupled from their underlying hardware instantiation, can be supported. In such an environment, the network state (forwarding and configuration) is decoupled from the underlying hardware, and thus, networks can be created, moved, cloned, and deleted just like VMs in the server world - all in software. A network hypervisor, FlowVisor[94], has been developed for OpenFlow-based SDN environments that allows carving out of virtualized *slices*, that are isolated from each other and controlled by a separate NOS, out of OpenFlow production networks[95].

There also has been work in exploiting SDN in addition to NFV to bulid service chains. OpenFlow has also been used for designing and prototyping high-speed networking by a reusable platform *OpenPipes*[199]. Using an OpenFlow network, new systems can be constructed quickly by OpenPipes, like the Click modular router, by 'plumbing' modules - be they implemented in CPU, FPGA, and ASIC - together in a pipeline. OpenPipes also allows flexible migration of modules (implemented in software or hardware, or both) from one subsystem to another, even in a running system.

It is pertinent here to clarify the connections between SDN and network virtualization (NV). Since both these technologies return some similar benefits, it is a common mistake to equate these two technologies[18]. The SDN architecture is characterized by its emphasis on the separation of the control plane and the data planes and the potential management of multiple data planes through the separated control plane. NV, on the other hand, is characterized by its emphasis on a new 'virtual network' (VN) abstraction that decouples the virtual network from the physical infrastructure. It is another myth that NV is just an application of SDN. It is worth stressing that NV is a *solution* while SDN is an *architecture* - while NV can be implemented more easily using the architectural flexibility offered by SDN, implementation of SDN is not a prerequisite for NV. It has been argued quite convincingly that NV is a distinct entity, important in its own right[18, 103, 200], which may turn out to be even bigger than the current SDN fad sweeping the networking community[103].

In summary, the cloud computing concepts, proposed originally for datacenters, are likely to play a big part in creating future programmable *wireless* networks. In a few years time, it is anticipated that mobile carrier and telecom networks will increasingly emulate datacenters and clouds in their reliance on commodity hardware, virtualization technology, and open software and interfaces in a break from the current scenario in which telecom service providers are full of proprietary vertically integrated hardware appliances. In particular, the combination of cloud computing/network virtualization (including network functions virtualization) allow programmability that leads to unprecedented flexibility in rapidly creating, deploying, and managing novel services in virtualized settings as per the demands of users. This can create a new service-oriented architecture for wireless networking where heterogeneous wireless access technologies may coexist and converge as *extended cloud infrastructure*[201]. For a more comprehensive discussion of wireless virtualization, the interested reader is referred to a specialized book on this topic[201].

In the remainder of this section, we will discuss applications of VWNs and will provide a network-specific characterization of VWN applications.

4.3.1 Applications of VWNs

- *Convergence of different technologies.* Wireless technologies are proliferating at a breakneck pace, and in such a dynamic ecosystem, technologies that facilitate multi-technology convergence will become increasingly important. VWNs are an important multi-faceted trend that promise to facilitate coexistence of diverse technologies.
- *Efficient resource sharing*. An important application pull and use-case of VWNs in general is the convenience of supporting multiple virtual owners (also known as *tenants*) on shared infrastructure. The trend of VWNs provides better support for multi-tenancy and multi-provider and infrastructure sharing, which is convenient both in terms of user experience and economics[202]. Each tenant is provided with a 'slice' - which is an virtualized abstraction of shared wireless infrastructural components - with a committed service-level agreement (SLA) to allow efficient resource sharing of expensive infrastructure. For instance, in the service provider domain, there is a need of supporting multiple virtual network operators (VNOs) - mobile VNOs (MVNOs) in the context of mobile carrier networks[203] - on shared infrastructure to lower the capital and operational expenditures (CAPEX and OPEX). The trend of VWNs can also be used with other trends such as SWNs to facilitate multi-tenancy as has been proposed in the MobileFlow framework[135] to build multi-tenant mobile networks.
- *Virtualization of network components.* There has been a lot of work done in virtualizing various aspects of wireless network components. In practice, the trend of VWNs encompasses many sub-solutions offering varying level of programmability and granularity of control. For example, VWNs encompass in the context of cellular networks, the virtualization of 4G/5G RAN[156], WiMAX BS[157], and LTE[158]; in a similar vein, virtualized abstractions have been proposed for WLAN environments such as the virtual NIC abstraction[159] and the virtual access point (AP) abstraction[160]. As we shall see later in section, virtualization has been applied in the context of WLANs, cellular networks, etc.

4.3.2 Network-specific categorization of VWNs

In the following, we shall discuss the application of virtualization in four environments: *i*) WLANs, *ii*) software-defined VWNs, *iii*) cellular mobile carrier networks, and *iv*) CRNs.

4.3.2.0 WLAN-based VWNs

With the widespread use of IEEE 802.11 WLANs (Wi-Fi) - and the pervasive commissioning of Wi-Fi hotspots in campuses, offices, business centers, airports, shopping centers, etc. - the wireless signal is almost ubiquitously available. There has been a lot of interest in exploiting this common infrastructure to support multi-tenant and multi-provider environments. The concept of 'slices' proposes to provide virtualized environment that runs on top of common shared infrastructure. The general area of network virtualization (NV) is explored in depth in[106], and the interested reader is referred to this paper, and the references therein, for more details. Some prominent contributions that have proposed virtualization for WLANs include wireless virtualization on commodity 802.11 hardware[159], the use of virtual access points (VAPs)[132, 160], virtual Wi-Fi[161], and building multi-purpose access point (MPAP) virtualization architecture[162].

4.3.2.0 Software-defined VWNs

There is a lot of interest in exploiting SDN technologies to enhance the functionality and utility of VWNs. With management protocols such as control and provisioning of wireless access points (CAPWAP)[204], it will be possible to implement infrastructure-wide virtualization using functions consolidated in a centralized software controller. In an SDN context, the FlowVisor[94] project has proposed mechanisms for ensuring isolation needed to support virtualization. In OpenRoads[123], OpenFlow has been extended to wireless APs with forwarding planes virtualized through FlowVisor. In[163], Sivaraman et al. have proposed to address the impasse on service quality in access networks by virtualizing the wireless access network via open APIs using SDN techniques. In a recent work, Schulz-Zander et al.[132] have proposed using SDN techniques using their Odin work for programmatic orchestration of Wi-Fi networks.

4.3.2.0 Cellular mobile VWNs

With the mobile traffic increasing exponentially[1], mobile carrier wireless networks are an attractive setting for wireless virtualization. Various works, focusing on mobile carrier VWNs, have exploited virtualization technology, with some sample works being Costa et al. work on RAN virtualization[156], Bhanage et al. work on WiMAX base station virtualization[157], and Zaki et al. work on LTE virtualization[158].

4.3.2.0 CRN-based VWNs

Relatively less work has been done on CRN-based VWNs. In[165], Tan et al. have presented a novel spectrum virtualization layer, that runs directly below the wireless PHY layer, that presents a seamless interface to the upper layers while allowing DSA. In[205], Nakauchi et al. have proposed AMPHIBIA as a cognitive virtualization platform that can be used to dynamically reconfigure wireless networks by integrated CR and network virtualization technology. More research needs to be done to explore the interplay between cognitive radio technology and virtualization technology.

4.4 Trend 4: cloud-based wireless networks (CbWNs)

The emergence of cloud computing is promising to revolutionize IT by democratizing and commoditizing computing. The main insight of cloud computing is to provide computing as a utility (i.e., using a pay-for-what-you-use pricing model) via a *centralized* setup exploiting economies of scale. The widespread popularity of cloud computing is being driven by advances in computing, virtualization, and web technology, along with support for 'programming the network' through SDN and web APIs. The ability to program the network through the high-level abstractions defined by cloud/SDN API allow unprecedented dynamism in service provisioning, network management, and control. The broad idea behind cloud-based management of wireless networking to implement CbWNs is illustrated in Figure6.

A recent trend in cloud computing is 'mobile cloud computing' (MCC) which integrates cloud computing in mobile environments to enable mobile wireless clients to perform computations in the cloud. A major challenge in MCC is to ensure that the mobile users are provisioned the necessary QoS even if the interfacing gateway changes due to node mobility. Misra et al. have proposed an auction-based QoS utility maximization-based approach for providing QoS-guaranteed bandwidth shifting and retribution in mobile clouds[170]. MCC can be used to overcome various performance impediments in mobile environments due to battery life, storage, bandwidth, and environment. These applications are discussed further in a detailed survey by Dinh et al.[171].

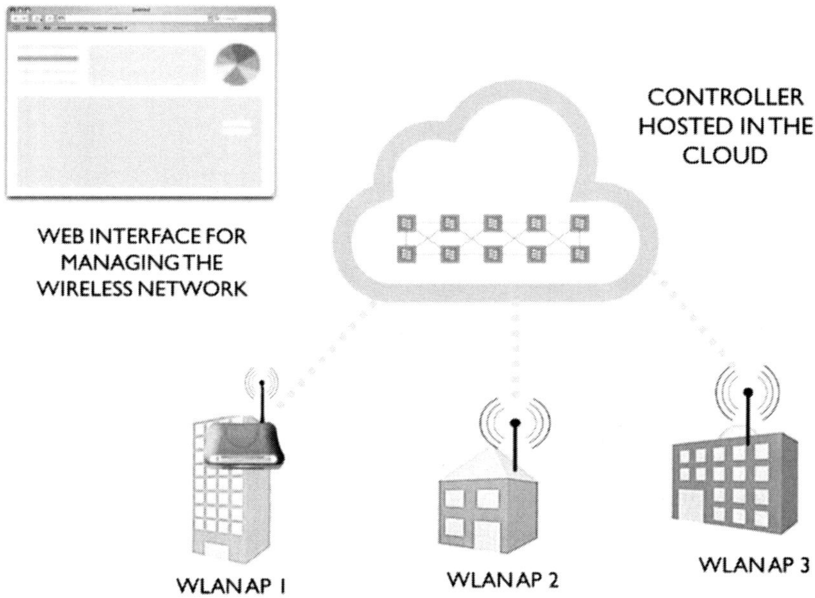

Figure 6. Cloud-based wireless networks (CbWNs).

In the following subsections, we will introduce applications of CbWNs and will provide a network-specific categorization of CbWNs.

4.4.1 Applications overview of CbWNs

- *Centralized management.* A major application of CbWNs is *centralized remote management* of wireless networks using which cloud-based provisioning of services can be performed scalably on the level of a large enterprise and/or service provider. The 'control and provisioning of wireless access points' (CAPWAP) protocol is a control and management (C&M) protocol defined by IETF in RFC 5415[204] which is aimed at relocating some important functionality from the hardware AP equipment to an external controller. The CAPWAP protocol allows a software-defined architecture and is amenable to implementation via the cloud. There exists significant interest in the research community[169] in proposing efficient approaches for central management of Wi-Fi networks. Various industrial solutions, such as Meraki Networks[206] and Aruba Networks[207], have also been proposed that perform cloud-based management of WLANs.

- *Zero-touch auto-configuration.* The centralized management paradigm of CbWNs can allow *zero-touch auto-configuration* of wireless APs regardless of the location. The centralized management of wireless networks also allow cloud-based performance management and the use of *advanced data analytics* for optimization performance including *real-time reconfiguration* of wireless parameters.
- *Context-aware mobile services.* The centralized management paradigm of CbWNs can also facilitate provisioning of context-aware mobile services. As an example, Papakos et al. have proposed Volare as a framework for context-aware adaptive cloud service discovery for mobile systems[208].
- *Optimized operations and management.* Apart from the ability to perform zero-touch auto-configuration, CbWNs can exploit their centralized management architecture to make dynamic changes in real-time in response to changing network conditions through the cloud-based management setup. As an example, Misra et al.[170] have proposed a framework for optimized QoS-guaranteed bandwidth redistribution among gateways in the settings of MCC. In addition, there has been work proposed for Papagianni et al.[209] for optimal allocation of VM resources in cloud computing environments with heterogeneous resources.
- *Computation offloading.* Another major application is computation offloading through which energy or computation-constrained devices can scalably handle complex tasks by offloading their computation to the cloud. The *cloudlet*approach[210] is to provision a trusted well-connected and resource-rich cluster of computers which is available for use by mobile devices that are close. When computation offload to the cloud is undesirable due to any reason (such as cost, delay, etc.), then mobile users can connect to a nearby cloudlet.
- Various other works have been proposed for offloading in a mobile cloud computing environment. Yang et al. have proposed techniques for effective offloading for resource-constrained mobile devices running resource-rich mobile Internet applications[166]. Chun et al. have proposed CloneCloud which allows unmodified mobile applications in an application-level VM to seamlessly offload part of their execution from the mobile devices onto device clones operating in the cloud[167].

Vestin et al. have proposed CloudMAC[175] architecture for managing enterprise WLANs through a virtual AP abstraction provided by the cloud. In the CloudMAC framework, OpenFlow is used to manage the network, and the physical CloudMAC APs are now simplified devices only tasked with relaying MAC frames between the virtual APs and mobile stations while the build of the processing and management functionality is now executed in the cloud.

- *Cloudifying mobile networks.* An extensive amount of work has focused on integrating cloud computing with mobile networks (including 4G[178] and 5G networks[177]) to improve the efficiency of next-generation mobile carrier networks[172]. In particular, there has been work done in extending cloud computing 'as-a-service' provisioning style to radio access networks[174, 177].

4.4.2 Interplay with other programmable wireless trends

In this section, we will describe the interplay of CbWNs with the other three trends of programmable wireless networking (SWNs, CWNs, and VWNs) that we have studied.

4.4.2.0 Cloud-based software-defined WNs

Cloud technology and SDN technology have been used in previous work to introduce programmability features in wireless networking. As part of the NFV drive, there is significant interest in the mobile industry for transforming fixed network functions typically implemented in proprietary hardware devices (such as radio network controller, LTE base stations, etc.) to software-based virtualized instances running in the cloud. In parallel as part of the 'software-defined datacenter' (SDDC) drive, datacenters are increasingly turning to SDN technology for better management.

4.4.2.0 Cloud-based cognitive WNs

There have been a few proposals for utilizing cloud technology to improve the performance in CWNs. In particular, cloud technology has been proposed for use with CWNs for cooperative spectrum sensing in the TV white spaces[176] and for a fast processing of vast volumes of data[211]. In the future, CWNs will perceivably leverage cloud technology increasingly to exploit its scalable computation capability along with its inherent programmability.

4.4.2.0 Cloud-based virtualable WNs

Cloud technology has recently been used with VWNs to provide its scalability and service benefits in such environments. The CloudMAC project has proposed virtualizing APs in a datacenter[175]. Vassilaras et al. present an approach in[172] to provide wireless networking as a service (in utility computing style associated with cloud computing). Other VWN projects that have utilized cloud computing include the 'wireless network cloud' (WNC) project[173]. Due to their popular usage and associated benefits, cloud technology has also been proposed for use with SDRs[212] and in CRs[211].

4.4.3 Network-specific categorization of CbWNs

As discussed in this paper, the use of cloud technology has been proposed in various wireless network settings. The trend of CbWNs is most prominent for WLANs, mobile cellular networks, and CRNs (as has been discussed already in this paper). In addition, cloud computing paradigm has also been applied to the setting of WSNs[213]. In WSNs, the network nodes are limited in their processing capability, battery life, and communication resources. WSNs can utilize the CbWN trend to offload its computation to the cloud and increase its efficiency.

5 OPEN RESEARCH ISSUES AND CHALLENGES

Research in programmable wireless networks has indeed gained momentum as highlighted in this paper. However, many issues still remain to be resolved in order to fully realize the potential benefits associated with this paradigm. We highlight a few important research issues in this domain.

5.1 Building software-defined cognitive wireless networks

An initial promise of SDR was seamless interworking with a plethora of technologies through software-defined adaptations. The vision of CRNs, on the other hand, has evolved from the foundations of SDRs and aims to provide users with seamless holistic experience that integrates potentially heterogeneous technologies. The interplay of cognitive radios with the SDN architecture appears to be a viable and promising hybrid technology that can be used to create programmable wireless networks. We refer to this hybrid technology as 'software-defined cognitive wireless networking'

(SCWN). Numerous interesting use-cases can plausibly emerge if we synergize the mainly centralized operational paradigm of SDNs with the mainly distributed operational paradigm of CRs. While the emphasis of SDN architecture has been on the separation of control and data planes, it is worth exploring if a combined SDN and CR architecture can help realize the vision of programmable wireless networks having a 'knowledge plane' as envisioned by Clark et al.[186].

5.2 Development of wireless-specific network APIs

The utility of any programming paradigm depends greatly on the abstractions available and the standardization of APIs. For the vision of programmable wireless networking to become established, it is important that there is progress in developing useful network APIs offering sophisticated high-level abstractions for wireless networking. In the SDN community, while numerous southbound APIs have been proposed, there is a lack of clarity and consensus about what abstractions and interface a northbound API should expose. While OpenFlow, an example southbound API, has become wildly popular, it is quite primitive in its functionality[17] - the metaphor of Assembly language programming is often used to describe direct OpenFlow programming. To increase the rate of innovation, it is important that higher level languages are developed that can be used by network programmers through a high-level standardized interface. In this regard, more work is required for both southbound and northbound API. Since a network application programmer will interface with the controller through a northbound API, quick consensus on the development of an effective northbound network API is of paramount importance.

5.3 Integrating wireless and cloud technologies

The paradigm of cloud computing - which is itself based on web technology, programmability through APIs and virtualization - is likely to play a big role in future wireless programmable networks. There is already a lot of work on integration of SDN and cloud technology[10]. Future designers of programmable wireless devices will be well-served by exploiting the performance and scalability advantages offered by cloud computing in their designs. Already, there has been significant work in incorporating cloud technology into existing frameworks (as has been discussed in Section 4.4), and this trend looks set to continue well into the future.

5.4 Wireless internet of things

While traditionally, the Internet communication paradigm has revolved around human consumption of Internet services, it is envisioned that in the future, networking will create many novel services through machine-to-machine communication by creating an Internet of things[214]. In particular, the convenience of untethered mobile communication facilitated by wireless communication can create a future wireless Internet of things. This is a potential future research area envisioned to have a significant impact on the community and the way stakeholders interact with the services provided by the Internet.

5.5 Balancing centralized and distributed paradigms

SDNs have proposed a separation of the control plane and data plane with the control logic placed on a separated controller. Pragmatic concerns about performance and security have dictated that this controller be implemented as a distributed system. Hence, although the controller is 'logically centralized', it is implemented as a distributed system - this has led to coining of the awkward term 'logically centralized control'. This draws our attention to the perennial tension between distributed and centralized control. The Internet's community has traditionally favored the distributed control paradigm due to its scalability. However, architectural ossification and inflexible network control has led through a rethink to the centralized SDN paradigm. Like the pre-SDN era, not all tasks can be, or should be, exclusively centralized or distributed. The modern shift to a centralized paradigm is sometimes codified in the mantra, 'centralize what you can, distribute what you must'. Finding the right balance between centralized and distributed control is an important fundamental design choice which needs careful evaluation. Also, it is important to address the scalability and performance concerns associated with the centralized control to make it a viable practical architecture. Future wireless networks will have to seamlessly manage the delicate balance between the centralized and distributed control paradigms of current technologies - such as WiMAX and Wi-Fi - and the centralized aspects and distributed aspects of future network architecture such as SWNs and CWNs, respectively.

6 CONCLUSIONS

In this paper, we have provided a general overview of architectural techniques useful for building next-generation programmable wireless networks. We have seen that the seemingly disparate schemes of software-defined radio, cognitive radio networking, software-defined networking, and programmable wireless processors are in fact themed on a common goal of creating flexible 'programmable wireless networks'. A self-contained tutorial is provided for these architectures followed by a detailed survey of their applications. We also proposed synergizing these technologies into newer hybrid technologies. In particular, we proposed a new framework of *software-defined cognitive wireless networking* which will employ both SDN and CRN principles to potentially open up new use cases. We have also highlighted important research issues in this field and identified future research work.

NOTES

[a] Tennenhouse et al. also made the sobering observation in[16] that object-oriented approaches to networking are proposed every 5 to 10 years with little impact on mainstream research. Time will tell how transformative the modern proposals (e.g., SDN languages approach) will be in the long run.

[b] James Hamilton, the architect of Amazon's cloud, made the now famous remark in 2010 that 'datacenter networks are in my way' bemoaning the lack of network programmability. See Hamilton's blog post at http://perspectives.mvdirona.com/2010/10/31/DatacenterNetworksAreInMyWay.aspxfor more perspective.

[c] David Clark, the former chair of the Internet Architecture Board (IAB), summarized the role of software implementations in the networking community ethos when he said, 'We reject kings, presidents and voting. We believe in rough consensus and *running code*'.

[d] The centralized SDN network controller can itself be built as a distributed system to be scalable and avoid a single point of failure.

[e]http://www.opendaylight.org.

[f] OpenFlow should not be confused with the overall SDN *architecture* since OpenFlow, popular as it is, is but only one *protocol* that implements the southbound API as envisioned in the SDN architecture.

AUTHORS' INFORMATION

Junaid Qadir is an Assistant Professor at the School of Electrical Engineering and Computer Sciences (SEECS), National University of Sciences and Technology (NUST), Pakistan. He is also the Director of the Cognet Lab at SEECS. He completed his BE in Electrical Engineering from UET, Lahore, Pakistan and his PhD from University of New South Wales, Australia in 2008. His research interests include cognitive radio networks, wireless networks, and software-defined networks.

Nadeem Ahmed received the BE degree from the University of Engineering and Technology, Lahore, Pakistan and the MS and PhD degrees in computer sciences from University of New South Wales (UNSW), Sydney, Australia in 2000 and 2007, respectively. He is currently an Assistant Professor at the School of Electrical Engineering and Computer Sciences (SEECS), National University of Sciences and Technology (NUST), Pakistan. His research interests include wireless sensor networks, software-defined networks and mobile ad-hoc networks.

Nauman Ahad is presently working as a software engineer with xFlow Research Inc. Nauman graduated from the School of Electrical Engineering and Computer Sciences (SEECS), National University of Sciences and Technology (NUST), Pakistan with a BE in Electrical Engineering in 2014. He was an internee in the Cognet research lab at SEECS, NUST, in the summer of 2013.

REFERENCES

1. Raychaudhuri D, Mandayam NB: Frontiers of wireless and mobile communications.Proc. IEEE 2012, 100(4):824-840.
2. Akamai State of Internet Report Q1 2013.http://www.akamai.com/stateoftheinternet/ . Accessed 30 Sept 2013
3. Koponen T, Shenker S, Balakrishnan H, Feamster N, Ganichev I, Ghodsi A, Godfrey P, McKeown N, Parulkar G, Raghavan B, Rexford J, Arianfar S, Kuptsov D: Architecting for innovation. Comput. Comm. Rev 2011, 41(3):24-36. 10.1145/2002250.2002256
4. Liskov B: The power of abstraction. In Distributed Computing. Springer, Berlin Heidelberg; 2010:3-3.
5. Rexford J: The networking philosopher's problem. Comput. Comm. Rev 2011, 41(3):5-9. 10.1145/2002250.2002252

6. Scott Shenker's talk at Open Networking Summit 2011.http://www.youtube.com/watch?v=YHeyuD89n1Y [Online]. Accessed 12 Sept 2013
7. Rexford J, Zave P: Report of the DIMACS working group on abstractions for network services, architecture, and implementation. Comput. Comm. Rev 2012, 43(1):56-59. 10.1145/2427036.2427047
8. McKeown N, Anderson T, Balakrishnan H, Parulkar G, Peterson L, Rexford J, Shenker S, Turner J: OpenFlow: enabling innovation in campus networks. Comput. Comm. Rev 2008, 38(2):69-74. 10.1145/1355734.1355746
9. Andreessen M: Why software is eating the world. Wall Street Journal 2011. [online], Accessed 22 Sept 2013 http://online.wsj.com/article/SB10001424053111903480904576512250915629460.html [online], Accessed 22 Sept 2013
10. Openstack's neutron plugin for "network-as-a-service" [online]. Accessed 22 Sept 2013 https://wiki.openstack.org/wiki/Neutron [online]. Accessed 22 Sept 2013
11. Nunes B, Mendonca M, Nguyen X, Obraczka K, Turletti T: A survey of software-defined networking: past, present, and future of programmable networks. IEEE Commun. Surv. Tutorials 2014, PP(99):1-18. IEEE, NY USA
12. Jain S, Kumar A, Mandal S, Ong J, Poutievski L, Singh A, Venkata S, Wanderer J, Zhou J, Zhu M, Zolla J, Hölzle U, Stuart S, Vahdat A: B4: Experience with a globally-deployed software defined WAN. In Proceedings of the ACM SIGCOMM 2013. ACM, NY USA; 2013:3-14.
13. Cisco's use of programmable networks for powering 'Internet of Everything' [Online]. Accessed 1 Oct 2013 http://tinyurl.com/o9y4mza [Online]. Accessed 1 Oct 2013
14. Campbell AT, De Meer HG, Kounavis ME, Miki K, Vicente JB, Villela D: A survey of programmable networks. Comput. Comm. Rev 1999, 29(2):7-23. 10.1145/505733.505735
15. Smith JM, Nettles SM: Active networking: one view of the past, present, and future. IEEE Trans. Syst. Man Cybern. C Appl. Rev 2004, 34(1):4-18.
16. Tennenhouse DL, Wetherall DJ: Towards an active network architecture. In DARPA Active Networks Conference and Exposition, 2002. Proceedings. IEEE, NY USA; 2002:2-15.
17. Foster N, Guha A, Reitblatt M, Story A, Freedman MJ, Katta NP, Monsanto C, Reich J, Rexford J, Schlesinger C: D Walker, R Harrison, Languages for software-defined networks. IEEE Comm. Mag 2013, 51(2):128-134.

18. Feamster N, Rexford J, Zegura E, The road to SDN: Queue. 2013, 11(12):20. 10.1145/2559899.2560327
19. Campbell AT, Katzela I, Miki K, Vicente J: Open signaling for ATM, internet and mobile networks (OPENSIG'98). Comput. Comm. Rev 1999, 29(1):97-108. 10.1145/505754.505762
20. Burgess DA, Samra HS: The OpenBTS Project. 2008.http://openbts.sourceforge.net [Online]. Accessed 22 Sept 2013
21. Reed JH: Software Radio: A Modern Approach to Radio Engineering. Prentice Hall Professional, USA; 2002.
22. Tuttlebee WH: Software Defined Radio: Enabling Technologies. Wiley, USA; 2003.
23. Ulversoy T: Software defined radio: challenges and opportunities. IEEE Commun. Surv. Tutorials 2010, 12(4):531-550. IEEE, NY USA
24. Partridge C: Realizing the future of wireless data communications. Commun. ACM 2011, 54(9):62-68. ACM, NY USA 10.1145/1995376.1995395
25. Ettus Research, acquired recently by National Instruments (NI) [Online]. Accessed 12 Sept 2013 http://www.ettus.com [Online]. Accessed 12 Sept 2013
26. GNU Radio [Online]. Accessed 12 Sept 2013 http://www.gnuradio.com [Online]. Accessed 12 Sept 2013
27. OSSIE Virginia Tech [Online]. Accessed 12 Sept 2013 http://ossie.wireless.vt.edu/ [Online]. Accessed 12 Sept 2013
28. Mackenzie P: Software and Reconfigurability for Software Radio Systems. PhD thesis, Ph. D dissertation, Trinity College Dublin, Ireland; 2004.
29. Ettus M: USRP User's and Developer's Guide. Ettus Research LLC [Online]Accessed 12 Sept 2013 http://www.olifantasia.com/gnuradio/usrp/files/usrp_guide.pdf [Online]Accessed 12 Sept 2013
30. Khattab A, Camp J, Hunter C, Murphy P, Sabharwal A, Knightly EW: WARP: a flexible platform for clean-slate wireless medium access protocol design. ACM SIGMOBILE Mobile Comput. Commun. Rev 2008, 12(1):56-58. ACM 10.1145/1374512.1374532
31. Ng MC, Fleming KE, Vutukuru M, Gross S, Balakrishnan H, Arvind: Airblue: a system for cross-layer wireless protocol development. Proceedings of the 6th ACM/IEEE Symposium on Architectures for Networking and Communications Systems 2010, 4-4. ACM
32. Tan K, Liu H, Zhang J, Zhang Y, Fang J, Voelker GM: Sora: high-performance software radio using general-purpose multi-core processors. Commun. ACM 2011, 54(1):99-107. ACM 10.1145/1866739.1866760

33. Bansal M, Mehlman J, Katti S, Levis P: Openradio: a programmable wireless dataplane. In Proceedings of the First Workshop on Hot Topics in Software Defined Networks. ACM; 2012:109-114.
34. Stewart G, Gowda M, Mainland G, Radunovic B, Vytiniotis D: Ziria: wireless programming for hardware dummies. Tech. Rep MSR-TR-2013-135, Microsoft Systems, November 2013
35. Stewart G, Gowda M, Mainland G, Radunovic B, Vytiniotis D, Patterson D: Ziria: language for rapid prototyping of wireless PHY. In Proceedings of the 2014 ACM conference on SIGCOMM. ACM; 2014:357-358.
36. Mishra SM, Cabric D, Chang C, Willkomm D, Van Schewick B, Wolisz S, Brodersen B: A real time cognitive radio testbed for physical and link layer experiments. In New Frontiers in Dynamic Spectrum Access Networks (DYSPAN), 2005. First IEEE International Symposium on. IEEE, NY USA; 2005:562-567.
37. Yuan Y, Bahl P, Chandra R, Chou PA, Ferrell JI, Moscibroda T, Narlanka S, Wu Y: KNOWS: cognitive radio networks over white spaces. In New Frontiers in Dynamic Spectrum Access Networks, 2007, DySPAN 2007. 2nd IEEE International Symposium on. IEEE; 2007:416-427.
38. Miljanic Z, Seskar I, Le K, Raychaudhuri D: The WINLAB network centric cognitive radio hardware platform: WiNC2R. Mobile Network Appl 2008, 13(5):533-541. 10.1007/s11036-008-0082-0
39. Lazar AA, Lim K-S, Marconcini F: xbind: The system programmer's manual. report # 452-96-16. tech. rep., Center for Telecommunications Research (CTR), Columbia University, 1996
40. Rooney S, van der Merwe JE, Crosby SA, Leslie IM: The tempest: a framework for safe, resource assured, programmable networks. IEEE Comm. Mag 1998, 36(10):42-53. 10.1109/35.722136
41. Angin O, Campbell AT, Kounavis ME, Liao R-F: The mobiware toolkit: programmable support for adaptive mobile networking. IEEE Pers. Comm 1998, 5(4):32-43. 10.1109/98.709367
42. Campbell A, De Meer H, Kounavis M, Miki K, Vicente J, Villela D: The genesis kernel: a virtual network operating system for spawning network architectures. In Open Architectures and Network Programming Proceedings, 1999. OPENARCH'99. 1999 IEEE Second Conference on. IEEE, NY USA; 1999:115-127.
43. Morris R, Kohler E, Jannotti J, Kaashoek MF: The click modular router. In ACM SIGOPS Operating Systems Review. ACM, NY USA; 1999:217-231.
44. Handley M, Hodson O, Kohler E: XORP: an open platform for network research. Comput. Comm. Rev 2003, 33(1):53-57. ACM, NY USA 10.1145/774763.774771

45. Neufeld M, Fifield J, Doerr C, Sheth A, Grunwald D: SoftMAC: flexible wireless research platform. In ACM Workshop on Hot Topics in Networks (HotNets). ACM, NY USA; 2005.
46. Doerr C, Neufeld M, Fifield J, Weingart T, Sicker DC, Grunwald D: MultiMAC - an adaptive MAC framework for dynamic radio networking. In New Frontiers in Dynamic Spectrum Access Networks (DYSPAN), 2005. First IEEE International Symposium on. IEEE, NY USA; 2005:548-555.
47. Lockwood JW, McKeown N, Watson G, Gibb G, Hartke P, Naous J, Raghuraman R, Luo J: NetFPGA–an open platform for gigabit-rate network switching and routing. In Microelectronic Systems Education, 2007. MSE'07. IEEE International Conference on. IEEE, NY USA; 2007:160-161.
48. Naous J, Gibb G, Bolcuki S, McKeown N: NetFPGA: reusable router architecture for experimental research. In Proceedings of the ACM Workshop on Programmable Routers for Extensible Services of Tomorrow. ACM, NY USA; 2008:1-7.
49. Dobrescu M, Egi N, Argyraki K, Chun B-G, Fall K, Iannaccone G, Knies A, Manesh M, Ratnasamy S: Routebricks: exploiting parallelism to scale software routers. In Proceedings of the 22nd ACM Symposium on Operating Systems Principles (SIGOPS). ACM, NY USA; 2009:15-28.
50. Anwer MB, Motiwala M, Tariq Mb, Feamster N: Switchblade: a platform for rapid deployment of network protocols on programmable hardware. Comput. Comm. Rev 2010, 40(4):183-194. ACM, NY USA 10.1145/1851275.1851206
51. Ansari J, Zhang X, Achtzehn A, Petrova M, Mahonen P: Decomposable MAC framework for highly flexible and adaptable MAC realizations. In New Frontiers in Dynamic Spectrum, 2010 IEEE Symposium on. IEEE, NY USA; 2010:1-2.
52. Zhang X, Ansari J, Yang G, Mahonen P: Trump: supporting efficient realization of protocols for cognitive radio networks. In New Frontiers in Dynamic Spectrum Access Networks (DySPAN) 2011 IEEE Symposium on. IEEE, NY USA; 2011:476-487.
53. Tinnirello I, Bianchi G, Gallo P, Garlisi D, Giuliano F, Gringoli F: Wireless MAC processors: programming MAC protocols on commodity hardware. In INFOCOM, 2012, Proceedings, IEEE. IEEE, NY USA; 2012:1269-1277.
54. Bianchi G, Gallo P, Garlisi D, Giuliano F, Gringoli F, Tinnirello I: MAClets: active MAC protocols over hard-coded devices. In Proceedings of the 8th International Conference on Emerging Networking Experiments and Technologies. ACM, NY USA; 2012:229-240.

55. Jondral FK: Software-defined radio: basics and evolution to cognitive radio. EURASIP J. Wireless Commun. Netw 2005, 2005(3):275-283. Springer Heidelberg
56. Mitola III J: Cognitive Radio Architecture: The Engineering Foundations of Radio XML. John Wiley & Sons, USA; 2006.
57. He A, Bae KK, Newman TR, Gaeddert J, Kim K, Menon R, Morales-Tirado L, Neel JJ, Zhao Y, Reed JH, Tranter W: A survey of artificial intelligence for cognitive radios. IEEE Trans. Veh. Tech 2010, 59(4):1578-1592.
58. Akyildiz I, Lee W, Vuran M, Mohanty S: Next generation/dynamic spectrum access/cognitive radio wireless networks: a survey. Comput. Network 2006, 50(13):2127-2159. 10.1016/j.comnet.2006.05.001
59. Haykin S: Cognitive radio: brain-empowered wireless communications. IEEE J. Sel. Area Comm 2005, 23(2):201-220.
60. Fette BA: Cognitive Radio Technology. Academic Press, Elsevier, USA; 2009.
61. Qadir J: Artificial intelligence based cognitive routing for cognitive radio networks. arXiv preprint arXiv:1309.0085 2013. [Online]. Accessed 12 Sept 2013 http://arxiv.org/abs/1309.0085 [Online]. Accessed 12 Sept 2013
62. Baldini G, Sturman T, Biswas AR, Leschhorn R, Gódor G, Street M: Security aspects in software defined radio and cognitive radio networks: a survey and a way ahead. IEEE Commun. Surv. Tutorials 2012, 14(2):355-379. IEEE, NY USA
63. Raychaudhuri D, Mandayam NB, Evans JB, Ewy BJ, Seshan S, Steenkiste P: Cognet: an architectural foundation for experimental cognitive radio networks within the future internet. In Proceedings of first ACM/IEEE International Workshop on Mobility in the Evolving Internet Architecture. ACM, NY USA; 2006:11-16.
64. Thomas RW, Friend DH, DaSilva LA, MacKenzie AB: Cognitive networks. In Cognitive Radio, Software Defined Radio, and Adaptive Wireless Systems. Edited by: Arslan H. Springer, Netherlands; 2007:17-41.
65. Nychis G, Hottelier T, Yang Z, Seshan S, Steenkiste P: Enabling MAC protocol implementations on software-defined radios. In USENIX NSDI. USENIX; 2009:91-105.
66. Gallo P, Krasilov A, Lyakhov A, Tinnirello I, Bianchi G: Breaking layer 2: a new architecture for programmable wireless interfaces. In ICT Convergence (ICTC), 2012 International Conference on. IEEE, NY USA; 2012:342-347.

67. Messerschmitt D: Rethinking components: from hardware and software to systems. Proc. IEEE 2007, 95(7):1473-1496. IEEE, NY USA
68. Gazis V, Patouni E, Alonistioti N, Merakos L: A survey of dynamically adaptable protocol stacks. IEEE Commun. Surv. Tutorials 2010, 12(1):3-23. IEEE, NY USA
69. Tennenhouse DL, Smith JM, Sincoskie WD, Wetherall DJ, Minden GJ: A survey of active network research. IEEE Commun. Mag 1997, 35(1):80-86. IEEE, NY USA
70. Lazar AA: Programming telecommunication networks. IEEE Netw 1997, 11(5):8-18. IEEE, NY USA 10.1109/65.620517
71. Wetherall D, Legedza D, Guttag J: Introducing new internet services: why and how. IEEE Netw 1998, 12(3):12-19. IEEE, NY USA 10.1109/65.690955
72. Wetherall D: Active network vision and reality: lessons from a capsule-based system. In DARPA Active NEtworks Conference and Exposition, 2002. Proceedings. IEEE, NY USA; 2002:25-40.
73. Psounis K: Active networks: applications, security, safety, and architectures. IEEE Commun. Surv. Tutorials 1999, 2(1):2-16. IEEE, NY USA
74. ActiveWare project at MIT . Accessed 12 Sept 2013 http://www.sds.lcs.mit.edu/activeware/
75. CANES project at Georgia Tech [Online]. Accessed 22 Sept 2013 http://www.cc.gatech.edu/projects/canes/ [Online]. Accessed 22 Sept 2013
76. SwitchWare project at University of Pennsylvania [Online]. Accessed 22 Sept 2013 http://www.cis.upenn.edu/switchware/ [Online]. Accessed 22 Sept 2013
77. Wetherall D, Guttag J, Tennenhouse D: Ants: network services without the red tape. Computer 1999, 32(4):42-48. IEEE, NY USA 10.1109/2.755004
78. ANTS project at the University of Washington [Online]. Accessed 22 Sept 2013 http://www.cs.washington.edu/research/networking/ants/ [Online]. Accessed 22 Sept 2013
79. Doria A, Hellstrand F, Sundell K, Worster T: General switch management protocol (GSMP) v3.IETF Request for Comments: 3292. 2002.https://www.ietf.org/rfc/rfc3292.txt [Online]. Accessed 12 Sept 2013
80. Yang L, Dantu R, Anderson T, Gopal R: RFC 3746: Forwarding and control element separation (FORCES) framework. Network Working Group 2004., 4: [Online]. Accessed 12 Sept 2013 http://tools.ietf.org/html/rfc3746 [Online]. Accessed 12 Sept 2013

81. Feamster N, Balakrishnan H, Rexford J, Shaikh A, Van Der Merwe J: The case for separating routing from routers. In Proceedings of the ACM SIGCOMM Workshop on Future Directions in Network Architecture. ACM, NY USA; 2004:5-12.
82. Lakshman T, Nandagopal T, Ramjee R, Sabnani K, Woo T: The softrouter architecture. In Proc. ACM SIGCOMM Workshop on Hot Topics in Networking. ACM, NY USA; 2004-2004.
83. Greenberg A, Hjalmtysson G, Maltz DA, Myers A, Rexford J, Xie G, Yan H, Zhan J, Zhang H: A clean slate 4d approach to network control and management. Comput. Comm. Rev 2005, 35(5):41-54. ACM, NY USA
84. Lakshminarayanan K, Stoica I, Shenker S, Rexford J: Routing as a Service. Computer Science Division, Technical Report No. UCB/EECS-2006-19. University of California Berkeley; 2004.
85. Farrel A, Vasseur J-P, Ash J: A path computation element (pce)-based architecture. tech. rep., RFC 2006, 4655.
86. Van der Merwe J, Cepleanu A, D'Souza K, Freeman B, Greenberg A, Knight D, McMillan R, Moloney D, Mulligan J, Nguyen H, Nguyen M, Ramarajan A, Saad S, Satterlee M, Spencer T, Toll D, Zelingher S: Dynamic connectivity management with an intelligent route service control point. In Proceedings of the 2006 SIGCOMM Workshop on Internet Network Management. ACM, NY USA; 2006:29-34.
87. Casado M, Garfinkel T, Akella A, Freedman MJ, Boneh D, McKeown N, Shenker S: SANE: a protection architecture for enterprise networks. In ACM USENIX Security Symposium. ACM; 2006.
88. Casado M, Freedman MJ, Pettit J, Luo J, McKeown N, Shenker S: Ethane: taking control of the enterprise. Comput. Commun. Rev 2007, 37: 1-12. ACM 10.1145/1282427.1282382
89. Juniper Networks . Accessed 12 Sept 2013 http://www.juniper.net
90. Cisco's One Platform Kit (onePK) . Accessed 12 Sept 2013 http://www.cisco.com/en/US/prod/iosswrel/onepk.html
91. OpenStack's Neutron Accessed 12 Sept 2013 https://wiki.openstack.org/wiki/Neutron?
92. Chun B, Culler D, Roscoe T, Bavier A, Peterson L, Wawrzoniak M, Bowman M: Planetlab: an overlay testbed for broad-coverage services. Comput. Comm. Rev 2003, 33(3):3-12. ACM, NY USA 10.1145/956993.956995
93. Anderson T, Peterson L, Shenker S, Turner J: Overcoming the internet impasse through virtualization. Computer 2005, 38(4):34-41. IEEE, NY USA
94. Sherwood R, Gibb G, Yap K-K, Appenzeller G, Casado M, McKeown N, Parulkar G: Flowvisor: a network virtualization layer. OpenFlow

Switch Consortium, Tech. Rep 2009. [Online]. Accessed 12 Sept 2013 http://archive.openflow.org/downloads/technicalreports/openflow-tr-2009-1-flowvisor.pdf [Online]. Accessed 12 Sept 2013
95. Sherwood R, Chan M, Covington A, Gibb G, Flajslik M, Handigol N, Huang T-Y, Kazemian P, Kobayashi M, Underhill D, Yap K, Appenzeller G, McKeown N: Carving research slices out of your production networks with Openflow. Comput. Comm. Rev 2010, 40(1):129-130. ACM, NY USA 10.1145/1672308.1672333
96. Guo C, Lu G, Wang HJ, Yang S, Kong C, Sun P, Wu W, Zhang Y: Secondnet: a data center network virtualization architecture with bandwidth guarantees. In Proceedings of the 6th ACM International Conference on emerging Networks EXperiments and Technologies (Co-NEXT'10). ACM, NY USA; 2010:1-12.
97. Koponen T, Amidon K, Balland P, Casado M, Chanda A, Fulton B, Ganichev I, Gross J, Gude N, Ingram P, Jackson E, Lambeth A, Lenglet R, Li S, Padmanabhan A, Pettit J, Pfaff B, Ramanathan R, Shenker S, Shieh A, Stribling J, Thakkar P, Wendlandt D, Yip A, Zhang R: Network virtualization in multi-tenant datacenters. In USENIX Networked Systems Design and Implementation (NSDI). USENIX; 2014.
98. Armbrust M, Fox A, Griffith R, Joseph AD, Katz R, Konwinski A, Lee G, Patterson D, Rabkin A, Stoica I, Zaharia M: A view of cloud computing. Commun. ACM 2010, 53(4):50-58. ACM, NY USA 10.1145/1721654.1721672
99. Dikaiakos MD, Katsaros D, Mehra P, Pallis G, Vakali A: Cloud computing: distributed internet computing for it and scientific research. IEEE Internet Comput 2009, 13(5):10-13. IEEE, NY USA
100. Casado M, Koponen T, Shenker S, Tootoonchian A: Fabric: a retrospective on evolving SDN. In Proceedings of the First Workshop on Hot topics in Software Defined Networks. ACM, NY USA; 2012:85-90.
101. What's Behind Network Downtime? - IBM www-05.ibm.com/uk/juniper/pdf/200249.pdf?. Accessed 30 Sept 2013
102. Statistics on "virtual server access port" vs. "physical server access port"; Source: Crehan Research Inc. and VMWare Estimates, March 2003 . Accessed 30 Sept 2013 http://www.crehan.com?
103. Scott Shenker's talk at Stanford 2013.http://www.youtube.com/watch?v=WabdXYzCAOU [Online]. Accessed 12 Sept 2013
104. Pfaff B, Pettit J, Amidon K, Casado M, Koponen T, Shenker S: Extending networking into the virtualization layer. In ACM Workshop on Hot Topics in Networks (Hotnets). ACM, NY USA; 2009-2009.

105. Wang A, Iyer M, Dutta R, Rouskas GN, Baldine I: Network virtualization: technologies, perspectives, and frontiers. J. Lightwave Technol 2013, 31(4):523-537.
106. Chowdhury N, Boutaba R: A survey of network virtualization. Comput. Network 2010, 54(5):862-876. 10.1016/j.comnet.2009.10.017
107. McKeown N: Software-defined networking. INFOCOM Keynote Talk, Apr 2009 [Online]. Accessed 12 Sept 2013 http://infocom2009.ieee-infocom.org/keynotes.html [Online]. Accessed 12 Sept 2013
108. Nicira [Online]. Accessed 12 Sept 2013 https://www.vmware.com/products/nsx [Online]. Accessed 12 Sept 2013
109. Contrail [Online]. Accessed 12 Sept 2013 http://www.contrail.com [Online]. Accessed 12 Sept 2013
110. Chung J, Gonzalez G, Armuelles I, Robles T, Alcarria R, Morales A: Experiences and challenges in deploying OpenFlow over a real wireless mesh network. In Communications (LATINCOM) 2012 IEEE Latin-America Conference on. IEEE, NY USA; 2012:1-5.
111. IW Group: Interface to the routing system (I2RS). Internet Engineering Task Force, 2014. 2014.http://datatracker.ietf.org/wg/i2rs/charter/ [Online] Available:. Accessed 12 Sept 2013
112. Alimi R, Penno R, Yang Y: Application-layer traffic optimization protocol. Internet Draft, Internet Engineering Task Force, March 2014. 2014.https://datatracker.ietf.org/doc/draft-ietf-alto-protocol/ [Online]. Available:. Accessed 12 Sept 2013
113. Lara A, Kolasani A, Ramamurthy B: Network innovation using OpenFlow: a survey. IEEE Commun. Surv. Tutorials 2013., 16: IEEE, NY USA
114. Gude N, Koponen T, Pettit J, Pfaff B, Casado M, McKeown N, Shenker S: NOX: towards an operating system for networks. Comput. Commun. Rev 2008, 38(3):105-110. ACM, NY USA 10.1145/1384609.1384625
115. Koponen T, Casado M, Gude N, Stribling J, Poutievski L, Zhu M, Ramanathan R, Iwata Y, Inoue H, Hama T, Shenker S: Onix: a distributed control platform for large-scale production networks. In USENIX OSDI. USENIX; 2010:1-6.
116. Open networking operating system . Accessed 12 Sept 2013 http://onlab.us/tools.html
117. Nascimento MR, Rothenberg CE, Salvador MR, Magalhães MF: Quagflow: partnering quagga with openflow. Comput. Commun. Rev 2010, 40: 441-442. ACM, NY USA 10.1145/1851275.1851252
118. Quagga Routing Suite [Online]. Accessed 12 Sept 2013 http://www.quagga.net [Online]. Accessed 12 Sept 2013

119. Nascimento MR, Rothenberg CE, Salvador MR, Magalhaes MF, Corrêa CN, de Lucena SC: The RouteFlow approach to IP routing services on software-defined networks. Brazilian Symposium on Computer Networks and Distributed Systems (SBRC) 2011. [Online]_{2}.pdf. Accessed 12 Sept 2013 http://sbrc2011.facom.ufms.br/files/workshops/wpeif/ST02 [Online]_{2}.pdf. Accessed 12 Sept 2013
120. Nascimento MR, Rothenberg CE, Salvador MR, Corrêa CN, de Lucena SC, Magalhães MF: Virtual routers as a service: the routeflow approach leveraging software-defined networks. In Proceedings of the 6th International Conference on Future Internet Technologies. ACM, NY USA; 2011:34-37.
121. Ghobadi M, Yeganeh SH, Ganjali Y: Rethinking end-to-end congestion control in software-defined networks. In Proceedings of the 11th ACM Workshop on Hot Topics in Networks. ACM, NY USA; 2012:61-66.
122. Egilmez HE, Dane ST, Bagci KT, Tekalp AM: OpenQoS: an OpenFlow controller design for multimedia delivery with end-to-end quality of service over software-defined networks. In Signal & Information Processing Association Annual Summit and Conference (APSIPA ASC), 2012 Asia-Pacific. IEEE, NY USA; 2012:1-8.
123. Yap K-K, Kobayashi M, Sherwood R, Huang T-Y, Chan M, Handigol N, McKeown N: Openroads: empowering research in mobile networks. 2010. ACM, NY USA
124. Metsch T, Edmonds A, Harsh P, Antonescu F: Real-time Analytics System for Service Operations & Management. ACM symposium applied computing. ACM, NY USA; 2014.
125. Ali-Ahmad H, Cicconetti C, De la Oliva A, Mancuso V, Sama MR, Seite P, Shanmugalingam S: An SDN-based network architecture for extremely dense wireless networks. In Future Networks and Services (SDN4FNS), 2013 IEEE SDN for. IEEE, NY USA; 2013:1-7.
126. Karimzadeh M, Valtulina L, Karagiannis G: Applying SDN/OpenFlow in Virtualized LTE to support distributed mobility management (DMM). In The 4th International Conference on Cloud Computing and Services Science (CLOSER). Citeseer; 2014.
127. Yang H, Kim Y: Routing optimization with SDN. 2013.http://tools.ietf.org/html/draft-yang-dmm-sdn-dmm-00 . Accessed 12 Sept 2013
128. Yap K-K, Sherwood R, Kobayashi M, Huang T-Y, Chan M, Handigol N, McKeown N, Parulkar G: Blueprint for introducing innovation into wireless mobile networks. In Proceedings of the second ACM SIGCOMM Workshop on Virtualized Infrastructure Systems and Architectures (VISA). ACM, NY USA; 2010:25-32.

129. Ding AY, Crowcroft J, Tarkoma S: Poster: Softoffload: a programmable approach toward collaborative mobile traffic offloading. In Proceedings of the 12th Annual International Conference on Mobile Systems, Applications, and Services. ACM, NY USA; 2014:368-368.
130. Akyildiz IF, Lee A, Wang P, Luo M, Chou W: A roadmap for traffic engineering in software defined networks. Comput. Network 2014, 71: 1-30. ACM, NY USA
131. Yi G, Lee S: Fully distributed handover based on SDN in heterogeneous wireless networks. In Proceedings of the 8th International Conference on Ubiquitous Information Management and Communication. ACM, NY USA; 2014:70-70.
132. Schulz-Zander J, Suresh L, Sarrar N, Feldmann A, Hühn T, Merz R: Programmatic orchestration of wifi networks. In 2014 USENIX Annual Technical Conference (USENIX ATC 14). USENIX Association; 2014.
133. Ding AY, Crowcroft J, Tarkoma S, Flinck H: Software defined networking for security enhancement in wireless mobile networks. ACM, NY USA; 2014:94-101.
134. Suresh L, Schulz-Zander J, Merz R, Feldmann A, Vazao T: Towards programmable enterprise WLANs with Odin. In Proceedings of the First Workshop on Hot Topics in Software Defined Networks. ACM, NY USA; 2012:115-120.
135. Pentikousis K, Wang Y, Hu W: Mobileflow: toward software-defined mobile networks. IEEE Commun. Mag 2013, 51(7):44-53. IEEE, NY USA
136. Gudipati A, Perry D, Li LE, Katti S: Softran: software defined radio access network. In Proceedings of the Second ACM SIGCOMM Workshop on Hot topics in Software Defined Networking. ACM, NY USA; 2013:25-30.
137. Jin X, Li LE, Vanbever L, Rexford J: SoftCell: scalable and flexible cellular core network architecture. ACM, NY USA; 2013.
138. Luo T, Tan H-P, Quek TQ: Sensor openflow: enabling software-defined wireless sensor networks. IEEE Commun. Lett 2012, 16(11):1896-1899. IEEE, NY USA
139. Costanzo S, Galluccio L, Morabito G, Palazzo S: Software defined wireless networks: unbridling SDNs. In European Workshop on Software Defined Networking (EWSDN). IEEE, NY USA; 2012:1-6.
140. Mahmoud QH: Cognitive Networks: Towards Self-Aware Networks. Wiley, USA; 2007.
141. Manoj B, Rao RR, Zorzi M: Architectures and protocols for next generation cognitive networking. In Cognitive Wireless Networks. Springer, Heidelberg; 2007:271-284.

142. Fortuna C, Mohorcic M: Trends in the development of communication networks: cognitive networks. Comput. Networks 2009, 53(9):1354-1376. 10.1016/j.comnet.2009.01.002
143. Cesana M, Cuomo F, Ekici E: Routing in cognitive radio networks: challenges and solutions. Ad Hoc Networks 2011, 9(3):228-248. 10.1016/j.adhoc.2010.06.009
144. Zhao Q, Sadler BM: A survey of dynamic spectrum access. IEEE Signal Process. Mag 2007, 24(3):79-89.
145. Zhang Z, Long K, Wang J: Self-organization paradigms and optimization approaches for cognitive radio technologies: a survey. IEEE Wireless Comm 2013, 20(2):36-42.
146. Cormio C, Chowdhury KR: A survey on MAC protocols for cognitive radio networks. Ad Hoc Networks 2009, 7(7):1315-1329. 10.1016/j.adhoc.2009.01.002
147. De Domenico A, Strinati EC, Di Benedetto M: A survey on MAC strategies for cognitive radio networks. IEEE Communicat. Surv. Tutorials 2012, 14(1):21-44.
148. Azarfar A, Frigon J-F, Sanso B: Improving the reliability of wireless networks using cognitive radios. IEEE Communicat. Surv. Tutorials 2012, 14(2):338-354.
149. Fragkiadakis AG, Tragos EZ, Askoxylakis IG: A survey on security threats and detection techniques in cognitive radio networks. IEEE Communicat. Surv. Tutorials 2013, 15(1):428-445.
150. Niyato D, Hossain E: Cognitive radio for next-generation wireless networks: an approach to opportunistic channel selection in IEEE 802.11-based wireless mesh. IEEE Wireless Commun 2009, 16(1):46-54.
151. Ahuja R, Corke R, Bok A: Cognitive radio system using IEEE 802.11 over UHF TVWS. In New Frontiers in Dynamic Spectrum Access Networks, 2008. DySPAN 2008. 3rd IEEE Symposium on. IEEE, NY USA; 2008:1-9.
152. Flores AB, Guerra RE, Knightly EW, Ecclesine P, Pandey S: "IEEE 802.11 af: a standard for tv white space spectrum sharing. IEEE Commun. Mag 2013, 51(10):92-100. IEEE, NY USA
153. Cordeiro C, Challapali K, Birru D, Sai Shankar N: IEEE 802.22: the first worldwide wireless standard based on cognitive radios. In New Frontiers in Dynamic Spectrum Access Networks, 2005. DySPAN 2005. IEEE, NY USA; 2005:328-337.
154. Akan OB, Karli O, Ergul O: Cognitive radio sensor networks. IEEE Netw IEEE, NY USA, 2009, 23(4):34-40.
155. Di Felice M, Doost-Mohammady R, Chowdhury KR, Bononi L: Smart radios for smart vehicles: cognitive vehicular networks. IEEE Veh. Tech. Mag 2012, 7(2):26-33. IEEE, NY USA

156. Costa-Perez X, Swetina J, Guo T, Mahindra R, Rangarajan S: Radio access network virtualization for future mobile carrier networks. IEEE Comm. Mag 2013, 51(7):27-35. IEEE, NY USA
157. Bhanage G, Seskar I, Mahindra R, Raychaudhuri D: Virtual basestation: architecture for an open shared WiMAX framework. In Proceedings of the Second ACM SIGCOMM Workshop on Virtualized Infrastructure Systems and Architectures (VISA). ACM, NY USA; 2010:1-8.
158. Zaki Y, Zhao L, Goerg C, Timm-Giel A: LTE wireless virtualization and spectrum management. In Wireless and Mobile Networking Conference (WMNC), 2010 Third Joint IFIP. IEEE, NY USA; 2010:1-6.
159. Smith G, Chaturvedi A, Mishra A, Banerjee S: Wireless virtualization on commodity 802.11 hardware. In Proceedings of the Second ACM International Workshop on Wireless Network Testbeds, Experimental Evaluation and Characterization. ACM, NY USA; 2007:75-82.
160. Hamaguchi T, Komata T, Nagai T, Shigeno H: A framework of better deployment for, WLAN access point using virtualization technique. In 24th International Conference on Advanced Information Networking and Applications Workshops (WAINA) 2010 IEEE. IEEE, NY USA; 2010:968-973.
161. Xia L, Kumar S, Yang X, Gopalakrishnan P, Liu Y, Schoenberg S, Guo X: Virtual WiFi: bring virtualization from wired to wireless. In ACM SIGPLAN Notices, vol 46. ACM, NY USA; 2011:181-192.
162. He Y, Fang J, Zhang J, Shen H, Tan K, Zhang Y: Mpap: virtualization architecture for heterogenous wireless aps. Comput. Comm. Rev 2010, 40: 475-476. ACM, NY USA 10.1145/1851275.1851271
163. Sivaraman V, Moors T, Habibi Gharakheili H, Ong D, Matthews J, Russell C: Virtualizing the access network via open apis. In Proceedings of the Ninth ACM Conference on Emerging Networking Experiments and Technologies. ACM, NY USA; 2013:31-42.
164. Philip VD, Gourhant Y, Zeghlache D: Openflow as an architecture for e-node b virtualization. In e-Infrastructure and e-Services for Developing Countries. Springer, Heidelberg; 2012:49-63.
165. Tan K, Shen H, Zhang J, Zhang Y: Enable flexible spectrum access with spectrum virtualization. In Dynamic Spectrum Access Networks (DYSPAN), 2012 IEEE International Symposium on. IEEE, NY USA; 2012:47-58.
166. Yang K, Ou S, Chen H-H: On effective offloading services for resource-constrained mobile devices running heavier mobile internet applications. IEEE Commun. Mag 2008, 46(1):56-63. IEEE, NY USA

167. Chun B-G, Ihm S, Maniatis P, Naik M, Patti A: Clonecloud: elastic execution between mobile device and cloud. In Proceedings of the Sixth Conference on Computer Systems, EuroSys '11. ACM, NY USA; 2011:301-314.

168. Calhoun P: Lightweight access point protocol. 2010.http://tools.ietf.org/html/rfc5412 . Accessed 12 Sept 2013

169. Dalvi A, Swamy P, Meshram B: Centralized management approach for, WLAN. In Computer Networks and Information Technologies. Springer; 2011:578-580.

170. Misra S, Das S, Khatua M, Obaidat M: QoS-guaranteed bandwidth shifting and redistribution in mobile cloud environment. IEEE Trans. Cloud Comput 2013, 99: 1-1. IEEE, NY USA

171. Dinh HT, Lee C, Niyato D, Wang P: A survey of mobile cloud computing: architecture, applications, and approaches. Wireless Comm. Mobile Comput Wiley, USA, 2011.

172. Vassilaras S, Yovanof GS: Wireless going in the cloud: a promising concept or just marketing hype? Wireless Pers. Comm 2011, 58(1):5-16. Springer 10.1007/s11277-011-0284-9

173. Lin Y, Shao L, Zhu Z, Wang Q, Sabhikhi RK: Wireless network cloud: architecture and system requirements. IBM J. Res. Dev 2010, 54(1):4-1. IBM

174. Ferreira L, Pichon D, Hatefi A, Gomes A, Dimitrova D, Braun T, Karagiannis G, Karimzadeh M, Branco M, Correia L: An architecture to offer cloud-based radio access network as a service. Networks and Communications (EuCNC), 2014 European Conference on (IEEE, 2014) pp. 1-5

175. Vestin J, Dely P, Kassler A, Bayer N, Einsiedler H, Peylo C: CloudMAC: towards software defined WLANs. ACM SIGMOBILE Mobile Comput. Commun. Rev 2013, 16(4):42-45. ACM, NY USA 10.1145/2436196.2436217

176. Ko C-H, Huang DH, Wu S-H: Cooperative spectrum sensing in TV white spaces: when cognitive radio meets cloud. In Computer Communications Workshops (INFOCOM WKSHPS), 2011 IEEE Conference on. IEEE, NY USA; 2011:672-677.

177. Rost P, Bernardos CJ, Domenico A, Girolamo M, Lalam M, Maeder A, Sabella D, Wübben D: Cloud technologies for flexible 5G radio access networks. IEEE Commun. Mag 2014, 52(5):68-76. IEEE, NY USA

178. Staring AJ: Applying the cloud computing model in LTE based cellular systems. In Proc. of the 17th Twente Student Conference on IT. University of Twente; 2012.

179. Anwer MB, Feamster N: Building a fast, virtualized data plane with programmable hardware. Comput. Commun. Rev 2010, 40(1):75-82. ACM, NY USA 10.1145/1672308.1672323
180. Chaudet C, Haddad Y: Wireless software defined networks: challenges and opportunities. In Microwaves, Communications, Antennas and Electronics Systems (COMCAS), 2013 IEEE International Conference on. IEEE, NY USA; 2013:1-5.
181. Zuniga JC, Bernardos CJ, de la Oliva A, Melia T, Costa R, Reznik A: Distributed mobility management: a standards landscape. IEEE Comm. Mag 2013, 51(3):80-87. IEEE, NY USA
182. Li LE, Mao ZM, Rexford J: Toward software-defined cellular networks. In European Workshop on Software Defined Networking (EWSDN), 2012. IEEE, NY USA; 2012:7-12.
183. Randall Schwartz MJ: White paper: Carrier Wi-Fi Offload. . Accessed 12 Sept 2013 http://tinyurl.com/carrierwifi
184. Zheng J, Lee MJ: A comprehensive performance study of IEEE 802.15.4. Sensor Network Operations, ISBN 0-471-71976-5, Chapter 4 218–237. IEEE Press, Wiley Interscience; 2006.
185. Singh CK, Kumar A, Ameer P: Performance evaluation of an IEEE 802.15.4 sensor network with a star topology. Wireless Network 2008, 14(4):543-568. Springer 10.1007/s11276-007-0043-8
186. Clark DD, Partridge C, Ramming JC, Wroclawski JT: A knowledge plane for the internet. In Proceedings of the 2003 Conference on Applications, Technologies, Architectures, and Protocols for Computer Communications. ACM; 2003:3-10.
187. Sherman M, Mody AN, Martinez R, Rodriguez C, Reddy R: IEEE standards supporting cognitive radio and networks, dynamic spectrum access, and coexistence. IEEE Commun. Mag IEEE, NY USA, 2008, 46(7):72-79.
188. Sarigiannidis A, Nicopolitidis P, Papadimitriou G, Sarigiannidis P, Louta M, Pomportsis A: On the use of learning automata in tuning the channel split ratio of wimax networks. IEEE Systems Journal 2013, PP(99):1-13. IEEE, NY USA
189. Misra S, Venkata P, Krishna Isaac K, Abraham Sasikumar N, Fredun S: An adaptive learning routing protocol for the prevention of distributed denial of service attacks in wireless mesh networks. Comput. Math. Appl 2010, 60(2):294-306. Elsevier, USA 10.1016/j.camwa.2009.12.035
190. Misra S, Krishna P, Kalaiselvan K, Saritha V, Obaidat MS: Learning automata-based QoS framework for cloud IaaS. IEEE Trans. Network Service Manage 2014, 11: 15-24. IEEE, NY USA
191. Kakali VL, Sarigiannidis PG, Papadimitriou GI, Pomportsis AS: A novel adaptive framework for wireless push systems based on

distributed learning automata. Wireless Pers. Comm 2011, 57(4):591-606. Springer 10.1007/s11277-009-9863-4

192. Misra S, Krishna PV, Saritha V: Lacav: an energy-efficient channel assignment mechanism for vehicular ad hoc networks. J. Supercomput 2012, 62(3):1241-1262. Springer 10.1007/s11227-011-0552-1

193. Network functions virtualization (NFV): Network functions virtualization: an introduction, benefits, enablers, challenges and call for action. Introductory white paper, SDN and OpenFlow world congress, Oct 2012 2013. . Accessed 12 Sept 2013 http://www.tid.es/es/Documents/NFV_White_PaperV2.pdf

194. John W, Pentikousis K, Agapiou G, Jacob E, Kind M, Manzalini A, Risso F, Staessens D, Steinert R, Meirosu C: Research directions in network service chaining. In Future Networks and Services (SDN4FNS), 2013 IEEE SDN for. IEEE, NY USA; 2013:1-7.

195. Bavier A, Feamster N, Huang M, Peterson L, Rexford J, VINI veritas: realistic and controlled network experimentation: Comput. Commun. Rev. 2006, 36: 3-14. ACM, NY USA

196. Hibler M, Ricci R, Stoller L, Duerig J, Guruprasad S, Stack T, Webb K, Lepreau J: Large-scale virtualization in the, Emulab network testbed. In USENIX Annual Technical Conference. USENIX; 2008:113-128.

197. Elliott C: GENI: Opening up new classes of experiments in global networking. IEEE Internet Comput 2010, 14(1):39-42. IEEE, NY USA

198. Casado M, Koponen T, Ramanathan R, Shenker S: Virtualizing the network forwarding plane. In Proceedings of the Workshop on Programmable Routers for Extensible Services of Tomorrow. ACM, NY USA; 2010:8-8.

199. Gibb G, Underhill D, Covington A, Yabe T, McKeown N: OpenPipes: prototyping high-speed networking systems. In Proc. ACM SIGCOMM Conference (Demo), Barcelona, Spain. ACM, NY USA; 2009.

200. Bruce Davie's talk in the Open Networking Summit on NV. [Online] 2013.http://www.opennetsummit.org/pdf/2013/presentations/bruce_davie.pdf . Accessed 12 Sept 2013

201. Wen H, Tiwary PK, Le-Ngoc T: Wireless Virtualization, Springer Briefs in Computer Science. (Springer; 2013.

202. Bosch P, Duminuco A, Pianese F, Wood TL: Telco clouds and virtual telco: consolidation, convergence, and beyond. In Integrated Network Management (IM), 2011 IFIP/IEEE International Symposium on. IEEE, NY USA; 2011:982-988.

203. Sachs S, Baucke J: Virtual radio: a framework for configurable radio networks. In Proceedings of the 4th Annual International Conference on Wireless Internet. ICST (Institute for Computer Sciences, Social-Informatics and Telecommunications Engineering). ACM, NY USA; 2008:61-61.
204. R. Control and Provisioning of Wireless Access Points (CAPWAP) Protocol Specifications 2009.https://tools.ietf.org/rfc/rfc5415.txt [Online]. Accessed 12 Sept 2013
205. Nakauchi K, Ishizu K, Murakami H, Nakao A, Harada H: Amphibia: a cognitive virtualization platform for end-to-end slicing. IEEE, NY USA; 2011.
206. Meraki Networks [Online]. Accessed 12 Sept 2013 http://meraki.cisco.com [Online]. Accessed 12 Sept 2013
207. Aruba Networks [Online]. Accessed 12 Sept 2013 http://cloud.arubanetworks.com/ [Online]. Accessed 12 Sept 2013
208. Papakos P, Capra L, Rosenblum DS: Volare: context-aware adaptive cloud service discovery for mobile systems. In Proceedings of the 9th International Workshop on Adaptive and Reflective Middleware. ACM, NY USA; 2010:32-38.
209. Papagianni C, Leivadeas A, Papavassiliou S, Maglaris V, Cervelló-Pastor C, Monje A: On the optimal allocation of virtual resources in cloud computing networks. IEEE Trans Comput IEEE, NY USA, 2013, 62(6):1060-1071.MathSciNet
210. Satyanarayanan M, Bahl P, Caceres R, Davies N: The case for VM-based cloudlets in mobile computing. IEEE Pervasive Comput 8(4):14-23. IEEE, NY USA
211. Ge F, Lin H, Khajeh A, Chiang CJ, Eltawil AM, Bostian CW, Feng W-C, Chadha R: Cognitive radio rides on the cloud. In Military Communications Conference, MILCOM 2010. IEEE, NY USA; 2010:1448-1453.
212. Miguelez IGomez, Marojevic V: A Gelonch Bosch, Resource management for software-defined radio clouds. IEEE Micro 2012, 32(1):44-53. IEEE, NY USA
213. Kurschl W, Beer W: Combining cloud computing and wireless sensor networks. In Proceedings of the 11th International Conference on Information Integration and Web-based Applications & Services. ACM, NY USA; 2009:512-518.
214. Atzori L, Iera A, Morabito G: The internet of things: a survey. Comput. Network 2010, 54(15):2787-2805. Elsevier 10.1016/j.comnet.2010.05.010

2

CHAPTER

ENABLING 5G MOBILE WIRELESS TECHNOLOGIES

Long Bao Le[1*], Vincent Lau[2], Eduard Jorswieck[3], Ngoc-Dung Dao[4], Afshin Haghighat[5], Dong In Kim[6] and Tho Le-Ngoc[7]

ABSTRACT

Research on 5G mobile wireless technologies has been very active in both academia and industry in the past few years. While there has been certain consensus on the overall requirements of 5G wireless systems (e.g., in data rate, network capacity, delay), various enabling wireless technologies have been considered and studied to achieve these performance targets. It has been quite clear, however, that there would be no single enabling technology that can achieve all diverse and even conflicting 5G requirements. In general, many fundamental changes and innovations to re-engineer the overall network architecture and algorithms in different layers and to exploit new system degrees of freedom would be needed for the future 5G wireless system. In particular, we may need to consider other potential waveform candidates that can overcome limitations of the orthogonal frequency multiple access (OFDM) waveform employed in the current 4G system, develop disruptive technologies to fulfill 5G rate and capacity requirements including network densification, employment of large-scale (massive) multiple input multiple output (MIMO), and exploitation of the millimeter wave (mmWave) spectrum to attain Gigabit communications. In addition, design tools from the computer networking domain including software defined networking, virtualization, and cloud computing are expected to play important roles in defining the more flexible, intelligent, and efficient 5G network architecture. This paper aims at describing key 5G enabling wireless mobile technologies and discussing their potentials and open research challenges. We also present how papers published in our special issue contribute to the developments of these disruptive 5G technologies.

Keywords 5G wireless 5G waveform Dense networks HetNets mmWave Massive MIMO C-RAN Software defined networking Virtualization

1 INTRODUCTION

Research and development activities on 5G mobile wireless technologies have attracted lots of interest in both academia and industry worldwide over the last several years. These initiatives have been driven by many evidences that the currently deployed 4G wireless network could not provide satisfactory support in the near future. In particular, the mobile data traffic has been exploding and this trend is expected to continue in the coming years [1, 2]. It is predicted that the 5G wireless system should be able to deliver as much as 1000 times of the capacity provided by today's mobile networks [3–5]. Moreover, many emerging broadband wireless applications such as high-resolution video streaming and tactile Internet demand much higher data rates, which could be 10–100 times of what is offered by 4G wireless networks [6, 7].

In addition, the next-generation 5G mobile wireless system is expected to accommodate considerably larger number of wireless connections to better support existing and emerging applications including machine-to-machine (M2M) applications many of which require more stringent quality-of-service (QoS) including better delay, reliability, and higher spectral and energy efficiency [2, 7]. For example, wireless communications supporting smart connected cars for road safety, remote monitoring, and real-time control must satisfy very stringent delay and reliability constraints. Furthermore, the vision of the hyperconnected world with billions of wireless connections where all the world of human and things can be inter-connected wirelessly in the so-called Internet of Things can become the reality soon. Importantly, future mobile technologies must be more cost-efficient since mobile operators have been experiencing reducing revenue per data unit.

To address the aforementioned challenges and requirements, various breakthroughs and advancements of wireless technologies at the physical and higher layers as well as system architecture are strongly needed to engineer the future 5G wireless system. In particular, waveform design must be rethought for 5G since the current orthogonal frequency multiple access (OFDM) technique may not be suitable for stringent delay-constrained wireless applications due to its large transmission time interval (TTI) [8, 9]. Moreover, OFDM may not be very efficient in terms of communication signaling overhead to support emerging M2M applications. In addition, novel solutions exploiting new network degrees of freedom must be sought to fundamentally increase the communication rate and network capacity. Since frequency spectrum below 6 GHz has been very crowded, exploitation of the millimeter wave (mmWave) frequency deems necessary to meet the aggressive requirements in terms of network capacity and from Gigabit broadband applications [10, 11]. However, mmWave spectrum with its unique characteristics including high attenuation, sensitivity to signal blockage, and fast channel variations requires innovative solutions from different design aspects ranging from communication algorithms, network protocols to network architecture engineering.

Ultra-dense wireless networks with multi-tier cells including dense small cells supporting multiple frequency bands (microwave and mmWave frequency), multiple radio access technologies (multi-RATs) for enhanced network capacity and energy efficiency through short-range communications, and dynamic load balancing play an important role in future 5G networks [3–5, 12–15]. Moreover, emerging massive multiple input multiple output (MIMO) and full-dimension MIMO technologies provide important paradigm shifts to push further the frontier of MIMO techniques [16–18]. Specifically, massive MIMO and full-dimension MIMO enable to significantly improve the network capacity and energy efficiency. Importantly, efficient utilization of massive MIMO techniques allows us to realize these benefits without relying on complicated and costly coordination techniques such as coordinated multi-point (CoMP) transmission/reception proposed in the 4G wireless cellular system.

To significantly improve the cost efficiency and flexibility in traffic and network management, enabling 5G technologies based on cloud computing, software defined networking (SDN), and network virtualization are expected to shape the way the 5G wireless network is engineered and managed [19–21]. In particular, cloud radio access networks (C-RAN) aims at realizing various network algorithms and network functionalities in different layers of the network protocol stack in the cloud. This allows to exploit the multiplexing gain due to centralized processing in the cloud. Moreover, SDN and virtualization provide mechanisms and tools to efficiently utilize the network resources with better traffic management, which results in improvements in the capital expenditures (CAPEX) and operational expenditures (OPEX) costs. All these technologies play vital roles in ensuring successful deployment of the 5G wireless systems, which lead to significant benefits in the mobile communications as well as many other application sectors.

Each aforementioned 5G enabling technology presents many interesting and challenging research issues, which deserve much more research in the coming years. Moreover, innovative solutions to efficiently combine their strengths in suitable network architecture and deployment scenarios must be developed [22]. In particular, massive MIMO techniques could play an important role in supporting the exploitation of mmWave spectrum in future dense wireless networks since they provide sufficient antenna gains to offset the high signal attenuation in these frequencies [23]. Moreover, novel solutions must be sought to effectively utilize the centralized C-RAN for dense network deployment. Furthermore, virtualization techniques can enable to exploit the cloud computational resources in the C-RAN and to shape the ways both wireless access and core networks are engineered in the 5G mobile wireless network. All together, these enabling technologies provide many fertile and exciting research areas for further research.

To build the future 5G wireless network that meet the projected requirements, many innovations from the architecture design concepts as well as algorithm development perspectives would be all needed. The paper [24] in our special issue discusses several potential enabling technologies, which are based on research experience and activities from the 5G METIS project. Moreover, estimation and feedback of channel state and other system information must be rethought to achieve scalable and efficient solutions in future wireless networks. These are the topics considered in three papers of our special issue [25–27]. While the work [25] develops a scalable channel state information (CSI) estimation with controllable resolution and complexity for future dense networks, the paper [26] studies the CSI estimation problem for a very wideband wireless channel using the sampling rate less than the Nyquist rate. In addition, an iterative estimation technique for total noise and interference in multi-carrier wireless systems is developed to assist the data detection in [27], which is relevant for dense wireless networks with aggressive frequency reuse. Future 5G wireless architecture should foster sophisticated spectrum-efficient resource sharing design such as the network-coding-based and relay-based cognitive spectrum sharing frameworks proposed respectively in two papers of our special issue [28, 29]. Other papers published in our special issue, which address research issues related to different enabling 5G wireless technologies, will be discussed in the next section.

2 PERSPECTIVES ON 5G ENABLING TECHNOLOGIES AND OPEN RESEARCH ISSUES

In this section, we provide our perspectives on the role and potential of some important 5G enabling technologies. We also describe in details some open research issues related to these technologies. Moreover, we will discuss how they can complement and be effectively combined to meet different performance requirements of the future 5G wireless network.

2.1 Waveform design for 5G wireless systems

It has been recognized that the OFDM waveform currently adopted in the long-term evolution (LTE) standard has several limitations in supporting the 5G requirements [8]. Specifically, 5G traffic is expected to have very diverse characteristics and requirements in data column, communications rate, delay, and reliability. For example, applications such as video streaming and augmented reality demand very high communication rates for real-time communications. Moreover, various emerging M2M applications such as those for smartgrids have sporadic and bursty traffic with small amount of data to be delivered in each transmission. Other wireless applications such as vehicular communications and robotics control require highly reliable and ultra low delay. Finally, the number of wireless connections supporting future M2M and Internet-of-Things applications is expected to be significantly larger than that due to mostly human-type communications in today's wireless networks.

The OFDM waveform possesses several desirable features including low-complexity implementation using fast Fourier transform (FFT) and inverse FFT (IFFT) blocks, simple equalization, flexibility in supporting per subcarrier adaptive modulation and MIMO communications [9]. However, strict synchronization, which involves large signaling overhead, is required to maintain communication orthogonality and satisfactory communication quality. Moreover, the OFDM signaling requires a cyclic prefix (CP) and null guard bands at the spectrum edges, which reduce the spectral efficiency. In general, development of more spectral and power efficient waveforms compared to the current state-of-the-art is of great interest to support the exponential growth of mobile traffic and operation cost reduction in the future wireless system. Finally, current frame structure and implementation of various protocols such as scheduling and retransmissions in the 4G LTE systems offers the latency of multiple 10 ms, which is far beyond the 5G strict delay requirement of 1 ms. In fact, this 5G delay requirement implies that the physical-layer delay budget is about 100 μs.

There have been several proposals on waveforms for 5G wireless networks [30–35]. In particular, faster-than-Nyquist (FTN) signaling has recently received renewed interest since it can potentially increase the system capacity by packing more data in the time and/or frequency domains [30–33]. In the time domain, this can be achieved by allowing data bearing pulses to be sent faster. While this design destroys the orthogonality, satisfactory detection performance can still be achieved by employing a sophisticated detection technique at the receiver to mitigate the intentional ISI introduced at the transmitter. In addition, employment of FTN signaling in both time and frequency domains results in multi-carrier FTN, which attains larger spectrum efficiency compared to FTN in time or frequency alone [31]. Although some significant progresses have been achieved in recent years on understanding FTN signaling, more research is needed to study and develop efficient designs to meet the performance and QoS requirements of 5G wireless systems. In particular, practical implementation aspects deserve more research and development efforts [36].

Filter bank multi-carrier (FBMC) is one of the 5G waveform candidates where per-subcarrier filtering is performed to suppress the sidelobes and the FFT/IFFT blocks can still be employed as in the OFDM modulation [9, 37]. The bandlimitedness of FBMC would enable it to offer higher spectrum efficiency than OFDM and more flexibility in exploiting white spaces in cognitive radio networks. Moreover, FBMC is more robust to synchronization and frequency misalignments compared to OFDM; therefore, FBMC would be more suitable for random access to support sporadic M2M traffic with loose synchronization requirements. Despite these advantages, FBMC is more difficult to realize the MIMO communications compared to OFDM. However, recent progresses on advanced physical techniques have led to efficient ways to implement MIMO-FBMC with comparable performance with respect to MIMO-OFDM systems [38, 39].

An enhanced version of FBMC, called universal filtered multi-carrier (UFMC), has been proposed recently [40, 41]. In UFMC, filtering is performed on the subband basis instead of per subcarrier as in FBMC. Therefore, UFMC can still acquire the advantages of FBMC including suppression of spectral sidelobes, robustness against time and frequency synchronization [8]. Moreover, the filter length of UFMC can be significantly shorter than that of FBMC since UFMC performs the filtering for each subband with larger bandwidth than subcarrier. Therefore, UFMC would be more suitable to support short-burst communications typically required by many M2M applications. UFMC does not require a CP, but using zero padding, which results in improved spectrum efficiency but becomes more sensitive to time misalignment compared to the existing CP-based OFDM waveform. On the other hand, the filtered OFDM (f-OFDM) technique in [42] allows the filter length to exceed the CP length of OFDM. The f-OFDM waveform can achieve a desirable frequency localization for bandwidths as narrow as a few tens of subcarriers, while keeping the inter-symbol interference/inter-carrier interference (ISI/ICI) within an acceptable limit.

Another waveform design referred to as generalized frequency division multiplexing (GFDM) has been proposed to address the low-power, broadband, and real-time challenges of future wireless cellular networks [34, 35]. The GFDM aims at achieving flexible resource and QoS management by performing modulation for independent blocks each of which comprises a number of subcarriers and subsymbols. Moreover, GFDM employs filters for individual subcarriers that are circularly shifted in time and frequency. While inter-symbol interference (ISI) and inter-carrier interference (ICI) might still exist in GFDM, these undesirable effects can be efficiently addressed by efficient detection techniques at the receiver side.

Several 5G requirements can be met by the GFDM via appropriate tuning of different waveform parameters. The per-subcarrier filtering adopted by GFDM leads to small out-of-band emission; therefore, dynamic spectrum access of fragmented spectrum in cognitive radio scenarios can be achieved. Moreover, GFDM design allowing to flexibly configure the time-frequency blocks, which enables us to meet the stringent delay constraints of various 5G applications such as tactile internet and many wireless control applications. In addition, it is possible to add small cyclic prefix and suffix for an entire block to relax the time synchronization requirement of emerging M2M applications while still maintaining improved spectral efficiency with respect to the OFDM waveform. In fact, most synchronization algorithms developed for OFDM remain applicable to the GFDM waveform. The paper [43] in our special issue demonstrates how space time coding can be combined with the GFDM waveform in the multi-antenna wireless system while still allowing flexible resource sharing among users by using wireless scheduling.

There are many open challenges to address regarding the 5G waveform, frame structure, and related design issues in the physical layer and higher layers of future 5G wireless systems. One such frame structure to support the 5G diverse QoS and traffic requirements has been proposed in [8]. To enhance the multiple access gain, non-orthogonal multiple access schemes have been proposed. For example, the NOMA framework [44] utilizes power-domain superposition multiplexing. In the code domain, the sparse code multiple access (SCMA) is presented in [45]. Data symbols of users are mapped to multi-dimensional sparse codewords. Codewords are non-orthogonal but are systematically designed to allow low-complexity detection [46]. Moreover, while a rich literature on adaptive resource allocation techniques has been developed for OFDM wireless systems over the past decades, significant more research on medium access, resource management, and higher layer design aspects to better understand the pros and cons of different waveform alternatives is expected in the coming years [47]. Such studies for relevant 5G deployment scenarios will be especially useful.

In general, there are several aspects one must consider in selection of a waveform for a communication system such as bandwidth efficiency, scalability, interoperability, implementation complexity and robustness to imperfections including non-linearity, phase noise. The final decision on a waveform hinges on a trade-off between these often contradictory requirements. Another important factor that must be considered for waveform design and selection is the operation frequency. As the current trend in 5G development indicates, there could be two different operating frequency regions for 5G systems, namely below 6 GHz and mmWave bands. As each band could be employed for different deployment scenarios and they would exhibit different propagation properties, 5G systems could rely on two different waveforms for operation in the two bands.

2.2 Ultra-dense wireless networks exploiting microwave and mmwave communications

Ultra-dense deployment of small cells, relays, distributed antenna systems operating on different frequency bands (e.g., microwave and mmWave) based on multiple RATs in a multi-tier heterogeneous network (HetNet) architecture provides a fundamental way to meet the capacity demand of future 5G wireless networks [48–50]. This heterogeneous network architecture also enables efficient traffic offloading among different network tiers and RATs to better support the growing mobile traffic with enhanced QoS, communication rates, and energy efficiency [51–56].

In fact, research on multi-tier HetNets where communications among different network tiers (e.g., macro, micro and small-cell tiers) share the same microwave spectrum has been one of the most active research topics over the past years [57]. One very important research theme in this direction is interference management, which can be performed using time, frequency, spatial, and power control approaches [49, 50, 58–62]. Three papers published in our special issue address different analysis and design issues for wireless HetNets. Specifically, the outage analysis for cognitive HetNets is performed by using the stochastic geometry technique in [63]. Moreover, the work [64] proposes enhanced inter-cell interference coordination (eICIC) techniques for the two-tier HetNet where the optimal cell range expansion bias and almost blank subframe rate are obtained to achieve the fair resource sharing and interference management between two network tiers. The paper [65] studies different interference management solutions for the multi-antenna moving network where it shows that the high-performance moving network can be indeed realized with minimal impacts on the regular outdoor communications.

Other important research issues for multi-tier HetNets include mobility management and load balancing via dynamic BS association and admission control [51–56]. To effectively manage cross-tier interference and perform load balancing in the dense HetNet, distributed solutions using the self-optimization network (SON) techniques are desirable, which typically require reliable and low-delay backhauls for a large number of small cells (i.e., pico and femto cells) High-speed backhauls are also critical to exchange the large volume of data between small cells and the core network. While low-cost backhauls using existing DSL can be used for indoor femtocells, this solution would not lead to a satisfactory QoS and performance. Indeed, backhauls remain to be one of the most critical issues, which hinders the successful widespread adoption of the dense small-cell network.

Future dense HetNets should support different types of nodes and communications. In particular, device-to-device (D2D) communications exploiting the cellular spectrum to establish direct high-speed and reliable communications among wireless devices is an important technology [66–69]. D2D communications can attain different benefits including enhancement of spectrum and energy efficiency, improving communication delay and reliability, reduction of traffic in the core network, and support of various emerging wireless applications such as public safety, proximity-based social networks, and vehicle-to-vehicle communications. To achieve these benefits, efficient radio resource management techniques for D2D communications must be developed. In [70], the interference alignment technique is employed jointly with user clustering and resource allocation for efficient spectrum sharing and interference management while the work [71] proposes a joint power control and resource allocation framework for spectrum sharing between D2D and uplink cellular communications in the LTE-based wireless network.

MmWave communications has received tremendous interest from both academia and wireless industry recently since mmWave spectrum would provide a unique solution to solve the capacity requirements of the future wireless network. Moreover, the ultra-dense wireless network is a perfect choice to exploit the huge spectrum available in the mmWave frequency since the high attention in these frequency bands limits the cell size within a few hundreds meters [72, 73]. Importantly, mmWave could potentially help solve the spectrum shortage problems for both wireless access and backhauls [74–76]. In fact, the distinction between wireless access and backhaul links would become very blur in the future ultra-dense wireless network where an extremely large number of wireless connections would be expected to support emerging wireless M2M and Internet-of-Things applications. It remains an open challenge to develop efficient physical layer techniques to achieve high-speed communications and advanced radio resource management solutions to support the joint wireless access and backhaul communications.

These potential benefits of mmWave communications for ultra-dense wireless deployment are not available without many challenges to overcome. In fact, mmWave communications require to employ large-scale antennas at both transmitter and receiver sides to provide the sufficient antenna gain to offset the high attention in this frequency regime. Toward this end, development of cost-efficient hardware architecture together with suitably designed channel estimation, beamforming, as well as other communications and signal processing techniques require much further research [77–83]. The cost-efficient large-scale MIMO architecture where the number of RF chains is smaller than the number of antennas has been considered very potential for mmWave communications [79, 80]. Here, the development of effective channel estimation and hybrid beamforming solutions for this architecture using phase shifters and analog-to-digital converters (ADCs) with limited resolution presents major research challenges [78]. Moreover, multi-user and multi-cell resource allocation and beamforming design for this hybrid architecture to achieve high-speed communications deserves much more research efforts. Preliminary results along this direction have been presented recently in [82] where the joint hybrid beamforming and scheduling solution is proposed.

As previously mentioned, mmWave could potentially provide unique solutions for Gigabit wireless access and backhaul communications. Since wireless backhaul links can be relatively static, adaptive algorithms for beam alignment to maintain reliable communications can be developed to compensate for the base station (BS) movements due to strong wind for example [81]. Engineering adaptive beam update algorithms for large-scale MIMO systems to support mobile users is obviously more challenging since the MIMO beamwidth is quite narrow with the large number of antennas. Furthermore, it is expected that both single-hop and multi-hop backhaul communication solutions would be needed to deal with different communication scenarios in the future dense wireless network. Toward this end, algorithms in different layers of the protocol stack including beamforming, resource allocation, and routing must be designed to support the wireless access and multi-hop backhaul communications.

Finally, novel connection and mobility management strategies must be proposed to effectively deal with the unique characteristics of the heterogeneous mmWave dense networks such as susceptibility to blockage of mmWave links, different user velocities, and QoS requirements. Given that the mmWave communications links are very sensitive to blockage, it is likely that mobile users must maintain dual communications with both macro-BS and small-cell BS using mmWave and microwave frequency bands, respectively, to maintain continuous and reliable communications. Here, communications with a nearby small-cell BS on the mmWave spectrum can be employed to achieve high-speed data delivery whenever possible but control information can be delivered through the macro-BS using more reliable microwave communications. The split of control and data planes deems necessary and presents the unique design choice for mmWave communications [84, 85]. Moreover, the research on efficient BS association to achieve load balancing and reliable and high-speed communications must be rethought. It is also obvious that mobility management and admission control design for the mmWave-based dense HetNet deserve much more research before one can indeed bring potential benefits of mmWave communications to practice.

2.3 Massive and full-dimension MIMO technologies

While the MIMO technology has played a very important role in current 3G/4G wireless systems, massive MIMO and full-dimension MIMO present significant innovations for this technology to fundamentally increase the spectrum and energy efficiency of wireless systems [86]. Moreover, massive MIMO can potentially simplify the ways we manage the radio resource allocation and co-channel interference in the multi-cell and multi-tier wireless network.

Many existing works design high energy-efficient massive MIMO systems. The number of users and transmit power were jointly optimized in [87]. The energy efficiency (EE) of massive MIMO systems depends greatly on the circuit power consumption. The number of antennas, users, and transmit power were respectively optimized for the single-cell massive MIMO system with zero-forcing beamforming (ZFBF) in [88]. The optimal transmit power increases with the number of antennas, which means that the transmit power is an important design parameter for high EE massive MIMO systems. In [89], the GEE considering both the uplink and downlink of a multi-cell massive MIMO system, subject to channel estimation errors, pilot contamination, and correlated channel paths, with either maximum ratio combining (MRC)/MRT or ZF transmission/reception is optimized.

Many existing designs for traditional MIMO systems must be re-thought for these new MIMO systems such as channel estimation, beamforming design, and signal detection [90–93]. Moreover, there are also various arising challenges to resolve such as the channel estimation and pilot contamination issues [86, 91, 92, 94, 95], suitable multi-cell coordination [90, 96], and hardware imperfection [97] to name a few. Three papers published in our special issue address some of these research issues. Specifically, the work [98] proposes a low-complexity subspace detection technique for the multi-stream MIMO communications. In [99], a novel semi-orthogonal pilot design for massive MIMO wireless systems is developed to reduce the pilot resource consumption where a successive interference cancelation (SIC)-based channel estimation is employed to manage the interference between data and pilot. Moreover, the work [100] develops a joint channel coding, modulation, and MIMO communication scheme by using the rotated modulation and space-time component interleaver.

Furthermore, the inter-cell coordination in multi-cell massive MIMO systems is addressed in various works, either suppressing the inter-cell interference and inter-user interference completely [101] by coordinated beamforming. Another technique introduced in [102, 103] allows interference suppression with smaller excess of base station antennas. In [104], a multi-cell MIMO downlink channel is studied and a distributed interference alignment (IA) algorithm is proposed to suppress or minimize the interference to non-intended users. Also, [105] develops an IA technique for a downlink cellular system with CSI exchange and feedback within each cell. In [106, 107], conditions for the feasibility of IA and degree of freedom (DoF) for MIMO cellular networks are investigated.

To reduce the complexity and CSI requirement, the concept of grouping-based IA (GIA) is proposed for a two-cell single-stream interfering MIMO-BC in [108]. This GIA is extended to a multi-cell interfering MIMO-BC in [109], where both the feasible condition on the GIA and a low complexity IA decoder design are studied. For the performance analysis of massive MIMO system, there exists a large set of tools from random matrix theory (see [110–116] and references therein).

2.4 C-RAN

Cloud radio access network (C-RAN) presents a major paradigm shift in the design of future wireless systems where various network functionalities in both access and core networks can be realized in the cloud [117]. Due to the high cost-efficiency, flexibility, and utilization efficiency, it is regarded as one of the most promising solutions to meet the huge capacity demand in 5G wireless networks. In a C-RAN, plenty of distributed remote radio heads (RRHs) are distributed within a specific geographical area and connected to a centralized centralized baseband unit (BBU) pool through high bandwidth fronthaul links. In such a system, enhanced energy efficiency can be achieved from shortened distances between RRHs and users as well as improved spectral efficiency per unit area. Moreover, since all baseband processings are performed at the BBU, it is possible to employ CoMP transmission [118] for capacity enhancement by exploiting the increased spatial degrees of freedom and beamforming gain [119, 120]. The paper [24] in our special issue proposes a general 5G network architecture based on the C-RAN technology.

There are numerous open research issues to address in this interesting research domain including efficient radio resource management (e.g., power control, bandwidth allocation, beamforming desing), fronthaul compression, efficient design, and utilization of computing resources in the cloud to name a few. Some key open research issues are elaborated below.

2.4.1 Joint RRH clustering and precoding

To perform full CoMP among all RRHs, the BBU must exchange the baseband signals with all RRHs and collect the global CSI from all RRHs to all users. As the number of RRHs increases, the fronthaul loading and CSI signaling overhead will become unacceptable. Moreover, from the system energy efficiency perspective, it is not necessary to turn on all the RRHs when the system loading is not heavy. This motivates dynamic clustering of RRHs into virtual base stations (VBSs) to serve users via partial CoMP. There are two different types of RRH clustering, namely the network centric clustering (NCC) and user centric clustering (UCC) [121]. In NCC, there is no overlap among VBSs (i.e., RRH can only belong to a single VBS). This can simplify the clutering and precoding design, but the performance of the cell edge users at the boundary of the clusters will be compromised. In UCC, there can be overlap among VBSs, and we have the freedom to choose a small number of "best" RRHs as a VBS for each user to avoid cell edge effect. Both NCC and UCC can significantly reduce the fronthaul loading and CSI signaling overhead by controlling the size of each VBS. However, the UCC usually outperforms the NCC but the associated clustering and precoding design is more challenging. One side effect of RRH clustering is the inter-cluster interferences among different VBS. As a result, efficient joint RRH clustering and interference mitigation (precoding) schemes are essential for practical deployment of C-RAN. The existing joint clustering and precoding schemes can be classified into two classes.

One-timescale schemes: In such schemes, both RRH clustering and precoding (beamforming) are optimized at each time slot based on instantaneous global CSI. In [122], the authors developed an efficient cell-clustering algorithm to maximize the sum rate of the users based on graph theory. In [123], a joint clustering and scheduling algorithm is proposed to maximize the weighted sum rate by greedily selecting the BS clusters from a set of predetermined candidate clusters. A major approach to avoid the combinatorial optimization of RRH clustering is group sparse beamforming (GSBF) [121, 124, 125], where the size of each RRH cluster (VBS) is reduced by imposing group sparsity constraints on the beamforming vectors such that the RRHs with zero beamforming vectors can be turned off. Various GSBF schemes have been proposed in [121, 124–126] based on the (weighted) mixed l_1/ l_p-norm techniques with different objective functions and constraints. However, all of the above one-timescale schemes suffer from huge CSI signaling overhead and large computation complexity due to global centralized optimization. Moreover, it is difficult to obtain instantaneous global CSI in a C-RAN with a large number of RRHs because of the signaling latency. Thus, these solutions are not scalable w.r.t. the network size.

Heuristic two-timescale schemes: In view of the challenges to implement the one-timescale schemes in practical C-RAN, some two-timescale schemes have been proposed where the RRH clustering is updated at slower timescale based on channel statistics, and the precoder is updated at each time slot based on instantaneous partial CSI from the active RRHs to the users. Since the number of active RRHs can be substantially smaller than the total number of RRHs in the network, such two-timescale design can significantly reduce the CSI signaling overhead and computatonal complexity. In [127], each user first chooses its serving BS cluster based on the channel statistics using the nearest base clustering and or the nearest interferer clustering. Then, the precoder is optimized using a soft interference nulling (SIN) precoding algorithm at each time slot. Liu and Lau [128] studied the joint optimization of RRH selection, regularization factor in RZF precoding, and power allocation, where the RRH selection is adaptive to the global path gain matrix using a heuristic algorithm based on the deterministic equivalent [129] of the average rate and power. In the existing two-timescale schemes, the RRH clustering and precoding solutions are obtained in a heuristic manner (i.e., the solution is not derived from a single joint optimiziation problem), and the performance gap of these heuristic solutions w.r.t. the optimal solution is still unknown.

One interesting future work in this direction is to develop an optimization based two-timescale joint RRH clustering and precoding scheme, where the long-term RRH clustering is adaptive to the channel statistics and the short-term precoding is adaptive to the instantaneous partial CSI from the active RRHs to the users. The optimization-based two-timescale solution gives implementation friendly and scalable solution because solutions of different timescales can be implemented at different hierarchies of C-RAN and they are robust to signaling latency. It is also expected to achieve better tradeoff between performance and implementation costs than the one-timescale scheme and heuristic two-timescale schemes. For example, similar two-timescale resource optimization schemes have been proposed in [130–132] for massive MIMO systems or heterogeneous networks and they are shown to achieve much better tradeoff between performance and various implementation costs. However, there are several technical challenges to achieve these good features of two-timescale resource optimization for C-RAN.

- *Two-stage stochastic non-convex optimization:* Due to the stochastic nature of the two-timescale optimization, there are usually expectation operators in the objective/constraint functions and the argument of the expectation operators involve the optimal short-term precoding solutions, which do not have closed form expressions. Moreover, the combinatorial nature of the RRH clustering optimization problem further complicates the problem.
- *Lack of Channel Statistics:* In practice, we might not even have the explicit knowledge of the channel statistics. The solution should be self-learning to the dynamic propagation environment.

2.4.2 Fronthaul loading compression/reduction

In conventional C-RAN, the BBU pool and RRHs exchange the baseband signals via low latency fronthaul links in the form of digitized IQ samples. Such design requires prohibitive fronthaul capacities in typical settings, and the capacity of fronthaul links has become a performance bottleneck in C-RAN [133]. Three major techniques have been proposed to reduce the fronthaul loading in the C-RAN: the RRH clustering which has already been reviewed above, the fronthaul compression, and the partially centralized C-RAN (PC-RAN). Various fronthaul compression schemes have been proposed based on vector quantization and possibly multivariate compression strategies [134–136]. In the uplink, one can also exploit the sparsity of the baseband signals (caused by bursty uplink traffic or random access) to further reduce the fronthaul loading [137] using compressive sensing (CS). The CS-based fronthaul compression method in [137] can only be applied to the case when the full CoMP is employed, and there is no inter-cluster interference. When partial CoMP is employed with RRH clustering, it is still possible to exploit the sparsity of the baseband signals and/or inter-cluster interference to do joint inter-cluster interference mitigation and fronthaul compression using the CS-based joint interference mitigation and detection algorithm in [138]. Recently, [139] proposed a PC-RAN architecture where part of the simple baseband processings such as modulation and precoding is shifted into RRHs. By doing so, precoder, data symbols, and channel information are separately transported instead of heavy IQ data. As a result, the fronthaul loading can be significantly reduced with slightly more complex RRHs.

However, the existing fronthaul loading compression/reduction schemes treat the information bits sent to users as random bits and do not exploit the fact that content (such as videos and musics), which are cachable at the RRH, contributes to a large portion of today's wireless traffic [1]. By caching some popular content at the RRHs, the RRHs may directly employ (partial) CoMP to transmit the cached content to the users without consuming the fronthaul. The BS-level PHY caching was first introduced in [140, 141] to exploit cached-induced opportunistic CoMP for cellular networks with limited backhaul capacity. In [140, 141], the BS-level PHY caching is designed for cellular networks where all the baseband processing is performed at the BS. Hence, the BS can directly cache content packets to exploit cached-induced opportunistic CoMP. However, in C-RAN where all baseband processing is implemented in the BBU, the RRH can only cache IQ samples. Since the IQ samples of the same content packet will be different under different modulation and coding schemes (MCS), it is challenging to design an efficient RRH-level PHY caching scheme to reduce the fronthaul loading, which could be an interesting research topic.

2.5 Software defined networking and wireless virtualization

Existing wireless cellular networks are based on standardized and inflexible hardware elements and network architecture, which hinder rapid deployment of new wireless applications, services, and adoption of emerging wireless technologies. It has been recognized that the SDN concept originally proposed for wired networks and data centers can be leveraged to engineer the 5G software defined wireless networks [142–146]. In general, SDN aims to separate the control and data forwarding planes, which enables to achieve quick provisioning and reconfiguration as well as more efficient utilization of the network infrastructure. An SDN-based network is realized by allowing network hardware elements (e.g., switches) to be programmed and controlled through a standardized interface such as OpenFlow [143] and by employing a network controller to make control decisions on the operations of the data forwarding infrastructure.

The network controller has a global view of the network by receiving updated traffic and network state information, which improves its decision-making efficiency. Moreover, network elements simply conduct required network functionality and protocols under the instruction of the centralized network controller. This saves a lot of time and resources and enables the operations and configuration of the multi-vendor network infrastructure to be realized more quickly and efficiently by software instead of custom configuration for many network elements deployed throughout the network.

Most existing research on SDN has been conducted for wired networks, and there are very few works addressing the SDN-based design for wireless cellular networks [147–149]. In fact, engineering the software defined wireless cellular network involves to develop the network and control architecture covering both wireless access and core networks as well as the corresponding interfaces [150, 151]. In addition, there is a strong demand in devising more flexible and efficient sharing of expensive wireless infrastructure among multiple (virtual) wireless service providers and operators using wireless virtualization technologies [152–158]. Sharing of the wireless cellular infrastructure among multiple parties can be applied for both wireless access and core networks [159]. This involves the sharing of different types of network resources including wireless spectrum and BSs, routers, and switches in the core network.

To realize such virtualization vision for wireless networks, network resources must be abstracted and partitioned into different "slices", which can be then allocated to different (virtual) wireless operators who provide services to the end users [156–158]. Development of such slicing mechanisms that guarantee efficient utilization of the network resources, satisfactory isolation among different slices to enable independent deployment of various network algorithms and protocols by the involved wireless operators and to provide QoS support for wireless users remains one major challenge to address in this interesting research theme. The SDN techniques can be leveraged to implement efficient virtualization of the future wireless cellular networks where the SDN controller can play an important role in the slicing and allocation operations of resource slices and in the control and activation of different network algorithms and protocols deployed by the corresponding wireless operators associated with different resource slices.

Efficient design of the software defined wireless cellular network is challenging because of several reasons. First, the wireless access network has become more and more complex to meet the exponential growth of the mobile traffic, the diverse QoS and throughput requirements of different wireless applications and services, and the need to achieve more cost-, energy-, and spectrum-efficient communications. Toward this end, engineering future wireless access networks follows some major trends including the evolution toward ultra-dense and multi-tiered deployment of different types of BSs (e.g., macro, micro, pico, and femto BSs), efficient utilization of the precious wireless spectrum in both microwave and mmWave bands using dynamic spectrum access and cognitive radio techniques, efficient coordination of radio resource management and transmission for better interference mitigation and spectrum utilization, development of advanced communication techniques such as massive and 3D MIMO, and employment of wireless access and backhaul communications. Therefore, optimization of the centralized access management for this complex wireless access presents a major challenge.

Development of an effective SDN architecture based on which control algorithms can be implemented for the software defined wireless cellular network is more challenging than that for the wired network. This is because the data forwarding plane in the wired network involves only routers, switches, and wired links with fixed capacity. In contrast, communication links in the wireless access network are strongly coupled since they share limited radio resources and simultaneous transmissions on one subband create co-channel interference for each other. Moreover, modern wireless cellular networks are based on sophisticated network algorithms and protocols, which adapt their operations to various system dynamics such as time-varying wireless channel state and traffic information.

Therefore, there would be a huge amount of control signaling between the SDN controller and wireless data forward planes (mobile users and BSs) if a fully centralized control architecture is adopted in which the SDN controller makes all control decisions. To achieve a scalable and efficient architecture, it is likely that different SDN controllers for the wireless access and core networks must be deployed, which closely interact with each other. This would lead to desirable tradeoff between network performance and control signaling overhead and provide more satisfactory end-to-end QoS performance for mobile users. Furthermore, the wireless access network may conduct various control functions which are based on highly dynamic system information such as wireless CSI in a semi-distributed manner using local controllers (e.g., employed at the BSs). Then, a well-designed network management framework can be developed so that these local controllers can efficiently interact with a centralized SDN controller, which makes more large-scale control decisions. For example, the CSI-driven wireless scheduling and radio resource management functions can be taken care by local controllers while load balancing and mobility control decisions can be realized by a centralized controller.

The control architecture for the core network of 4G and beyond wireless network could be established by leveraging significant advances on the SDN technology, which has been developed for the packet-switched networks. The fact that the core network of 4G wireless systems is based, the Evolved Packet Core (EPC) architecture established by the 3GPP [148] with well-separated control and data planes provides a suitable foundation for engineering this SDN architecture. Moreover, the SDN controller for the core network must be designed considering unique characteristics and requirements of different elements in the data forwarding plane such as mobility management, different service, and internet gateways. Furthermore, this design should be done so that it can support efficient sharing of the core network facility among multiple wireless operators in the future virtualized wireless mobile network.

The overall control architecture and interfaces for the next-generation SDN-based wireless network should also be established to enable efficient integration of many advanced physical-layer wireless technologies, to allow effective QoS provisioning for diverse set of wireless applications with very different characteristics and requirements, to support very large numbers of access nodes and wireless connections, and to foster the emerging active infrastructure sharing among multiple wireless operators via virtualization techniques. Specifically, the northbound interfaces should be appropriately defined to allow active sharing of wireless access and/or core networks among wireless operators and service providers where service providers can have sufficient power to control their traffic and optimize the assigned network resources. Moreover, the southbound interfaces should be engineered to achieve efficient utilization of network infrastructure and flexible programmable capability to provision and control the resource management and sharing operations under different RAN and wireless technologies.

An SDN architecture for the heterogeneous wireless dense network is proposed in [160], and another proposal of an SDN architecture for LTE-evolved packet core with design and analysis of five different procedures is discussed in [161] of our special issue. Development of efficient resource allocation solutions for future SDN-based wireless networks is the topic considered in two other papers of the special issue [162, 163]. Specifically, a three-step optimization framework to optimize the energy efficiency of the wireless network via adaptation of the user and control planes of the SDN-based wireless network is proposed in [162]. Moreover, an SDN-based resource allocation and load balancing algorithm is developed to enhance the spectrum efficiency and users' QoS of the multi-tier and multi-RAT wireless network in [163].

There are still many open research challenges related to the development of different network and resource management and optimization tools to be deployed on the SDN control and interface architecture. Research on developing these important tools also has strong impacts on the design and tuning of the overall architecture since they are closely interacting elements of the software defined wireless cellular network. Examples of the tools to be developed are network and radio resource management, mobility and load balancing, traffic engineering, spectrum allocation and management, radio resource, and networking slicing tools. It is also worth mentioning that there are strong interactions and synergy between research on SDN and C-RAN for future 5G mobile wireless systems since both technologies are based on the centralized optimization and management concepts of network resources and functionalities. In particular, data centers can be deployed to implement the SDN controller, which realizes various critical network management functions such as baseband processing, radio resource, mobility, and spectrum management. Obviously, there is much room for further research innovations as we evolve toward a unified, flexible, and efficient wireless ecosystem.

3 CONCLUSION

There have been some significant research advances in different 5G enabling wireless technologies and these research outcomes will form strong foundations for the 5G standardization activities, which are expected to start soon. Until then, it is desirable to develop a unified 5G architecture for wireless access and core networks that enable to integrate many advanced wireless technologies to meet the 5G requirements: realizing Gigabit, real-time, and ultra-reliable communications, supporting an extremely large number of wireless connections for both human and machine wireless applications, diverse set of mobile traffic with different characteristics, and QoS, and achieving more spectrum, energy, and cost-efficiency. There would be no single technology that can deliver these performance expectations. Therefore, a number of enabling 5G wireless technologies must be developed and jointly implemented on the newly developed 5G architecture. In this paper, we have reviewed the 5G requirements and discussed several key enabling 5G wireless technologies with their open research challenges. Moreover, we have also described research papers in our special issue in the corresponding 5G network and technology contexts.

ACKNOWLEDGEMENTS

The guest editors wish to thank all the authors for their contributions to this special issue. We would also like to thank all reviewers for their detailed and constructive review comments for all papers. We would like to further thank Prof. Luc Vandendorpe, Editor-in-Chief, EURASIP Journal on Wireless Communications and Networking as well as the whole Editorial Board for their support. Finally, we wish to express our sincere thanks to the whole editorial and publications staff of EURASIP Journal on Wireless Communications and Networking for their assistance and support of the special issue.

REFERENCES

1. Cisco, Cisco visual network index: Global mobile traffic forecast update 2014-2019. White paper (2015).
2. G Wu, S Talwar, K Johnsson, N Himayat, M2M: from mobile to embedded Internet. IEEE Commun. Mag. 49(4), 36–43 (2011)
3. N Bhushan, J Li, D Malladi, R Gilmore, D Brenner, A Damnjanovic, R Sukhavasi, C Patel, S Geirhofer, Network densification: the dominant theme for wireless evolution into 5G. IEEE Commun. Mag. 52(2), 82–89 (2014).
4. F Boccardi, RW Heath, A Lozano, TL Marzetta, P Popovski, Five disruptive technology directions for 5G. IEEE Commun. Mag. 52(2), 74–80 (2014).
5. R Baldemair, T Irnich, K Balachandran, E Dahlman, G Mildh, Y Seleń, S Parkvall, M Meyer, A Osseiran, Ultra-dense networks in

millimeter-wave frequencies. IEEE Commun. Mag. 53(1), 202–208 (2015).
6. GP Fettweis, The tactile internet: applications and challenges. IEEE Veh. Technol. Mag. 9(1), 64–70 (2014).
7. A Osseiran, F Boccardi, V Braun, K Kusume, P Marsch, M Maternia, O Queseth, M Schellmann, H Schotten, H Taoka, H Tullberg, MA Uusitalo, B Timus, M Fallgren, Scenarios for 5G mobile and wireless communications: the vision of the METIS project. IEEE Commun. Mag. 52(5), 26–35 (2014).
8. G Wunder, P Jung, M Kasparick, T Wild, F Schaich, Y Chen, S Brink, I Gaspar, N Michailow, A Festag, L Mendes, N Cassiau, D Ktenas, M Dryjanski, S Pietrzyk, B Eged, P Vago, F Wiedmann, 5GNOW: non-orthogonal, asynchronous waveforms for future mobile applications. IEEE Commun. Mag. 52(2), 97–105 (2014).
9. B Farhang-Boroujeny, OFDM versus filter bank multicarrier. IEEE Signal Process Mag. 28(3), 92–112 (2011).
10. Z Pi, F Khan, An introduction to millimeter-wave mobile broadband systems. IEEE Commun. Mag. 49(6), 101–107 (2011).
11. S TS Rappaport, R Sun, H Mayzus, Y Zhao, K Azar, GN Wang, JK Wong, M Schulz, F Samimi, Gutierrez, Millimeter wave mobile communications for 5G cellular: it will work. IEEE Access. 1, 335–349 (2013).
12. JG Andrews, What will 5G be. IEEE J. Sel. Areas Commun. 32(6), 1065–1082 (2014).
13. V Chandrasekhar, JG Andrews, A Gatherer, Femtocell networks: a survey. IEEE Commun. Mag. 46(9), 59–67 (2008).
14. MN Tehrani, M Uysal, H Yanikomeroglu, Device-to-device communication in 5G cellular networks: challenges, solutions, and future directions. IEEE Commun. Mag. 52(5), 86–92 (2014).
15. N NaderiAlizadeh, S Avestimehr, ITLinQ: a new approach for spectrum sharing in device-to-device communication systems. IEEE J. Sel. Areas Commun. 32(6), 1573–1577 (2014).
16. F Rusek, D Persson, BK Lau, E Larsson, T Marzetta, O Edfors, F Tufvesson, Scaling up MIMO: opportunities and challenges with very large arrays. IEEE Signal Process. Mag. 30(1), 40–60 (2013).
17. Y-H Nam, BL Ng, K Sayana, Y Li, J Zhang, Y Kim, J Lee, Full-dimension, MIMO (FD-MIMO) for next generation cellular technology. IEEE Commun. Mag. 51(6), 172–179 (2013).
18. A Kammoun, H Khanfir, Z Altman, M Debbah, M Kamoun, Preliminary results on 3D channel modeling: from theory to standardization. IEEE J. Sel. Areas Commun. 32(6), 1219–1229 (2014).

19. P Rost, CJ Bernardos, AD Domenico, MD Girolamo, M Lalam, A Maeder, D Sabella, D Wübben, Cloud technologies for flexible 5G radio access networks. IEEE Commun. Mag. 52(5), 68–76 (2014).
20. S Sezer, Are we ready for SDN? Implementation challenges for software-defined networks. IEEE Commun. Mag, 36–43 (2013).
21. C Liang, FR Yu, Wireless network virtualization: a survey, some research issues and challenges. IEEE Commun. Surveys & Tutorials. 17(1), 358–380 (2014).
22. E Bjornson, E Jorswieck, M Debbah, B Ottersten, Multi-objective signal processing optimization: the way to balance conflicting metrics in 5G systems. IEEE Signal Process. Mag. 31(6), 14–23 (2014).
23. S Sun, TS Rappaport, RW Heath Jr., A Nix, Rangan S, MIMO for millimeter-wave wireless communications: beamforming, spatial multiplexing, or both?IEEE Commun. Mag. 52(12), 110–121 (2014).
24. JF Monserrat, G Mange, V Braun, H Tullberg, G Zimmermann, 'O Bulakci, METIS research advances towards the 5G mobile and wireless mobile system definition. EURASIP J. Wirel. Commun. Netw, 53 (2015).
25. A Chiumento, M Bennis, C Desset, L der Perre, S Pollin, Adaptive CSI and feedback estimation in LTE and beyond: a Gaussian process regression approach. EURASIP J. Wirel. Commun. Netw. 2015, 168 (2015).
26. X Yin, C Ling, M-D Kim, H Chung, Parameter estimation using the sliding-correlator's output for wideband propagation channels. EURASIP J. Wirel. Commun. Netw. 2015, 165 (2015).
27. Y Shibata, N Yoda, T Ohtsuki, J Mashino, T Sugiyama, Iterative estimation of undesired signal power for superposed multicarrier transmission with channel estimation error. EURASIP J. Wirel. Commun. Netw. 2015, 23 (2015).
28. Z Liu, W Xu, S Li, J Lin, Network-coded primary-secondary cooperation in OFDM-based cognitive multicast networks. EURASIP J. Wirel. Commun. Netw. 2015, 144 (2015).
29. C Ling, X Yin, S Boqué, M García-Lozano, Power minimization of cooperative beamforming networks with spectrum sharing. EURASIP J. Wirel. Commun. Netw. 2015, 72 (2015).
30. A Barbieri, D Fertonani, G Colavolpe, Time-frequency packing for linear modulations: spectral efficiency and practical detection schemes. IEEE Trans. Commun. 57(10), 2951–2959 (2009).
31. F Rusek, J Anderson, Multistream faster than Nyquist signaling. IEEE Trans. Commun. 57(5), 1329–1340 (2009).
32. J Anderson, F Rusek, V Owall, Faster-than-Nyquist signaling. Proc. IEEE. 101(8), 1817–1830 (2013).

33. MJ Abdoli, M Jia, in Proc. IEEE VTC-Fall. Trellis decoding for multi-user faster-than-Nyquist transmission, (2014).
34. N Michailow, Matthé M, IS Gaspar, AN Caldevilla, LL Mendes, A Festag, G Fettweis, Generalized frequency division multiplexing for 5th generation cellular networks. IEEE Trans. Commun. 62(9), 3045–3061 (2010).
35. G Fettweis, M Krondorf, S Bittner, in Proc. IEEE Vehicular Technology Conference. GFDM: generalized frequency division multiplexing, (2009), pp. 1–4.
36. D Dasalukunte, F Rusek, V wall, Multicarrier faster-than-Nyquist signaling transceivers: hardware architecture and performance analysis. IEEE Trans. Circ. Syst. I, Reg. Papers. 58(4), 827–838 (2011).
37. MJ Abdoli, M Jia, J Ma, in Proc. IEEE PIMRC. Weighted circularly convolved filtering in OFDM/OQAM, (2013).
38. M Caus, AI Perez-Neira, Multi-stream transmission for highly frequency selective channels in MIMO-FBMC/OQAM systems. IEEE Trans. Signal Process. 62(4), 786–796 (2014).
39. M Caus, AI Perez-Neira, Transmitter-receiver designs for highly frequency selective channels in MIMO FBMC systems. IEEE Trans. Signal Process. 60(12), 6519–6532 (2012).
40. V Vakilian, T Wild, F Schaich, Brink St, J-F Frigon, in Proc. 9th International Workshop on Broadband Wireless Access, IEEE Globecom. Universal filtered multi-carrier technique for wireless systems beyond LTE, (2013).
41. T Wild, F Schaich, Y Chen, in Proc. Int. Conf. Dig. Signal Process. (DSP). 5G air interface design based on universal filtered (UF-) OFDM, (2014), pp. 699–704.
42. J Abdoli, M Jia, J Ma, in Proc. IEEE Int'l Workshop on Signal Processing Advances in Wireless Communications (SPAWC). Filtered OFDM: a new waveform for future wireless systems, (2015).
43. M Matthe, LL Mendes, I Gaspar, N Michailow, D Zhang, G Fettweis, Multi-user time-reversal STC-GFDMA for future wireless networks. EURASIP J. Wirel. Commun. Netw. 2015, 132 (2015).
44. K Higuchi, Y Kishiyama, in Proc. APWCS 2012. Non-orthogonal access with successive interference cancellation for future radio access, (2012).
45. H Nikopour, H Baligh, in Proc. IEEE PIMRC. Sparse code multiple access, (2013).
46. M Taherzadeh, H Nikopour, A Bayesteh, H Baligh, in Proc. IEEE VTC-Fall. CMA codebook design, (2014).

47. SN Premnath, D Wasden, SK Kasera, N Patwari, B Farhang-Boroujeny, Beyond OFDM: best-effort dynamic spectrum access using filterbank multicarrier. IEEE/ACM Trans. Netw. 21(3), 869–882 (2013).
48. HS Dhillon, RK Ganti, F Baccelli, JG Andrews, Modeling and analysis of K-tier downlink heterogeneous cellular networks. IEEE J. Sel. Areas Commun. 30(3), 55060 (2012).
49. V Chandrasekhar, JG Andrews, A gatherer femtocell networks: a survey. IEEE Commun. Mag. 46(9), 59–67 (2008).
50. N Saquib, E Hossain, LB Le, DI Kim, Interference management in OFDMA femtocell networks: issues and approaches. IEEE Wirel. Commun. 19(3), 86–95 (2012).
51. LB Le, D Niyato, E Hossain, DI Kim, DT Hoang, QoS-aware and energy-efficient resource management in OFDMA femtocells. IEEE Trans. Wirel. Commun. 12(1), 180–194 (2013).
52. Q Ye, B Rong, Y Chen, M Al-Shalash, C Caramanis, JG Andrews, User association for load balancing in heterogeneous cellular networks. IEEE Trans. Wirel. Commun. 12(6), 2706–2716 (2013).
53. K Shen, W Yu, Distributed pricing-based user association for downlink heterogeneous cellular networks. IEEE J. Sel. Areas Commun. 32(6), 1100–1113 (2014).
54. VN Ha, LB Le, Distributed base station association and power control for heterogeneous cellular networks. IEEE Trans. Vehicular Technol. 63(1), 282–296 (2014).
55. S Singh, JG Andrews, Joint resource partitioning and offloading in heterogeneous cellular networks. IEEE Trans. Wirel. Commun. 13(2), 888–901 (2014).
56. S Deb, P Monogioudis, J Miernik, JP Seymour, Algorithms for enhanced inter-cell interference coordination (eICIC) in LTE HetNets. IEEE/ACM Trans. Netw. 22(1), 13750 (2014).
57. E Hossain, LB Le, D Niyato, Radio Resource Management in Multi-Tier Cellular Wireless Networks (Wiley, New York, NY, USA, 2013).
58. D Lopez-Perez, D Lopez-Perez, I Guvenc, G de la Roche, M Kountouris, TQS Quek, J Zhang, Enhanced intercell interference coordination challenges in heterogeneous networks. IEEE Wirel. Commun. 18(3), 22–30 (2011).
59. D Nguyen, LB Le, T Le-Ngoc, Multiuser admission control and beamforming optimization algorithms for MISO heterogeneous networks. IEEE Access. 3, 759–773 (2015).
60. VN Ha, LB Le, Fair resource allocation for OFDMA femtocell networks with macrocell protection. IEEE Trans. Vehicular Technol. 63(3), 1388–1401 (2014).

61. DT Ngo, LB Le, T Le-Ngoc, Distributed Pareto-optimal power control for utility maximization in femtocell networks. IEEE Trans. Wirel. Commun. 11(10), 3434–3446 (2012).
62. DT Ngo, LB Le, T Le-Ngoc, E Hossain, DI Kim, Distributed interference management in two-tier CDMA femtocell networks. IEEE Trans. Wirel. Commun. 11(3), 979–989 (2012).
63. FH Panahi, T Ohtsuki, Stochastic geometry modeling and analysis of cognitive heterogeneous cellular networks. EURASIP J. Wirel. Commun. Networking. 2015, 141 (2015).
64. Q Zhang, T Yang, Y Zhang, Z Feng, Fairness guaranteed novel eICIC technology for capacity enhancement in multi-tier heterogeneous cellular networks. EURASIP J. Wirel. Commun. Networking. 2015, 62 (2015).
65. Y Sui, I Guvenc, T Svensson, Interference management for moving networks in ultra-dense urban scenarios. EURASIP J. Wirel. Commun. Networking. 2015, 111 (2015).
66. TD Hoang, LB Le, T Le-Ngoc, in Proc. IEEE GLOBECOM. Dual decomposition method for energy-efficient resource allocation in D2D communications underlying cellular networks, (2015).
67. TD Hoang, LB Le, T Le-Ngoc, in Proc. IEEE ICC. Energy-efficient resource allocation for D2D communications in cellular networks, (2015).
68. TD Hoang, LB Le, T Le-Ngoc, in Proc. IEEE GLOBECOM. Resource allocation for D2D communications under proportional fairness, (2014).
69. LB Le, in Proc. IEEE GLOBECOM. Fair resource allocation for device-to-device communications in wireless cellular networks, (2012).
70. Q Wang, C Lai, Y Dong, XXu Y Shu, Joint user clustering and resource allocation for device-to-device communication underlaying MU-MIMO cellular networks. EURASIP J. Wirel. Commun. Networking. 2015, 145 (2015).
71. S Shah, J Gu, S Hasan, M Chung, SC-FDMA-based resource allocation and power control scheme for D2D communication using LTE-A uplink resource. EURASIP J. Wirel. Commun. Networking. 2015, 137 (2015).
72. W Roh, J-Y Seol, J Park, B Lee, J Lee, Y Kim, J Cho, K Cheun, F Aryanfar, Millimeter-wave beamforming as an enabling technology for 5G cellular communications: theoretical feasibility and prototype results. IEEE Commun. Mag. 52(2), 106–113 (2014).
73. T Rappaport, F Gutierrez, E Ben-Dor, J Murdock, Y Qiao, J Tamir, Broadband millimeter-wave propagation measurements and models using adaptive-beam antennas for outdoor urban cellular

communications. IEEE Trans. Antennas Propag. 61(4), 1850–1859 (2013).
74. Mathias C, Fixing the cellular network: backhaul is the key (2008). http://www.networkworld.com/community/print/35920. Accessed Dec 2008.
75. It all comes back to backhaul, Ericsson white paper (2014). http://www.ericsson.com/res/docs/whitepapers/WP-Heterogeneous-Networks-Backhaul.pdf. Accessed Sep 2015.
76. S Chia, M Gasparroni, P Brick, The next challenge for cellular networks: backhaul. IEEE Microwave Mag. 10(5), 54–66 (2009).
77. TS Rappaport, JN Murdock, F Gutierrez, State of the art in 60-GHz integrated circuits and systems for wireless communications. Proc. IEEE. 99(8), 1390–1436 (2011).
78. H Zhang, S Venkateswaran, U Madhow, in Proc. IEEE Globecom. Analog multitone with interference suppression: relieving the ADC bottleneck for wideband 60 Ghz systems, (2012).
79. S O El Ayach, S Rajagopal, Z Abu-Surra, R Pi, Heath, Spatially sparse precoding in millimeter wave MIMO systems. IEEE Trans. Wirel. Commun. 13(3), 1499–1513 (2014).
80. A Alkhateeb, O El Ayach, G Leus, R Heath, Channel estimation and hybrid precoding for millimeter wave cellular systems. IEEE J. Sel. Topics Signal Process. 8(5), 831–846 (2014).
81. S Hur, T Kim, DJ Love, JV Krogmeier, TA Thomas, A Ghosh, Millimeter wave beamforming for wireless backhaul and access in small cell networks. IEEE Trans. Commun. 61(10), 4391–4403 (2013).
82. T Bogale, LB Le, A Haghighat, in Proc. IEEE ICC. User scheduling for massive MIMO OFDMA systems with hybrid analog-digital beamforming, (2015).
83. T Bogale, LB Le, in Proc. IEEE GLOBECOM. Beamforming for multiuser massive MIMO systems: digital versus hybrid analog-digital, (2014).
84. H Ishii, Y Kishiyama, H Takahashi, in Proc. Globecom Workshops. A novel architecture for LTE-B: C-plane/U-plane split and phantom cell concept, (2012).
85. Q Li, H Niu, G Wu, RQ Hu, in Proc. IEEE Globecom Workshop. Anchor-booster based heterogeneous networks with mmWave capable booster cells, (2013).
86. T Marzetta, Noncooperative cellular wireless with unlimited numbers of base station antennas. IEEE Trans. Wirel. Commun. 9(11), 3590–600 (2010).

87. H Yang, TL Marzetta, in Proc. IEEE GreenCom. Total energy efficiency of cellular large scale antenna system multiple access mobile networks, (2013).
88. E Björnson, L Sanguinetti, J Hoydis, M Debbah, Optimal design of energy-efficient multi-user MIMO systems: is massive MIMO the answer?IEEE Trans. Wirel. Commun. 14(2), 3059–3075 (2015).
89. W Liu, A Zappone, C Yang, E Jorswieck, in Proc IEEE SPAWC. Global EE optimization of massive MIMO systems, (2015).
90. J Hoydis, S ten Brink, M Debbah, Massive MIMO in the UL/DL of cellular networks: how many antennas do we need. IEEE J.Sel. Areas Commun. 31(2), 160–171 (2013).
91. H Yang, TL Marzetta, Performance of conjugate and zero-forcing beamforming in large-scale antenna systems. IEEE J. Sel. Areas Commun. 31(2), 172–179 (2013).
92. X Rao, VKN Lau, Distributed compressive CSIT estimation and feedback for FDD multi-user massive MIMO systems. IEEE Trans. Signal Process. 62(12), 3261–3271 (2014).
93. J Chen, VKN Lau, Two-tier precoding for FDD multi-cell massive MIMO time-varying interference networks. IEEE J. Sel. Areas Commun. 32(6), 1230–1238 (2014).
94. T Bogale, LB Le, X Wang, L Vandendorpe, in Proc. IEEE GLOBECOM. Pilot contamination mitigation for wideband massive MMO: number of cells vs multipath, (2015).
95. H Yin, D Gesbert, M Filippou, Y Liu, A coordinated approach to channel estimation in large-scale multiple-antenna systems. IEEE J. Sel. Areas Commun. 31(2), 264–273 (2013).
96. T Nguyen, VN Ha, LB Le, Resource allocation optimization in multi-user multi-cell massive MIMO networks considering pilot contamination. IEEE Access. 3, 1272–1287 (2015).
97. J E Björnson, M Hoydis, M Kountouris, Debbah, Massive MIMO systems with non-ideal hardware: energy efficiency, estimation, and capacity limits. IEEE Trans. Inform. Theory. 60(11), 7112–7139 (2014).
98. MM Mansour, A low-complexity MIMO subspace detection algorithm. EURASIP J. Wirel. Commun. Networking. 2015, 95 (2015).
99. H Zhang, X Zheng, W Xu, X You, On massive MIMO performance with semi-orthogonal pilot-assisted channel estimation. EURASIP J. Wirel. Commun. Networking. 2014, 220 (2014).
100. Z Wu, X Gao, An efficient MIMO scheme with signal space diversity for future mobile communications. EURASIP J Wirel. Commun. Networking. 2015, 87 (2015).

101. J Kim, SH Park, H Sung, I Lee, in Proc IEEE ICC. Sum rate analysis of two-cell MIMO broadcast channels: spatial multiplexing gain, (2010).
102. SA Jafar, Interference alignment—a new look at signal dimensions in a communication network. Found. Trends in Commun. Inf. Theory. 7(1), 1–134 (2011).
103. VR Cadambe, SA Jafar, Interference alignment and degrees of freedom of the K-user interference channel. IEEE Trans. Inf. Theory. 54(5), 3425–3441 (2008).
104. V Nagarajan, B Ramamurthi, Distributed cooperative precoder selection for interference alignment. IEEE Trans. Veh. Technol. 59(9), 4368–4376 (2010).
105. C Suh, M Ho, DNC Tse, Downlink interference alignment. IEEE Trans. Commun. 59(9), 2616–2626 (2011).
106. CM Yetis, T Gou, SA Jafar, AH Kayran, Feasibility conditions for interference alignment. IEEE Trans. Signal Process. 58, 4771–4782 (2010).
107. T Kim, DJ Love, B Clerckx, D Hwang, in Proc. IEEE Globecom. Spatial degrees of freedom of the multi-cell multiple access channel, (2011).
108. W Shin, N Lee, J-B Lim, C-Y Shin, K Jang, On the design of interference alignment scheme for two-cell MIMO interfering broadcast channels. IEEE Trans. Wirel. Commun. 10(2), 437–442 (2011).
109. J Tang, S Lambotharan, Interference alignment techniques for MIMO multi-cell interfering broadcast channels. IEEE Trans. Commun. 61(1), 4907–4922 (2013).
110. S Wagner, R Couillet, M Debbah, DTM Slock, Large system analysis of linear precoding in correlated MISO broadcast channels under limited feedback. IEEE Trans. Inform. Theory. 58(7), 4509–4537 (2012).
111. C-N Chuah, DNC Tse, JM Kahn, RA Valenzuela, Capacity scaling in MIMO wireless systems under correlated fading. IEEE Trans. Inform. Theory. 48(2), 637–650 (2002).
112. A Moustakas, S Simon, A Sengupta, MIMO capacity through correlated channels in the presence of correlated interferers and noise: a (not so) large N analysis. IEEE Trans Inform. Theory. 49(10), 2545–2561 (2003).
113. B Hochwald, T Marzetta, V Tarokh, Multi-antenna channel hardening and its implications for rate feedback and scheduling. IEEE Trans. Inform. Theory. 50(9), 1893–1909 (2004).

114. HQ Ngo, EG Larsson, TL Marzetta, The multicell multiuser MIMO uplink with very large antenna arrays and a finite-dimensional channel. IEEE Trans. Commun. 61(6), 2350–2361 (2013).
115. C Guthy, W Utschick, ML Honig, Large system analysis of sum capacity in the Gaussian MIMO broadcast channel. IEEE J. Sel. Areas Commun. 31(2), 149–159 (2013).
116. H Huh, AM Tulino, G Caire, Network MIMO with linear zero-forcing beamforming: large system analysis, impact of channel estimation and reduced-complexity scheduling. IEEE Trans. Inform. Theory. 58(5), 2911–2934 (2012).
117. A Checko, HL Christiansen, Y Yan, L Scolari, G Kardaras, MS Berger, L Dittmann, Cloud RAN for mobile networks—a technology overview. IEEE Comm. Surveys and Tutorials. 17(1), 405–426 (2014).
118. R Irmer, H Droste, P Marsch, M Grieger, G Fettweis, S Brueck, H-P Mayer, L Thiele, V Jungnickel, Coordinated multipoint: concepts, performance, and field trial results. IEEE Commun. Mag. 49(2), 102–111 (2011).
119. VN Ha, LB Le, N-D Dao, in Proc. IEEE CISS. Energy-efficient coordinated transmission for Cloud-RANs: algorithm design and tradeoff, (2014).
120. VN Ha, LB Le, in Proc IEEE GLOBECOM. Joint coordinated beamforming and admission control for fronthaul constrained Cloud-RANs, (2014).
121. B Dai, W Yu, Sparse beamforming and user-centric clustering for downlink cloud radio access network. IEEE Access. 2, 1326–1339 (2014).
122. J-M Moon, D-H Cho, in Proc. IEEE 73rd Vehicular Technology Conference (VTC Spring). Inter-cluster interference management based on cell-clustering in network MIMO systems, (2011), pp. 1–6.
123. P Baracca, F Boccardi, V Braun, in Proc International Symposium on Wireless Communication Systems (ISWCS). A dynamic joint clustering scheduling algorithm for downlink CoMP systems with limited CSI, (2012), pp. 830–834.
124. M Hong, R Sun, H Baligh, Z-Q Luo, Joint base station clustering and beamformer design for partial coordinated transmission in heterogeneous networks. IEEE J. Select. Areas Commun. 31(2), 226–240 (2013).
125. Y Shi, J Zhang, K Letaief, Group sparse beamforming for green cloud-ran. IEEE Trans. Wirel. Commun. 13(5), 2809–2823 (2014).
126. B Dai, W Yu, in IEEE GLOBECOM. Sparse beamforming for limited-backhaul network MIMO system via reweighted power minimization, (2013), pp. 1962–1967.

127. C Ng, H Huang, Linear precoding in cooperative MIMO cellular networks with limited coordination clusters. IEEE J. Select. Areas Commun. 28(9), 1446–1454 (2010).MathSciNetView Article
128. A Liu, V Lau, Joint power and antenna selection optimization in large cloud radio access networks. IEEE Trans. Signal Process. 62(5), 1319–1328 (2014).
129. S Wagner, R Couillet, M Debbah, DTM Slock, Large system analysis of linear precoding in correlated MISO broadcast channels under limited feedback. IEEE Trans. Inform. Theory. 58(7), 4509–4537 (2012).
130. A Liu, V Lau, Hierarchical interference mitigation for massive MIMO cellular networks. IEEE Trans. Signal Process. 62(18), 4786–4797 (2014).
131. A Liu, V Lau, L Ruan, J Chen, D Xiao, Hierarchical radio resource optimization for heterogeneous networks with enhanced inter-cell interference coordination (eICIC). IEEE Trans. Signal Process. 62(7), 1684–1693 (2014).
132. A Liu, VKN Lau, Phase only RF precoding for massive MIMO systems with limited RF chains. IEEE Trans. Signal Process. 62(17), 4505–4515 (2014).
133. M Peng, C Wang, V Lau, H Poor, Fronthaul-constrained cloud radio access networks: insights and challenges. IEEE Wirel. Commun. 22(2), 152–160 (2015).
134. S-H Park, O Simeone, O Sahin, S Shamai, Joint precoding and multivariate backhaul compression for the downlink of cloud radio access networks. IEEE Trans. Signal. Process. 61(22), 5646–5658 (2013).
135. SH Park, O Simeone, O Sahin, S Shamai, Joint decompression and decoding for cloud radio access networks. IEEE Signal Process. Letters. 20(5), 503–506 (2013).
136. P Patil, W Yu, in Proc. Information Theory and Applications Workshop (ITA). Hybrid compression and message-sharing strategy for the downlink cloud radio-access network, (2014), pp. 1–6.
137. X Rao, V Lau, Distributed fronthaul compression and joint signal recovery in Cloud-RAN. IEEE Trans. Signal Process. 63(4), 1056–1065 (2015).
138. A Liu, V Lau, Joint interference mitigation and data recovery in compressive domain: a sparse MLE approach. IEEE Trans. Signal Process. 62(19), 5184–5195 (2014).
139. S Park, C-B Chae, S Bahk, Large-scale antenna operation in heterogeneous cloud radio access networks: a partial centralization

approach (2015). [Online] Available: http://arxiv.org/abs/1405.2295. Accessed Sep 2015. Accessed Sep 2015.

140. A Liu, V Lau, Cache-enabled opportunistic cooperative MIMO for video streaming in wireless systems. IEEE Trans. Signal Process. 62(2), 390–402 (2014).
141. A Lui, UKN Lau, Mixed-timescale precoding and cache control in cached MIMO interference network. IEEE Trans. Signal Process. 61(24), 6320–6332 (2013).
142. BAA Nunes, M Mendonca, X-N Nguyen, K Obraczka, T Turletti, A survey of software-defined networking: past, present, and future of programmable networks. IEEE Commun. Surveys & Tutorials. 16(3), 1617–1634 (2014).
143. N McKeown, Openflow: enabling innovation in campus networks. SIGCOMM Comput. Commun. Rev. 38(2), 69–74 (2008).
144. H Kim, N Feamster, Improving network management with software defined networking. IEEE Commun. Mag. 51(2), 114–119 (2013).
145. F Hu, Q Hao, K Bao, A survey on software-defined network and openFlow: from concept to implementation. IEEE Commun. Surveys & Tutorials. 16(4), 2181–2206 (2014).
146. Y Jarraya, T Madi, M Debbabi, A survey and a layered taxonomy of software-defined networking. IEEE Commun. Surveys & Tutorials. 16(4), 1955–1980 (2014).
147. A Gudipati, D Perry, LE Li, S Katti, in Proc. ACM SIGCOMM workshop on Hot topics in software defined networking. SoftRAN: software defined radio access network, (2013).
148. K Pentikousis, Y Wang, W Hu, Mobileflow: toward software-defined mobile networks. IEEE Commun. Mag. 51(7), 44–53 (2013).
149. CJ Bernardos, A De La Oliva, P Serrano, A Banchs, LM Contreras, H Jin, JC Zuńiga, An architecture for software defined wireless networking. IEEE Wirel. Commun. 21(3), 52–61 (2014).View Article
150. H Farmanbar, H Zhang, in Prof. IEEE Network Operations and Management Symposium (NOMS). Traffic engineering for software-defined radio access networks, (2014).
151. N-D Dao, H Zhang, X Li, P Leroux, in Proc. IEEE Int'l Conf. Computing, Networking and Communications (ICNC). Radio access network coordination framework toward 5G mobile wireless networks, (2015).
152. NMMK Chowdhury, R Boutaba, A survey of network virtualization. Comput. Net. 54(5), 862–876 (2010).
153. R Kokku, R Mahindra, H Zhang, S Rangarajan, NVS: a substrate for virtualizing wireless resources in cellular networks. IEEE/ACM Trans. Networking. 20(5), 1333–1346 (2012).

154. F Fu, UC Kozat, in Proc. INFOCOM. Wireless network virtualization as a sequential auction game, (2010).
155. X Costa-Pérez, J Swetina, T Guo, R Mahindra, S Rangarajan, Radio access network virtualization for future mobile carrier networks. IEEE Commun. Mag. 51(7), 27–35 (2013).
156. M Kamel, LB Le, A Girard, in IEEE VTC-Fall. LTE wireless network virtualization: dynamic slicing via flexible scheduling, (2014).
157. M Kamel, LB Le, A Girard, in Proc. IEEE WCNC. LTE multi-cell dynamic resource allocation for wireless network virtualization, (2015).
158. JS Panchal, RD Yates, MM Buddhikot, Mobile network resource sharing options: performance comparisons. IEEE Trans. Wirel. Commun. 12(9), 4470–4482 (2013).
159. H Zhang, S Vrzic, G Senarath, ND Dao, H Farmanbar, J Rao, C Peng, H Zhuang, 5G wireless network: MyNET and SONAC. IEEE Netw. 29(4), 14–23 (2015).
160. VM Borges, K Cardoso, E Cerqueira, M Nogueira, A Santos, Aspirations, challenges, and open issues for software-based 5G networks in extremely dense and heterogeneous scenarios. EURASIP J. Wirel. Commun. Networking. 2015, 164 (2015).
161. VG Nguyen, Y Kim, Proposal and evaluation of SDN-based mobile packet core networks. EURASIP J. Wirel. Commun. Networking. 2015, 172 (2015).
162. X Xu, X Dai, Y Liu, R Gao, X Tao, Energy efficiency optimization-oriented control plane and user plane adaptation with a frameless network architecture for 5G. EURASIP J. Wirel. Commun. Networking. 159(2015) (2015).
163. X Duan, AM Akhtar, X Wang, Software-defined networking-based resource management: data offloading with load balancing in 5G HetNet. EURASIP J. Wirel. Commun. Networking. 2015, 181 (2015).

3
CHAPTER

COMPARISON OF INDOOR LOCALIZATION SYSTEMS BASED ON WIRELESS COMMUNICATIONS

Rejane Dalce, Thierry Val, Adrien Van Den Bossche

ABSTRACT

Localization using a Wireless Sensor Network (WSN) has become a field of interest for researchers in the past years. This information is expected to aid in routing, systems maintenance and health monitoring. For example, many projects aiming to monitor the elderly at home include a personal area network (PAN) which can provide current location of the patient to the medical staff. This article presents an overview of the current trends in this domain. We introduce the mathematical tools used to determine position then we introduce a selection of range-free and range-based proposals. Finally, we provide a comparison of these techniques and suggest possible areas of improvement.

Keywords: Wireless Sensor Networks (WSNs), Localization, Range-Based, Range-Free

1. INTRODUCTION

The modern world is centered on communication and speed. We want to be informed as quickly and reliably as possible about our health, the state of our environment, our current safety. Wireless Sensor Networks (WSNs) are a type of network that has been developed as a response to this need for control. Their nodes aim to become part of our environment and, using embedded sensors, help us monitor the status of relevant parameters.

As the needs evolved, the attention has turned to mobile sensors, introducing a new constraint: the collected data must be mapped to a specific location in order to ensure a correct interpretation of the values. Let's consider a mobile temperature sensor. When a measurement exceeds a set threshold, the system's reaction will depend on the current location of the node. Each predefined zone will be associated to an acceptable temperature range according to the usual activities taking place in this space. For example, during winter, detecting a temperature below 4°C in a bedroom hints a possible failure of the heater. On the other hand, the same value in a garage is perfectly normal. For the system to choose between alerting the system manager and ignoring the new value, the sensor's location must be available.

Position aware network nodes can also help in routing [1]. The benefits of localization being shown, the question is: How do we determine it?

We are not the first group to find this topic interesting. Therefore, we plan on beginning with a description of the tools and methods generally used to solve the localization problem. We will focus on indoor solutions and ones that require minimum participation from the target. We will then introduce propositions that represent the current trends in location determination before identifying possible research areas and concluding this article.

2. CLASSICAL LOCALIZATION ALGORITHMS

When trying to localize an item, the first step is to define references. In this document, our references will be a subset of the network nodes. They will be aware of their location and will be referred to as "beacons". Another term frequently used for this purpose in the literature is "anchor".

The network's deployment area can be either indoor or outdoor. In the first case, the radio signal mainly suffers from attenuation and reflections. In the latter, setting up links on high distances and maintaining them becomes a great challenge.

Once the references have been established, relative location determination becomes possible. According tc the implementation choices, this operation can be executed either by the mobile node or by a central node with greater processing power. When the mobile node is responsible for calculating its own position, an efficient algorithm must be used. The other solution requires the installation of at least one localization server who will process requests from all nodes in the network.

The following subsections will introduce mathematical tools for positioning using 3 anchors as a reference.

2.1. Trilateration

This method requires the knowledge of the distance between the mobile and each beacon.

Let the anchors be B1, B2 and B3. M is the node we want to localize. The positions (xi, yi) of the beacons as well as the distances di are known. The relationship between M, B1 and B2 can be written as:

$$(x-x_1)^2+(y-y_1)^2=d_1^2 \tag{1}$$

$$(x-x_2)^2+(y-y_2)^2=d_2^2 \tag{2}$$

M is therefore located at the intersection of circles C1(B1,d1) and C2(B2, d2). A third circle generated using the information of B3 enables us to choose the correct position between the two solutions of the general case of two circles intersecting (Figure 1).

2.2. Triangulation

In order to apply this method, the mobile must be aware of the distance between two beacons and the angles α and β as shown in Figure 2. The law of sines yields:

$$a=\frac{d*\sin\alpha}{\sin(\pi-\alpha-\beta)} \tag{3}$$

$$b = \frac{d * \sin \alpha}{\sin(\pi - \alpha - \beta)} \tag{4}$$

Once the characteristics of the triangle are known, the coordinates of the third summit can be calculated. As stated while describing trilateration, the third beacon or an additional hypothesis will allow identification of the correct solution.

2.3. Fingerprinting

This method is a pattern-recognition technique. It comprises to steps: in the first step or learning stage, some characteristics of the environment are measured at different locations and the data is stored along with a spatial reference information. This step is also qualified as offline phase since it is usually performed before the activation of the localization service provided by the network.

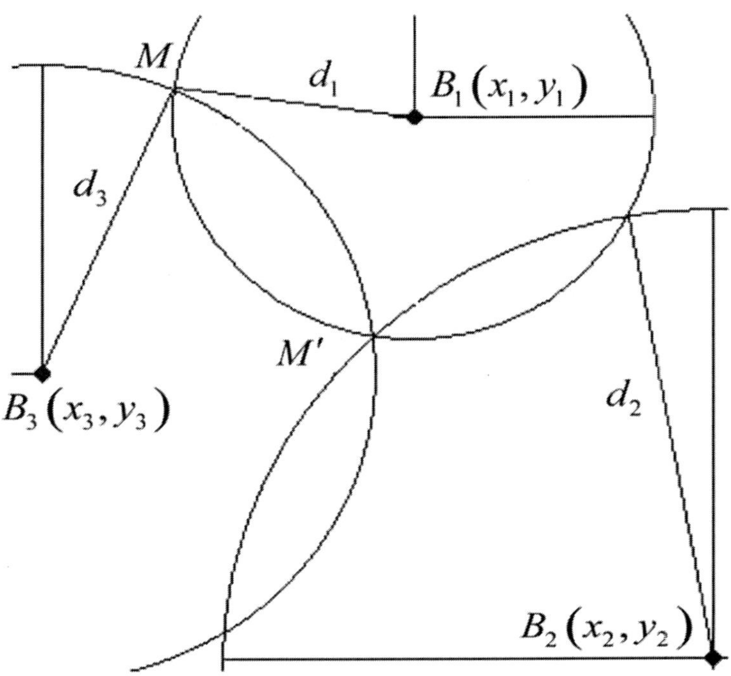

Figure 1. Trilateration.

The second step or online phase begins when the network is up and a mobile wants to determine its position. The same parameters are measured and the results are compared to the stored values.

The inherent modularity of fingerprinting enables researchers to work specifically on one phase or the other: hence the various propositions for selecting a position by comparing online and offline data.

According to the behavior of the measured characteristic through time, this method can be very interesting. Unfortunately, its cost in terms of setup time and data volume is high. Furthermore, any change in the configuration such as moving a beacon or modifying the environment, will imply creating a new database. In the case of a mobile WSN, the need for partial to complete updates will happen frequently.

2.4. Multilateration

Multilateration requires time information rather than distance. The time difference of arrival of a signal (TDOA) is used to determine position.

Figure 3 presents the case with the mobile being able to reach 4 beacons which are synchronized. The mobile node broadcasts a frame. Every beacon receiving the frame stores the local time of reception then calculates the difference between this time tag and the one provided by a set reference beacon. The collected time differences

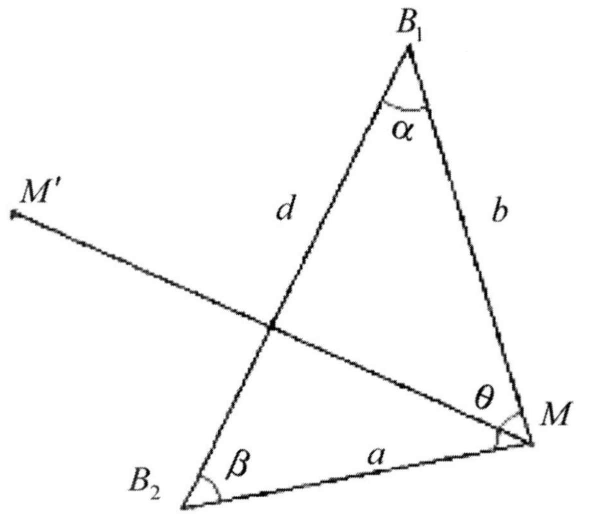

Figure 2. Triangulation.

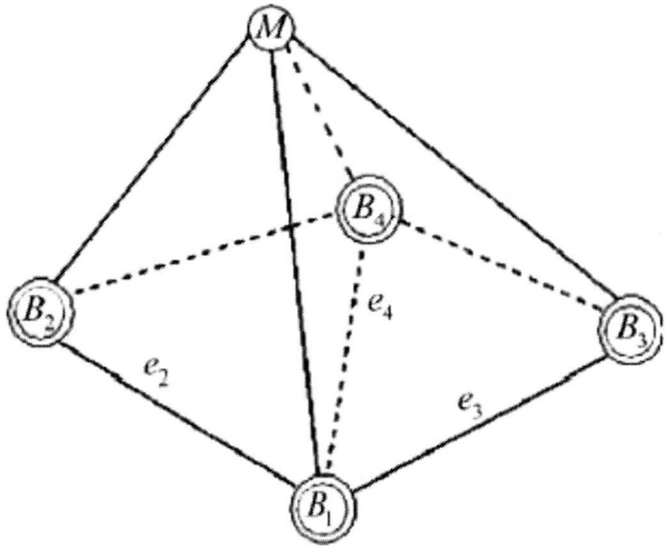

Figure 3. Multilateration.

are used to write the following equations:

$$0 = c\left(t_2 - t_4\right) - \frac{e_2^2}{ct_2} + \frac{e_4^2}{ct_4} + 2x\left(\frac{x_2}{ct_2} - \frac{x_4}{ct_4}\right) + 2y\left(\frac{y_2}{ct_2} - \frac{y_4}{ct_4}\right) + 2z\left(\frac{z_2}{ct_2} - \frac{z_4}{ct_4}\right) \quad (5)$$

$$0 = c\left(t_2 - t_3\right) - \frac{e_2^2}{ct_2} + \frac{e_3^2}{ct_3} + 2x\left(\frac{x_2}{ct_2} - \frac{x_3}{ct_3}\right) + 2y\left(\frac{y_2}{ct_2} - \frac{y_3}{ct_3}\right) + 2z\left(\frac{z_2}{ct_2} - \frac{z_3}{ct_3}\right) \quad (6)$$

$$0 = c(t_3 - t_4) - \frac{e_3^2}{ct_3} + \frac{e_4^2}{ct_4} + 2x\left(\frac{x_3}{ct_3} - \frac{x_4}{ct_4}\right) + 2y\left(\frac{y_3}{ct_3} - \frac{y_4}{ct_4}\right) + 2z\left(\frac{z_3}{ct_3} - \frac{z_4}{ct_4}\right) \quad (7)$$

where :

c: propagation speed

(xi, yi): beacon i coordinates

ei: distance from the chosen reference to node Bi. For the situation depicted in Figure 3, B1 is the reference beacon.

From a mathematical point of view, this equation system is easy to solve.

So far, we have browsed through methods that take as input a geometric relationship between the nodes and produce as an output the position of a node on a plane. Adding a beacon or a new measurement tool to the system would enable a 3D position to be identified.

We will now investigate the inputs which can be used in the context of radio communications.

3. RADIO SIGNAL AND DISTANCE MEASUREMENT

Having presented the main classical methods used to determine position using fixed reference nodes, we will now analyze the various radio signal characteristics that can be used for location determination purposes.

3.1. Signal Power and Received Signal Strength Indication (RSSI)

Due to interactions with the environment, a radio wave's energy decreases as it travels a certain distance. Many models describe the relationship between energy-loss and distance. Therefore they can be used to retrieve the spatial information.

3.1.1. Free Space Model (Friis' Equation)

$$\frac{P_r}{P_t} = G_t \times G_r \times \frac{1}{L} \times \left(\frac{\lambda}{4\pi R}\right)^2 \tag{8}$$

Widely used in link budget estimation, the Free Space Model (Equation (8)) [2] expresses the impact of travelled distance R, gains Gt and Gr of the transmit and receive antennas, free-space losses L and signal frequency f on the received power Pr, Pt being the transmitted power.

3.1.2. Two-Way Ground Reflection Model

Whereas Friis' equation considers two antennas in Line of Sight (LOS) with only one signal path, Two Way ground reflection model gets closer to reality by taking into account a path due to signal reflection on the ground (Figure 4).

Deciding between the two models is done by examining the antenna heights and the distance R between them. Let Dc (Equation (10)), be the critical distance. When the spacing R is below Dc, the Free Space model is used. Otherwise, the two-way ground reflection model (Equation (9)) is employed.

$$\frac{P_R}{P_T} = \frac{G_R \times G_T}{L}\left(\frac{h_T \times h_R}{R^2}\right)^2 \tag{9}$$

$$D_c = \frac{4\pi h_R h_T}{\lambda} \tag{10}$$

3.1.3. Shadowing Model [3]

This model comprises two parts:

1) Attenuation model: let d0 be the reference distance and $P_r(d_0)$ the received power d0. We use the following (Equation (11)) to calculate the received power at an arbitrary distance d:

$$\frac{P_r(d_0)}{P_r(d)} = \left(\frac{d}{d_0}\right)^{\beta} \qquad (11)$$

The attenuation factor β is determined empirically.

2) Changes in power level at a fixed distance Observation has revealed that a radio wave power's varies through time even when the measurements are taken at the same location. A log-normal probability law is typically used to represent these changes.

Considering these two aspects, the complete model is written as:

$$\left[\frac{P_r(d_0)}{P_r(d)}\right]_{dB} = -10\beta\log\left(\frac{d}{d_0}\right) + X_{dB} \qquad (12)$$

Although the Shadowing model is more complex, it is

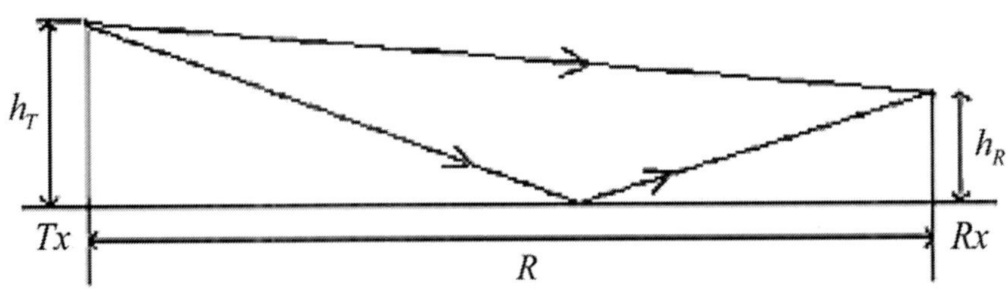

Figure 4. Propagation and reflection.

the closest to reality as it does not consider the communication range as a circle or sphere but rather as a surface (volume) which shape changes through time, even without moving the source.

3.1.4. RSSI

Some physical layers give access to the raw power level information but almost all provide the RSSI. This parameter, usually a positive byte, informs about the received power level but a standard relationship between power and RSSI has not been defined. This function depends on the implementation. Therefore, a model based on data produced by platform P might not be valid for platform P'.

RSSI can also be used as an input to a fingerprinting method. In the case of beacons transmitting frames, the RSSI values are stored in a database along with the location where the measurement was taken and used during the online phase as a reference for position estimation.

RSSI-based localization is one of the most prolific themes since radio signal reception is the main requirement for its exploitation. Moreover, in the case where the mobile receives frames from beacons and processes them to extract its position, there is no Multi-User Interference problem to consider: two nodes receiving data do not disturb each other.

On the downside, the environment's effect is not null: the human body, due to the large proportion of water, attenuates the radio signal [4].

Using a model to match received power to distance requires configuration of the parameters. To perform this profiling, a measurement campaign is necessary. This step not only is an additional cost but is contrary to the concept and implementation of a spontaneous network.

3.2. Time of Flight

Two typical usages of Time of Flight (TOF) exist: use the TOF directly to calculate the distance or generate Time Difference of Arrival (TDOA) pairs.

Regardless of the location determination activities, message exchanges are to be expected in a network: taking advantage of this when designing the positioning system is a cost effective decision in terms of message generation. The time references needed will correspond to the reception and transmission time of messages. Let us illustrate this with a message exchange between M and beacon B (Figure 5). In the following description; M and B are synchronized.

Let ta be the Tx instant from B and tb the Rx instant at M. Considering the fact that they share a common clock, the TOF is:

$$t_{vol} = t_b - t_a \tag{13}$$

Two typical implementations are often encountered:

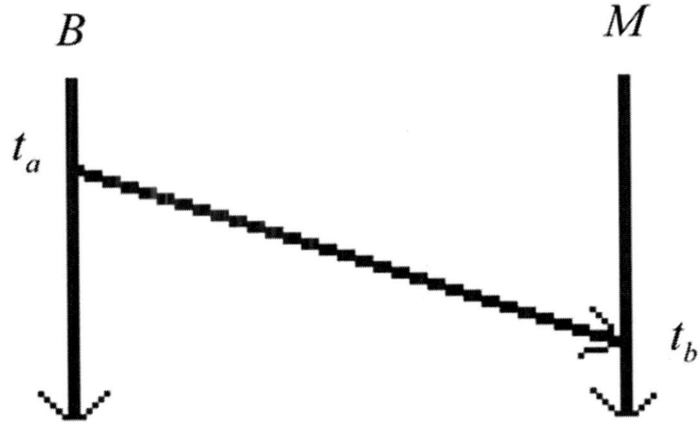

Figure 5. Time of Flight in a synchronized network.

3.2.1. Single Signal

In this case, distance is calculated as:

$$\text{distance} = vitesse * temps \tag{14}$$

The speed being known, we only need to measure the duration. To obtain this value, we either consider a synchronized network or let each node measure local time information then combine them to obtain the TOF.

For example, in the situation depicted in Figure 5, the distance between B and M can be obtained by c * tvol.

3.2.2. Combined Signals

Sometimes, nodes in the network are equipped with another device which generates a signal which speed is different from that of the radio signal. It is then possible to obtain the distance from the difference in arrival time of the two signals. Assuming the two signals' emission started at the same time, the distance/time equation can be rewritten as (Equation (15)).

$$\text{distance} = \tau * \left(\frac{v_1 * v_2}{v_1 - v_2} \right) \tag{15}$$

Parameters vi represent the signal speed of the two signals involved and τ corresponds to the measured TDOA.

Typically, a signal with lower speed is associated to the radio signal. Often, ultrasound is selected (vultrasound = 344 m/s in air at 20°C).

In any case, the nodes have to measure time. The results are influenced by the accuracy and precision of the measuring tools and also by the environment.

The first source of errors can easily be understood: if the smallest measurable time interval is a microsecond and the radio wave's speed is approximated at 3 × 108 m/s, the resolution of the distance measurement is 300 m. This could be fine for measuring distance between very very large objects but is not acceptable for an indoor environment.

The indoor aspect is closely related to the second source of errors which is multipath. This term is used to designate the fact that, in an environment where signal reflection and diffraction exist, the receiver will detect multiple copies of the sent message with different delays and attenuation factors. In this context, identifying the first occurrence, hence the shortest path, can become a very complicated task.

Depending on whether the nodes are synchronized, the choice between the following TOF extraction procedures is straightforward:

1) In a shared clock configuration, the emission time of the signal is known to the receiver which can begin to generate at the same time its local version of the sequence. Correlating the two signals will allow estmation of the TOF as illustrated in Figure 6. Receiver and sender agreed on t0 as the emission time of S(t). Since the two nodes are separated by a certain distance, the transmitted version will reach the receiver with a delay τ, which is proportional to the said distance. The value of τ maximize the correlation between the two signals.

2) In the case where each node has its own clock, a time interval will be associated with each node. This node in turn will be responsible for measuring the length of the interval and forwarding the data to the node requesting localization. We will explain this in much more detail as we present the location determination provisions of IEEE 802.15.4a [5].

3.3. Angle of Arrival

As introduced in 2.2, the angle of arrival (AOA) of the signal can also be used for location determination. Let's consider the mobile M as the source of the signal received by beacons B1(x1, y1) and B2(x2, y2). Angles α and β are measured between the MBi axis and the perpendicular to the B1B2 direction.

In order to find M's position, let's define a coordinate system with B1 as the origin and (B1, B2) as the y-axis.

The angles being known, finding the intersection of (B1, M) and (B2, M) is straightforward. M's coordinates (Figure 7) can be written as Equation (16).

$$M\left(\frac{y_2}{\tan\alpha+\tan\beta};\frac{y_2\tan\alpha}{\tan\alpha+\tan\beta}\right) \qquad (16)$$

Solving the problem is therefore very simple. The only complicated matter is finding the angles. Usually sophisticated means like motorized antennas or antenna array are used. Care should also be taken during beacon place ment phase as reflected signals introduce errors in the angles, hence the position

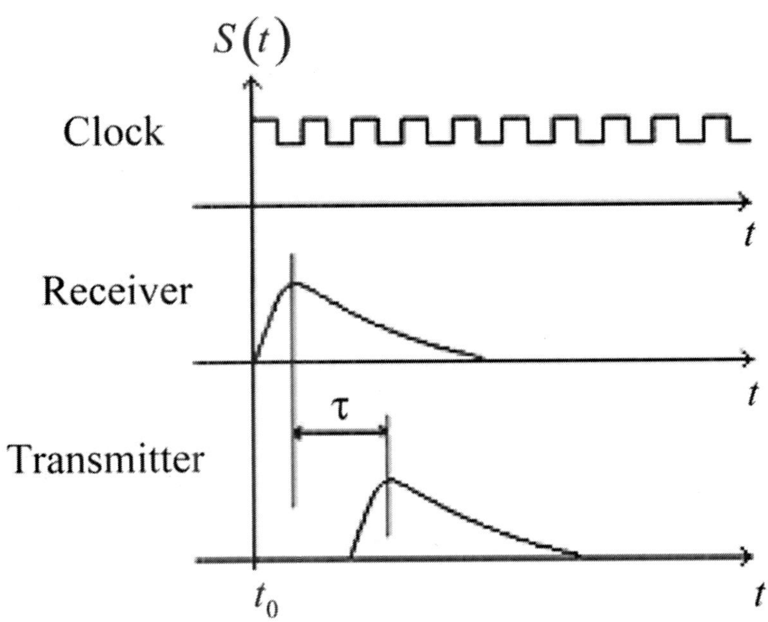

Figure 6. Time difference between local and received version.

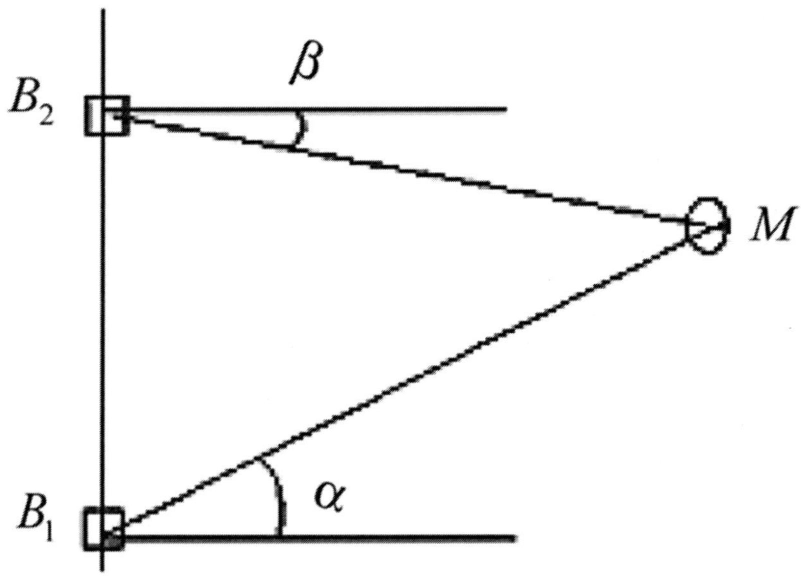

Figure 7. Angle of arrival and mobile position.

3.4. Signal Phase

Signal phase is one of the most accurate means of measuring the distance. Let's consider our signal as a simple sine wave with wavelength λ. Let's also restrict our description to one cycle. When the measuring tool's output is θ, it means the distance between receiver and transmitter is λ^*:

$$\lambda^* = \frac{\lambda\theta}{2\pi} \tag{17}$$

Therefore, if the signal frequency is 1 GHz, λ will be 30 cm: θ ranging from 0 to 2π, λ^*'s limits will be zero and 30 cm.

Measuring this angle means crosscorelating the received signal and a local version of the same signal. Therefore, at least during this part of the process, the nodes must be synchronized.

We must also keep in mind that this technique can only be used without an additional tool for distances below λ. Since we cannot differentiate θ and $\theta + 2k\pi$, $k \in N$, we will either need a radio technology that implies a large λ (as in low frequency techniques [6]) or add another method/set of tools that will allow us to solve the uncertainty.

4. LOCALIZATION IN WSNS

After exposing the mathematical tools and exploitable signal characteristics, we will present the additionnal constraints introduced by targeting a WSN and the various proposals that have been made.

4.1. The Challenge

A WSN is made of low-cost nodes with small processing power. A subset of nodes are aware of their location. The rest can be attached to a person or device and is used to collect information about this entity's motion and/or location.

In some situations, no maintenance plan is defined: Once deployed, new nodes can be added to the network but failing nodes are not repaired. Specifically, battery replacement is not an option: a node's lifetime is then equal to the battery's lifetime. Making the best out of this constrained resource implies energy-efficient components as well as energy-aware protocols and algorithms. However, we ought to mention studies in the field of energy-harvesting ([7,8]) that could help lengthen the network's lifetime.

4.2. The Constraints

Calculating node position in a WSN means using the resources of the network for this purpose. As these resources are limited, the localisation algorithm will need to abide by the following constraints:

1) Energy: A node's lifetime can often be considered equal to that of its embedded battery. A protocol running on such module should be designed with this in mind: particularly, reducing the number of frames used and taking advantage of sleep modes are two common ways for the software to help conserve energy. Short-range technologies are another means of reducing power consumption as a longer range has a higher cost.

2) Environment: Most of the time, WSNs are deployed indoor. This implies a good resolution for the localisation system: the environment being made of rooms of size n × p square meters (10 m < n, p < 100 m), the system must be able to at least select one of the rooms and associate it with the mobile. The expected precision is also influenced as the results must render the decision possible.

3) Simplicity: Among the characteristics of WSN nodes is the low processing power. Therefore, the algorithms that will be implemented on these platforms must not require complex computations. This becomes a major concern when a fully distributed solution is envisioned.

4) Minimal intrusion: One field of application for localisation using WSNs is the monitoring of the elderly. In this case, the system must not depend on a specific action from the person carrying the module.

After this brief listing of the constraints imposed by the use of a WSN, we will present a few propositions that were made by the scientific community. This is not an extensive description but we believe it identifies the current trends in the location-determination field.

4.3. Related Work

Location determination techniques for WSNs can be classified in two groups: range-based techniques and range-free methods. This chapter will present the charac- -teristics of each category and present a few propositions of each type.

4.3.1. Range-Based Techniques

As their name implies, range-based techniques use as input measurements taken by the system. In a WSN, it means signal characteristics being measured in real-time and fed to an entity that transforms it into a distance then a position.

These methods usually produce accurate results: if we recall the principles of signal phase-based distance measurement, in the appropriate situation, the accuracy is below the signal wavelength. The downside is the need to equip the nodes with specialized hardware, augmenting the cost and often size of the unit.

4.3.1.1. UWB, 802.15.4a and TOF

One of the basic characteristics of an UWB signal is having a bandwidth of at least 500 MHz or a fractional bandwith (Equation (18)) greater than 20%.

$$B_{fract} = 2 \times \frac{(f_h - f_l)}{(f_h + f_l)} \tag{18}$$

with fh and fl being respectively the highest/lowest frequency in the band The Spectral Power Density (SPD) being low, the disturbance caused to other systems using the same band is minimal. As a comparison, 802.11a has a DSP of 0.05 W/MHz while UBW's is 6.67 × 10–8 W/MHz. More information can be found in [9].

In Impulse-Radio UWB (IR-UWB), the pulse characteristic allow the existence of a wideband signal. The pulses (Figure 8) are typically Gaussian doublets (Equation (19)): being very narrow, they enhance signal resolution and render it immune to multipath (§3.2). If the delay between the first signal and the copies is large enough, the first symbol can be correctly detected. In the context of localisation, it implies that measuring time by correlating the received signal to a locally generated sequence will produce a peak that can be clearly identified and used to indicate the TOF.

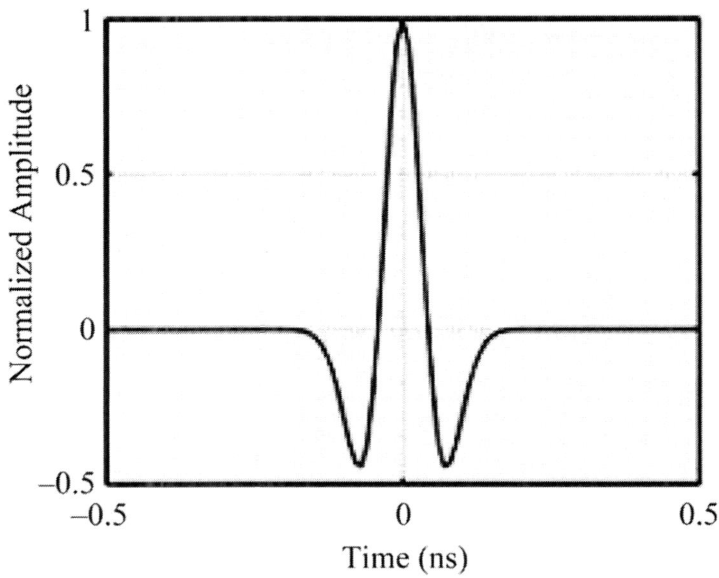

Figure 8. Gaussian doublet.

This spacing can be achieved either by spacing the pulses or by reducing the pulse width. The first solution has a negative impact on the radio link's bitrate. Therefore, the second has been chosen by IR-UWB.

$$G(x) = \frac{1}{\sqrt{2\pi\sigma^2}} e^{\left(\frac{-x^2}{\sqrt{2\sigma^2}}\right)} \tag{19}$$

UWB technology is also promising in terms of low cost [10] and high bandwidth. The expected rates range from 110 Mbps to 480 Mbps and vary according to the emitter-receiver distance [9].

IEEE has recently updated its communication standard for low-power low-rate WSNs: one uses Chirp Spread Spectrum and the other one UWB. The latter's bands are 250 - 750 MHz, 3.244 - 4.742 GHz and 5.944 - 10.234 GHz. Just as in previous versions of 802.15.4, these bands are divided in channels [11]. Three channels are more than 1 GHz wide and would be suitable for very accurate distance measurement.

A complete distance measurement using TOF is included in the standard. Network nodes that have the ability to measure TOF are designated as Ranging-capable DEVices (RDEVs). They must support a ranging counter, the ranging bit and the Ranging MARKER (RMARKER). These system functionalities will be described in detail as we explain the distance measurement process.

Common frames are used to measure time. The PHY header includes a special bit, the ranging bit, which purpose is to indicate that the current frame is a ranging frame (RFRAME) and is part of a measurement process. The actual time measurement uses the first pulse of the PHY header or RMARKER as a reference. No synchronization between the nodes is required. Figure 9 describes the message exchange.

The process begins with RDEV1 sending a frame to RDEV2 and storing the emission time, T0, of the RMARKER of frame1 (Figure 9). Upon reception, RDEV2 stores T1 and replies to RDEV1, storing T2. RDEV1 retrieves the fourth time information as it receives RDEV2's response. At this point, all relevant pieces of information have been collected but half is kept locally with each node since the frame is not generated

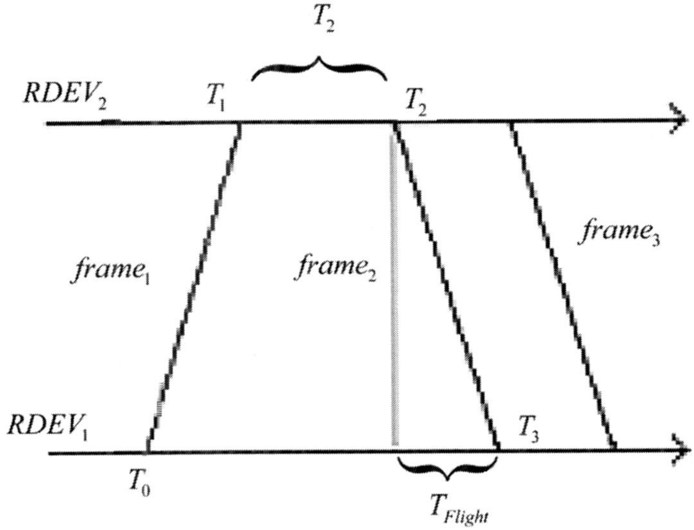

Figure 9. Two-way ranging (IEEE 802.15.4a).

"on the fly". Since RDEV1 initiated the process, RDEV2 has to forward the data to RDEV1. The node will then be able to compute the TOF using (Equation (20)).

$$T_{vol} = \frac{T_3 - T_0 - (T_2 - T_1)}{2} \tag{20}$$

The ranging counter is the entity responsible for producing the Ti values.

Although the clocks of both nodes are not synchronized, a certain level of accuracy is necessary. As the processing time of a frame (Ta in Figure 9) is in the order of the millisecond and the TOF for a 10 m distance is 30 ns, errors in measuring Ta will certainly and easily corrupt the results: at computation time, the nodes will combine time measurements that are not really similar (a second of time measured by node i would be too different from a second of time measure by node j).

One means of solving this problem is finding the difference between the two clocks and using it to correct the results. The IEEE standard suggests extracting this information from the PHY layer; since a node needs to synchronize its clock to the sender's clock when decoding a frame, it means that the difference between the clocks is an available information.

This implies that the PHY layer must be able to forward this type of information to the upper layers. As this functionality might not be available, another technique is also proposed, which is Symmetric Double-Sided TwoWay Ranging (SDS-TWR) (Figure 10). In this case, the frame exchanges of Figure 9 are duplicated: the first part corresponds to Figure 9 and in the second part, the roles are reversed. This uses a special kind of frame, the acknowledgement for an acknowledgement: this enables the use of three frames for the two-measurements. Please note that the final data forwarding to RDEV1 is not illustrated.

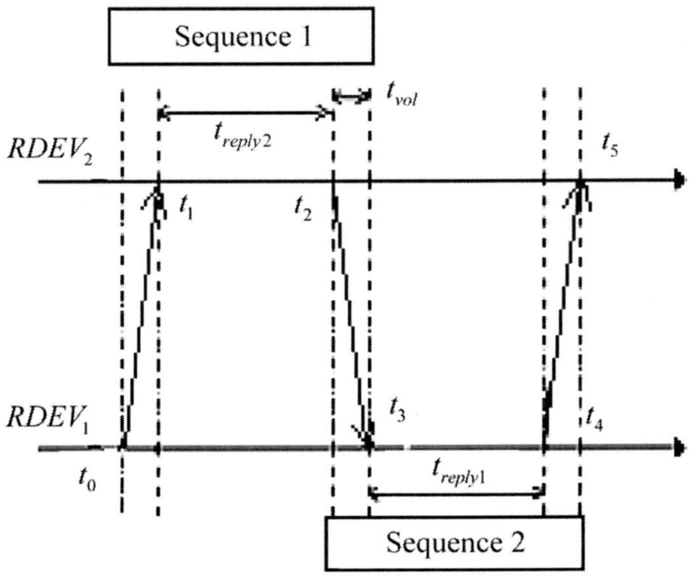

Figure 10. Symmetric double-sided two-way ranging (SDSTWR).

Applying SDS-TWR reduces the difference between the estimated value and the true TOF. Let treply1 be the difference t4 – t3, treply2 be the difference t2 – t1, Δreply the difference between treply2 and treply1, tvol the real value and $\hat{t}_{vol}$ the estimated value. The accuracy goes from:

$$\hat{t}_{vol} - t_{vol} \approx \frac{1}{2} \times t_{reply2} \times (e_1 - e_2) \tag{21}$$

to:

$$\hat{t}_{vol} - t_{vol} \approx \frac{1}{4} \times \Delta_{reply} \times (e_1 - e_2) \qquad (22)$$

Accuracy can be further improved by taking into account the propagation delay induced by circuits between buffers and the antenna. The standard also suggest adding an auto-calibration feature to the node which would be responsible for evaluating this delay.

The IEEE 802.15.4a is a well detailed proposition but nevertheless, we must keep in mind that the localisation service is an optional functionality: not all 802.15.4acompatible devices will implement them.

4.3.1.2. Fingerprinting

One of the earliest propositions in the field of signal power fingerprinting is RADAR [12]. This system uses signal power and SNR to calculate position.

The first step, as described in section 2.3, is recording the power level of frames sent periodically by the mobile. These frames contain the current position and orientation of the mobile unit. During this offline procedure, the beacon nodes map each frame with the measured power and the time of reception. Since all the nodes have been synchronized, the time values are valid throughout the network.

As the offline phase ends, a database is built containing, for each mobile position and orientation (North, South, East, West), the measurements taken by each beacon. These measurements constitute the Signal Strength (SS) vector.

During the online phase, the beacon nodes keep a mean value of the measured signal level and use either empirical values contained in the database or a propagation model to determine position.

Nearest Neighbour in Signal Space (NNSS) method is used to search for the position using empirical data: this method selects as the solution the position for which the distance between the stored SS vector and the real-time SS vector is minimal. The results have been improved by only choosing the k NNSS and setting their centre of gravity as the position. This can be further enhanced by ignoring the orientation differences when choosing the k NNSS as this ensures physically different nodes.

Whereas RADAR used averaged RSSI values from the mobile, the work presented in [13] followed a different path and chose to consider the difference between RSSI values as their system's input. Since their solution is based on 802.11, we will use the term Access Point (AP) to refer to the beacon nodes.

The starting point of this work was the observation that the RSSI value returned was highly dependent on the measuring hardware (see 3.1.3). Therefore, the team first conducted an experiment with APs responsible for emitting frames and two different 802.11-compatible mobile nodes recording the RSSI values in a finite set of predefined positions. AP pairs (Api, APj) were then formed with the constraint of i < j. For each pair, the difference between RSSI values was computed. This hypothesis has been validated using real-world data: even though the values returned by the mobiles were not identical, the graphs showed the same global shape.

The impact of this hypothesis on localisation was determined by including it as a component of two wellknown methods: Nearest Neighbour (NN) and Bayesian Inference.

NN method uses a measurement database during the online phase as its reference. Each location in the database is associated to a point in the Signal Space: choosing a location as the estimated position is equivalent to finding the closest point in Signal Space to the measured RSSI. Through some adjustments, this method is made suitable to compute distance between RSSI differences rather than RSSI points.

Bayesian Inference is a means of choosing the most probable position given the current values. In order to become applicable to the problem at hand, an observation vector o must be defined and mapped to a position l. The contents of vector ok will be (AP pair ID, RSSI difference for the pair). Combining these values and the stored measurement yields:

$$\Pr\left(o \mid l_k\right) = \prod_{1 \le i < j \le m} \Pr\left(d\left(AP_i, AP_j\right) \mid AP_i \times AP_j, l_k\right) \tag{23}$$

The probability of being in the position referenced by lk givent the current measured values is:

$$\mathrm{P_r}\left(l_k \mid o^*\right) = \frac{P_r\left(o^* \mid l_k\right) P_r\left(l_k\right)}{\sum_{k=1}^{n} P_r\left(o^* \mid l_k\right) P_r\left(l_k\right)} \tag{24}$$

Maximizing this probability with respect to lk yields the true position.

The experiment conducted using off-the-shelf hardware has showed improved performance of the RSSIaugmented versions against the traditional implementations.

4.3.1.3. TOF-Based Solutions

The basic exploitation of TOF consist of measuring distances from the mobile to the beacons. As a result a vector containing the various measured distances is obtained, along with the coordinates of each beacon. Position computation shows two main trends: using mathematical tools such as Maximum Likelihood (ML) methods to obtain the position and improving the performance of the PHY layer in order to achieve better accuracy.

ML methods' aim is to obtain the parameter value which maximizes the probability that the measurements are correct, given some system characteristics. The problem at hand can be written as:

$$\hat{u} = \arg\max P_u\left(\hat{d} \mid u\right) \tag{25}$$

u : matrix containing beacon coordinates

û : estimated mobile position

$\hat{d}$: measured mobile to beacon distances Another method makes use of a cost function [14] such as Equation (26).

$$F(x) = \sum_{i=1}^{N} \alpha_i^2 f_i^2(x) \tag{26}$$

with:

$$f_i(x, y) = c\Delta_{t_i} - \sqrt{(x - x_i)^2 + (y - y_i)^2} \tag{27}$$

In these cases, a seed is required which should not be too far from the real solution in order for the algorithm to converge. Unfortunately, this type of computation requires a processing unit more powerful than a typical WSN node and therefore calls for a centralized solution.

In [15], the authors aim to calculate TOF between synchronized nodes without adding complex components or trying to counter clock phase differences. In order to mitigate multipath effect, the team proposed the use of the whole 2.4 GHz band.

The method is comprised of two steps: unlike the fingerprinting methods, the first step is the online phase. The nodes use this phase to communicate then they turn off their transceivers and the offline phase begins. The sequence in which the actions are executed is regulated by the Event Clock

(EC) which is present in each module. This clock's period, TEC, serves as an upper limit to the desynchronization between nodes.

The online phase can take place once synchronization has been attained: node A starts transmission of k copies of an n-symbol long ranging signal. Emission starts at the next rising edge of EC and the signal's duration is chosen equal to TEC. Node B turns on its receiver circuit two rising edges later and captures k-2 copies of the signal. After demodulation, B combines them in order to obtain one complete copy. Then B emits the received signal k times. When A has received the k-2 versions, both modules turn off their transceiver.

The offline phase can then begin with each node performing a crosscorrelation between the received copy of the signal and a locally generated version. The ranging signal being a pseudonoise code (PN code) helps obtain a peak that indicates the time travelled between the nodes.

The prototype version of this system has been tested both indoor and outdoor and the accuracy was 1 - 3 meters depending on the environment.

4.3.1.4. TDOA

In [16], phase measurements and TDOA are combined in order to determine location in a synchronized network. The whole 2.4 GHz ISM band is used to enhance accuracy: the signal's band increases from that of a single channel, 5 MHz, to 80 MHz for the 16 channels defined: as the bandwidth increases, path separation and distance measurement are improved.

Four nodes are involved in the localisation process: two emitters, T1 and T2, and two receivers, R1 and R2. All nodes are aligned (Figure 11) and the receivers are included in both transmitters' range. Since the emitters are IEEE 802.15.4-based, their range will be 30 m and $\max\left(d\left(R_1,T_2\right),D\left(R_2,T_1\right)\right)\leq 30$ m, where $d(A,B)$ is the distance between two nodes A and B.

The Ti transmit simultaneously in two different channels chosen among the 16 802.15.4 channels.

The TDOA for each receiver is then computed. These values are referred to as Δt1 and Δt2 and written as:

$$\Delta t_1 = \tau_{21} - \tau_{11} + \tau_{t_0} = \tau_0 + \tau_{t_0} \tag{28}$$

$$\Delta t_2 = \tau_{22} - \tau_{12} + \tau_{t_0} = -\tau_0 + \tau_{t_0} \tag{29}$$

The difference between emission time is termed τt0 while τ0 is the time taken to travel the distance between T1 and T2. This distance, d0, can be computed using a combination of these time information:

$$d_0 = c_0\tau_0 = \frac{c_0}{2}\Delta\tau \tag{30}$$

with:

$$\Delta\tau = \Delta t_1 - \Delta t_2 \tag{31}$$

From this we can see that as soon as the position of one of the transmitters is known, the other can easily be calculated.

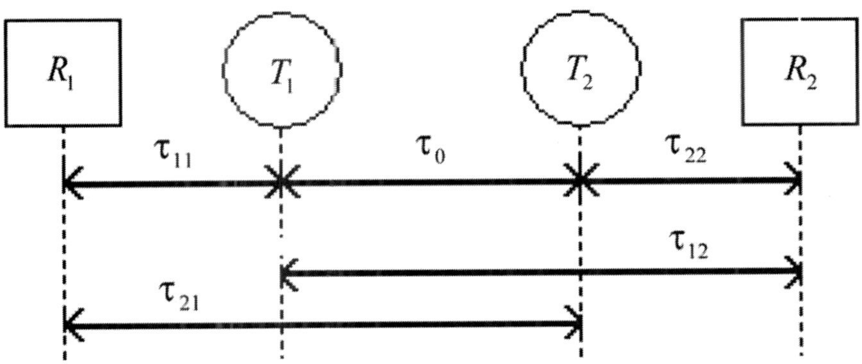

Figure 11. Node placement and time measurements.

Phase measurements are also used in conjunction with the time measurements: as mentioned in 3.4, phase measurements are precise on one signal cycle. The longrange technology is then used to select a point and from then on, the phase measurements are used to refine the position/distance estimate.

4.3.1.5. RSSI and Accelerometers

The work presented in [17] combines RSSI-based fingerprinting to accelerometers values. The network is made of access points, beacons and mobiles. The APs receive broadcast frames from the beacons and the mobiles. A map is then generated, an RSSI vector being linked to a mobile at a certain position. This database is updated regularly, during the online phase. This compensates for the changes in the environment which might have an impact on the stored values.

When location determination is requested, the most probable position with respect to the APs' values is selected. Two search methods have been proposed: Maximum Search (MxS) and Minimum Search (MnS). The following points can help distinguish them from each other:

1) Search space: MxS determines the solution for each AP then computes the intersection. MnS, on the other hand, sets an initial solution as the possible points according to the AP which measured the maximum RSSI value. The subsequent tests use this set as their starting point, reducing the computations but augmenting the risk of a bad estimation, as the correct position might not be included in the initial solution.

2) Results: MxS yields better results than MnS but at a higher computational cost.

An interesting aspect of this proposition is the addition of accelerometers to the localisation system. They are used to record the number of paces executed by the user. Knowing the initial position, the user's motion characteristics and the accessible zones of the environment, the application can identify a possible zone of arrival.

When combined to MnS and MxS, the accelerometerbased subsystem enhances the precision by 8%. Nevertheless, the accelerometer information had to be ignored sometimes as their results did not match those of the RSSI.

4.3.2. Range-Free Techniques

The previous section introduced a handful range-based propositions. In the simplest case, access to the RSSI values was all it took to calculate the position. In a more complex example, a new physical layer had to be designed, synchronization between nodes was required before phase and time difference measurement could be taken.

Range-free methods aim to eliminate the need for additional hardware by relying on hypothesis about network connectivity. In this case, being cost effective takes a toll on system accuracy which will often be expressed as a percentage of the radio range.

4.3.2.1. APIT

APIT method [18] is based on the Point In Triangulation test. It enables a node to determine whether or not it is located inside a given triangle. If any displacement of the node gets it closer to/farther from at least one of the triangle summits, then the node is inside the triangle (Figure 12(a)). If it gets farther from/closer to all summits at the same time, then the node is outside the triangle (Figure 12(b)).

This description works in the mathematical world but in real-life, some approximations must be done. The node's movement is not always a parameter that can be controlled by the system: as we stated earlier, the solution should avoid depending on a specific user behaviour. Hence, the node uses the relationships between it's neighbours and the beacons that act as triangle summits to investigate the situation. The test can then be written as: if all the node's neighbours are closer to at least one of the beacons, then the node is inside the triangle (Figure 12(c)). This version of the test is likely to fail when the node placement is irregular or part of the neighbours are outside de triangle (Figure 12(d)).

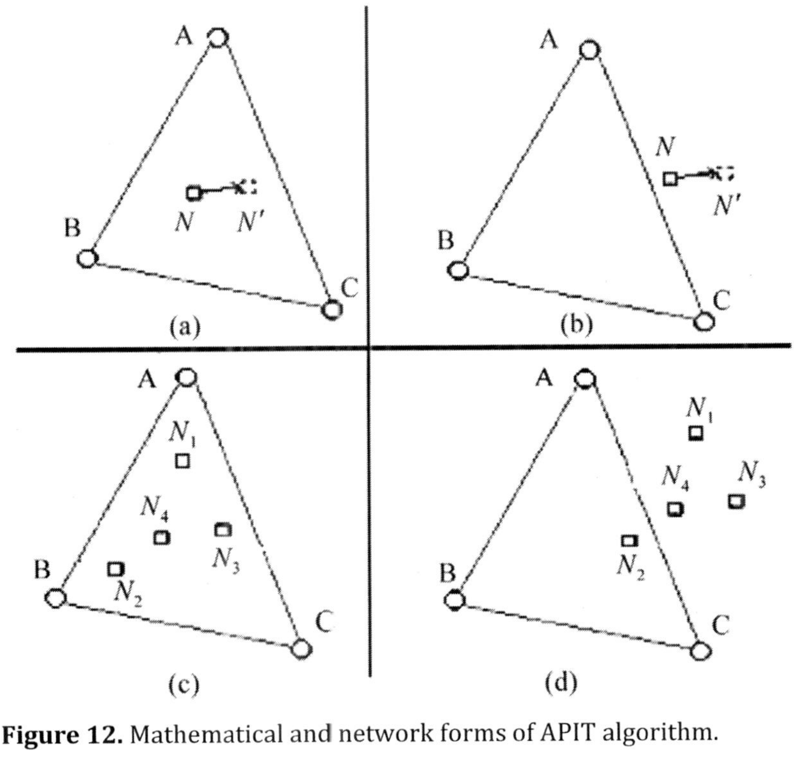

Figure 12. Mathematical and network forms of APIT algorithm.

The network is comprised of beacons or anchor nodes which are aware of their position and have a powerful transceiver, and mobile nodes which use APIT to determine their position. APIT is an area-based algorithm because its aim is to obtain a surface where the node is most likely to be.

The mobile nodes collect the frames transmitted by the anchors in order to build a list of in-range beacons. The node then selects groups of three beacons and runs the test. This process is repeated until no more groups can be formed or a certain surface-size threshold is reached, whichever happens first.

The node also uses a matrix that represents the environment: each cell corresponds to a zone and a zone is part of one of the triangles. Every successful run of the test leads to the incrementation of the value contained in all the cells corresponding to the surface of the selected triangle. Upon failure, the value is decreased by one.

In the end, the surface with the highest associated values is selected as the area and its centre of gravity is assumed to be the node's true position..

4.3.2.2. DV-Hop Suite

DV-Hop is a popular range-free algorithm. Its original version along with a few proposed enhancements are presented in this subsection.

4.3.2.2.1. Original DV-Hop

DV-Hop [19] is a method often cited as a reference in range-free propositions. It considers mobile nodes as unknowns and uses a logic similar to that of distance vector routing protocols to find the position. Each beacon broadcasts a frame containing its true position and a field dedicated to hop counting. Each node that receives this frame stores the hop count value, increments the field and forward the frame if the hop count is lower that the previously known value for the specified link. Once the beacons have collected these data, they compute a mean hop size. For example, in Figure 13, beacon A1 can mathematically calculate its true distance to A2 and A3. The hop count for each link is also obtained via flooding. Estimating hop size makes use of the following equation:

$$hopsize_1 = \frac{d_{1,2} + d_{1,3}}{hop_{1,2} + hop_{1,3}} \tag{32}$$

with di,j being true distance between beacons i and j, hopi,j representing the number of hops between beacons i and j This hop size is flooded to the network. The mobile nodes will use it to estimate the distance between itself and the anchor (Equation (33)).

$$d_{i,x} = hopsize_i \times hopcount_{i,x} \tag{33}$$

where hopcounti,x is the number of hops between mobile x and beacon i. Once the distances to three beacons are known, trilateration can be applied.

By using the hop count mechanism, DV-Hop aims to enable location determination on mobile nodes that are out of range of beacons.

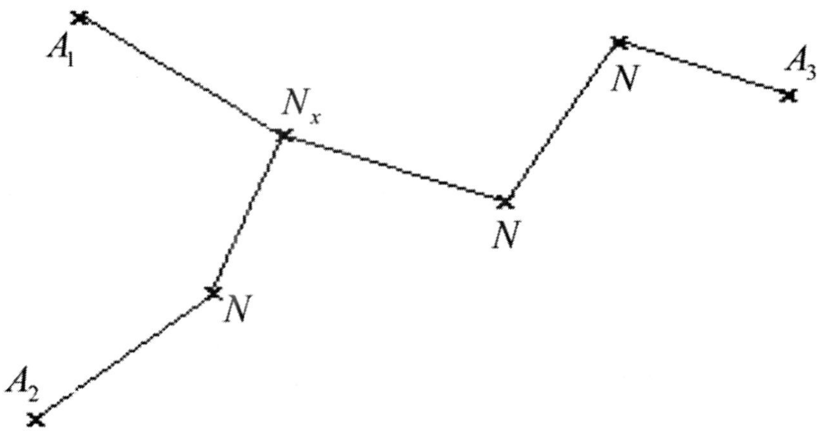

Figure 13. DV-Hop.

4.3.2.2.2. Constraint DV-Hop

Many groups have tried to improve DV-Hop. We will first mention Constrained DV-Hop or CDV-Hop, which is described in [20].

The proposal stemmed from the fact that an implicit relationship exists between the communication range D and the distance calculated by DV-Hop. When a node's hop count to a beacon is 1, it implies that the distance is lower than 1 * D. If the spacing is 2 hops, the distance is then bounded by 2 * D. The new system will therefore take these constraints into account and it can be rewritten as:

$$\min \sum_{i=1}^{M} \left((x_i - x_u)^2 + (y_i - y_u)^2 - d_i^2 \right)^2 \tag{34}$$

with

$$\sqrt{\left(x_{v_j}-x_u\right)^2+\left(y_{v_j}-y_u\right)^2} \le D,\ L=1 \tag{35}$$

where

$$D \le \sqrt{\left(x_{v_j}-x_u\right)^2+\left(y_{v_j}-y_u\right)^2} \le DL,$$
$$L \ne 1,\ j=1,2,\cdots,M_1 \tag{36}$$

This proposal was validated using simulations of a hundred nodes, 82 out of which were mobiles. The deployment area was 50 × 50 square meters. Various tests were conducted with different mobile to beacons rations and values of D. In the end, they all indicated that CDV-Hop was a definite improvement to DV-Hop.

4.3.2.2.3. Checkout DV-Hop

Also aiming to improve DV-Hop is the work described in [21]. Mobile nodes in this case are divided in two classes: level 1 mobiles can communicate with 3 beacons or more while level 2 nodes do not. Depending on its class, a node will choose which algorithm to use in order to obtain its position.

Level 1 nodes will refer to the Mid-Perpendicular Method: In the case of three anchors, we will consider the mid-perpendicular of each side of the beacon triangle. The node will assume their intersection to be its position. this solution only requires knowledge of the coordinates of two beacons.

The improvement is more noticeable when we consider the case of level 2 nodes. The Checkout DV-Hop, as its name implies, integrates a new step to the classical algorithm. The closest anchor, Anear, in terms of hopcount, corresponds to the link for which the smallest error accumulation will occur during path length estimation. The node can then use the distance between itself and Anear to improve the position obtained with DV-Hop.

Figure 14 shows the effect of Checkout DV-Hop. True node position is Nx but according to DV-Hop, the node is located in NDV-Hop. Checkout DV-Hop modifies the estimated position by substracting the Anear to Nx distance from the Anear-NDV-Hop axis, leading to a new estimated position, Ncheckout.

The new method, 2RaLA, has been compared to other methods such as Centroid (§4.3.2.3) + DV-Hop and Convex Position Estimation (CPE) [22] + DV-Hop using MATLAB. Under different configurations, 2RaLA has produces better results than the other algorithm combinations.

4.3.2.3. Centroid and Weighted Centroid

Centroid [23] assumes spherical isotropic propagation of the radio signal as well as identical range for all nodes in the network. The beacon nodes form a regular mesh and broadcasts frames every T seconds without collisions.

Each mobile receiving these frames computes a connectivity metric using (Equation (37)).

$$CM_i = \frac{N_{recv}(i,t)}{N_{sent}(i,t)} \times 100 \tag{37}$$

where:

$N_{recv}(i,t)$: number of frames emitted by beacon i and received during interval t $N_{sent}(i,t)$: total number of frames emitted by beacon i during interval t According to a predefined threshold, the mobile node selects a subset of beacons to take into account when calculating its position. The coordinates of this group of beacons will then be used to compute their centre of gravity, which the mobile node will assume to be its position. Figure 15 is an example of the results of the process for a non regular-mesh of beacons. The estimated position is represented by a plus sign while the dots correspond to the beacon nodes.

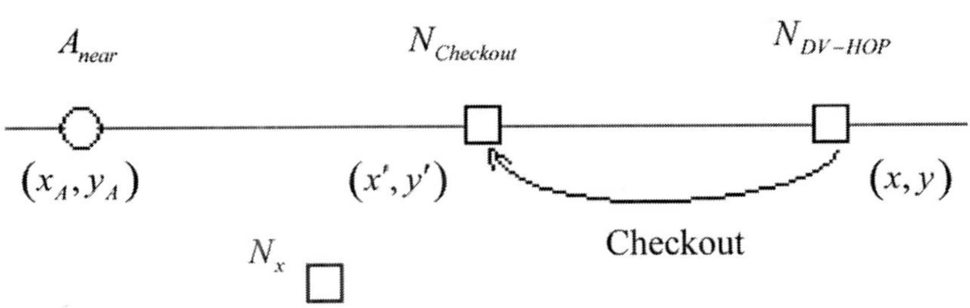

Figure 14. Checkout DV-Hop.

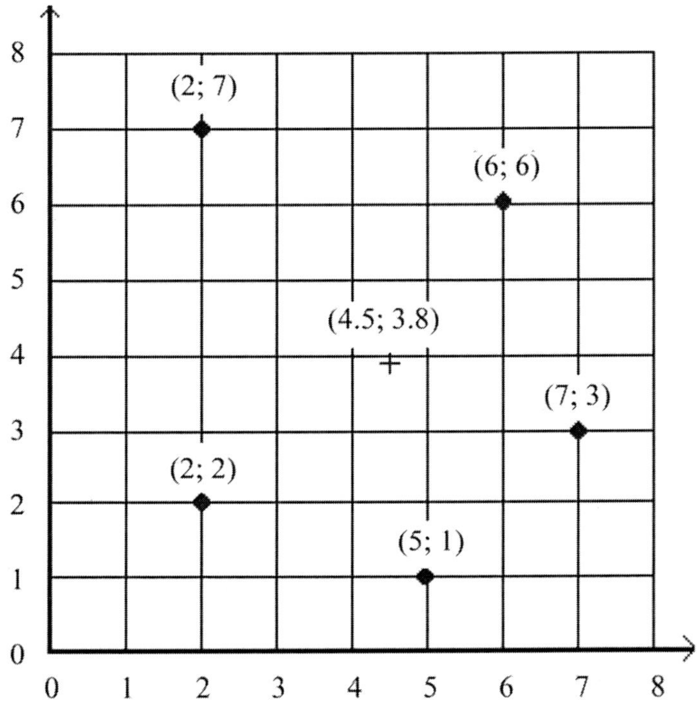

Figure 15. Localizing with Centroid.

Centroid has been validated in an outdoor setup: four beacons have been used to localize in a 10 × 10 square meters, leading to a mean error of 2 m.

Improvements such as Weighted Centroid Localization (WCL) [24] exist. WCL adds the RSSI information to the process by associating weights to the links between the mobile node and the beacons. This estimated position is then written as:

$$P = \left(\frac{\sum_{j=1}^{n} w_{ij}(d)x}{\sum_{j=1}^{n} w_{ij}(d)}, \frac{\sum_{j=1}^{n} w_{ij}(d)y}{\sum_{j=1}^{n} w_{ij}(d)} \right) \tag{38}$$

where $P = P(x, y)$ is the estimated position, (x_{Bi}, y_{Bi}) are the coordinates of beacon i and $w_{ij}(d)$ (Equation (39)) is the weight associated with the link.

$$w_{ij}(d) = \frac{1}{RSSI^{q}_{\text{stand}}} \tag{39}$$

The weights are calculated as follows; the mean RSSI value for each beacon is determined then, as the prototype returned a negative value in the range [–110 dB, –50 dB], translated to positive values using (Equation (40)).

$$RSSI_{strand} = -(RSSI + 49) \tag{40}$$

This RSSIstand then takes its values in the interval [1 dB, 51 dB].

As expected, the results have shown that, with the addition of weights, the estimated position moved closer to beacons with higher weight.

Although RSSI is exploited in this proposition, we did not consider it as range-based since its usage here corresponds neither to a power to distance model, nor to a power to position mapping.

4.3.3. Hybrid Solutions

Range-free and range-based methods do not always oppose each other. In fact, they sometimes combine their strenghts to bring novel solutions combining real-time measurements to network-wide hypothesis.

4.3.3.1. HexNet

In Hexnet [25], the environment was subdivided in hexagons (Figure 16). Two types of nodes are available and three behaviours can be implemented. The network begins with a fixed node called anchor which has two emitters: one is very directional while the other is omnidirectional. Every anchor is aware of its position and together, the anchors of the network form a regular grid. This type of node is not energy constrained and its communication range R is greater than that of the sensor nodes.

The second category comprises the sensor nodes. They can act as backbone sensors or as simple sensors depending on their relative position to the anchor.

The anchor's position defines the centre of the first hexagon. The subsequent hexagons are defined in a pattern that ensures all polygons are adjacent.

For each new hk, a backbone sensor (BS) is selected. The sensors that do not implement the BS behavior assume their own position to be that of the centre of gravity of all heard BSs.

The anchor is responsible for selecting the sensor that will act as the BS for the hexagons adjacent ot its own. From then on, the peripheral BSs will manage the BS election process.

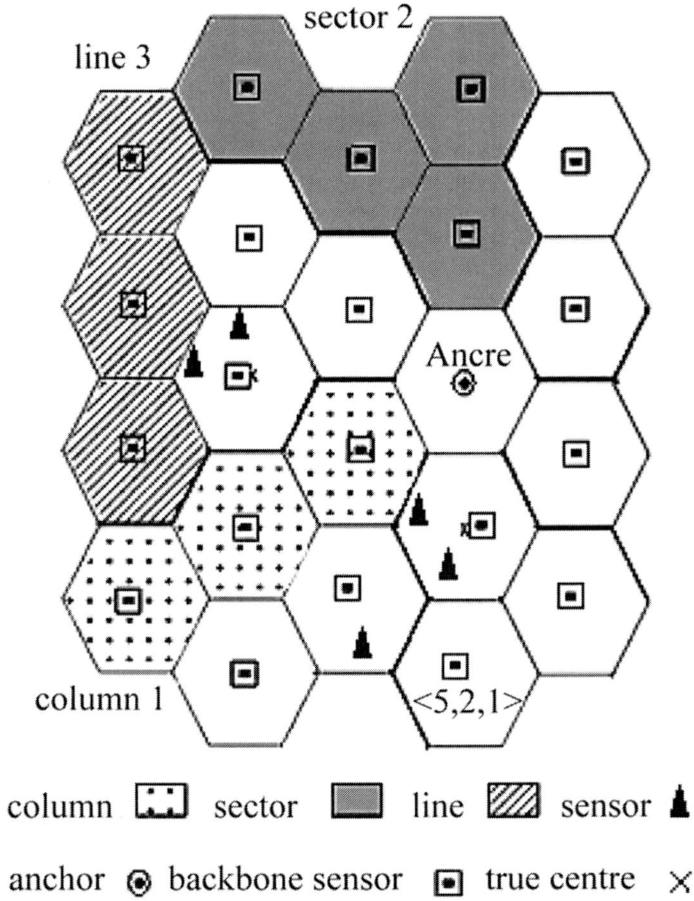

Figure 16. HexNet network.

First, nodes must be aware of their angular position in the coordinate system defined by the anchor and an arbitrary direction that will serve as the x-axis. The anchor uses its directional antenna and, while rotating it slowly, broadcasts the current angle. When BSk has to choose the BSi for one of its adjacent hexagons, it can reduce the number of sensors involved by first announcing the angle corresponding to the hexagon. Along with the angle of interest, the BSk sends data about the geometrical relationship between itself, the target and the anchor.

Once the candidates have received these informations, they estimate the distance between themselves and the centre of the hexagon. This result serves as an initial value for a local timer: the node with the smallest result will consider itself as the new BS upon expiration of the timer.

Hexnet therefore enables localization using a single anchor, given the specific hardware requirements are met.

4.3.3.2. GPS and Mobility Pattern

The idea in [26] is to combine GPS-enabled nodes to a mobility pattern in order to localize nodes in an outdoor environment. The beacons would be equipped with GPS receivers and would be synchronized. Their position can change through time.

These beacons are responsible for broadcasting regularly the time of their last GPS update, their position and the time between reception of GPS information and current emission.

These data are stored by the mobiles along with the local reception timestamp in the Anchor History Table (AHT). The GPS coordinates (Longitude, latitude) are converted to seconds and their size is reduced to ease computations. The mobile then uses this to create its own mobility pattern, rewriting its longitude and latitude as functions of time. Once the mobility pattern has been defined, the mobile will estimate its position periodically, regardless of reception of frames from the beacons.

In this localisation system, GPS is used to synchronize the beacons and also to enable for position correction after the mobile has left the network for some time. While out of coverage, the mobile will keep estimating its position using a model which is probably not valid. As soon as the mobile re-enters the network, the received frames will help rectify the estimations.

This proposition has been evaluated through both simulation and prototype. The simulation involved 400 nodes placed on a 500 × 500 square meters and was used to examine the effects of node speed, beacon ration and AHT size.

In the prototype configuration, the mobile was connected to a laptop which acted as a data sink while the node moved on a 100 × 100 square meters surface.The experiment revealed the impact of weather, through GPS, on the prototype. It also indicated that sudden change in movement would induce erroneous predictions. The results also showed a localization around of roughly 10m with 5 beacons seen by the mobile per minute.

5. SUMMARY AND COMPARISON

The previous sections were devoted to general methods and specific propositions from the research community. The current chapter will focus on comparisons, first of the means of measuring offered by WSNs then on the various propositions presented in section 4.

5.1. Signal Characteristics Comparison

We will begin with the comparison of the signal characteristics usable for localisation in a WSN.

Table 1 lists six means of obtaining positioning information using radio communications. The best with respect to accuracy are TOF-UWB and signal phase. Their performance make them really interesting for location determination purpose but the hardware is not always simple to obtain. In the case of UWB, very few companies offer chips that feature distance measurement functionality. For example, Decawave [27] intends to launch its 802.15.4a compatible chip but mass production is planned for 2012. Until then, the per chip price keeps testing out of the reach of many.

Phase measurement imply adding expensive modules to the system. Because of the angular uncertainty, it must be coupled with another long range technique. Furthermore, the node placement must abide by the LOS constraint or the measured phase will corrupt the computation results. This dependence upon node placement is also a factor to consider when designing an AOA-based localisation system.

Although the time information retrieved using narrow band technologies cannot compare to the data UWB is expected to produce, it still ranks higher than RSSI. Distance estimation using signal power indication suffers from environment obstructions and its own variations through time. In order to attain precise measurements with RSSI, a fingerprint database with an automatic and cost-effective update system must be created. Regardless of its insufficiencies, RSSI will still spawn new propositions as the information is available on almost every hardware implementation, unlike TOF.

Just as easy to obtain is hopcount. Originally, this was a metric used by Distance-Vector routing protocols but its ease of use has enabled its usage in the field of localisation. Unfotunately, in the case of nodes which do not form a regular grid, the results of distance estimation through hopcount are most unreliable.

5.2. Propositions Comparison

Before diving in the comparison of the various rangebased and range-free propositions, let us explain the criteria that have been chosen. We will evaluate the localisation algorithms based on environment constraints, whether they are decentralized, the computational requirements and the accuracy.

One of the first aspects to consider is whether the system operates indoor or outdoor: it indicates how well the proposed system manages in case of obstructions.

Some solutions assume there will be a special node with high processing power which will be responsible for all calculations. This node will be termed the localisation server. This type of centralized architecture allows for very low-cost mobile nodes but the failure of the server implies the unavailability of the localisation service. Also, the strain on the nodes closest to the server cannot be neglected as there is a need for data forwarding to and from the server. A decentralized solution will ensure computation is not dependant on a single node but will require all the nodes to be upgraded.

The fourth criterion is accuracy but sometimes, we will take into account the improvements brought by the proposition. These benefits will have to be considered in conjunction with the computational constraints, i.e. the conditions that must be satisfied for the proposed method to be applied.

Table 1. Comparison of ranging means.

Criterion / Signal	Accuracy	Cost	Environment (constraints, degradations...)
TOF UWB	centimeter	Synchronization Hardware	High frequency attenuation
TOF Narrow band technology	Meter	Hardware Synchronization	Multipath
Signal phase	<wavelenth on one cycle	Hardware	Line Of Sight
RSSI/Power	Meter and varies through time	X	Human
AOA	--	Hardware	Multipath
Hopcount	Radio range	X	X

Since the goal is location determination, a critical parameter will be its performance. In the cases where a prototype has been built, real-world data is available and make the comparison easier. For example, the results for [15] are far better than those of RADAR. [15] being based on TOF, the impact of the measuring tool is not to be overlooked. On the other hand, since both of them are centralized methods, [15] is more cost-effective than RADAR since the amount of data necessary to execute localisation is minimal: [15] calculates a crosscorrelation whereas RADAR operates with an extensive RSSI-locantion database. Finally, [15] has been tested both in indoor and outdoor environments.

In [16], a method for calculating position in a 1D problem has been provided using TDOA and phase measurement in a synchronized network. Since the underlying measuring tools are among the bests according to Table 2, the performance of this technique is quite impressive. Nevertheless, we find it rather complicated for a 1D problem, especially when all nodes can hear each other. Generalization to the 3D situation is expected to involve a dense network or invoque an astute solution.

The comparison becomes more difficult when we turn to range-free methods, especially when evaluating the accuracy. Due to the assumption of "hardware independency", these propositions are most of the time only simulated. Centroid stands out as real-world performance information is provided. The attained precision is similar to that of RADAR while requiring 4 anchors.

An interesting characteristic is the fact that most of these solutions are distributed: each node is responsible for collecting the data related to its situation and estimate its position. Using simple operations makes attaining this goal easier.

However, the assumptions on the network configuration are not trivial: even if it is not always clearly mentioned, having an isotropic network is not

Table 2. Propositions' summary.

		Indoor VS outdoor	Centralized VS Decentralized	Computational constraints	Accuracy/improvements
RANGE-BASED	**RADAR**	Indoor	Centralized	Synchronized beacons	3 - 4.3 m (50%)
	TOA [15]	Tested in both environments	Centralized	Crosscorrelation Sensor synchronization	$1m_{RMS}$ - $3m_{RMS}$
	TDOA phase [16] 1D solution	Indoor	Distributed	Sensors synchronization Sensors placed on a line New PHY layer	16 cm (phase) 79 cm (TDOA)
RANGE-FREE	**DV-HOP**		Distributed	Isotropic node placement	
	CENTROID	Outdoor	Distributed	Isotropic beacon placement	2 m mean error with 4 beacons
	APIT	Unspecified	Distributed	Isotropic node placement High beacon number Border effect	Similar to DV-Hop
HYBRID	**HEXNET**	unspecified	Centralized	Rotating antenna (anchor) Isotropic placement of anchors	Single anchor

to be overlooked. It implies either accepting non negligible errors or ensuring a regular grid of beacons.

If we consider the whole table, we can identify two tendencies: in the range-based context, both data collection and processing are costly. On the other hand, rangefree methods take advantage of simple computations while assuming a very constrained node deployment.

6. CONCLUSIONS

The ideal WSN is expected to allow fast node deployment and propose a light infrastructure. Combining properties from range-based and range-free techniques would enable the production of solutions closer to this goal: Range-based's ability to obtain a localization information without placing tight constraints on the deployment pattern of the beacons would make the network more spontaneous while a distributed approach would benefit from the simplicity of the processings proposed in range-based techniques. At the same time, range-based methods could enhance the reliability of the estimates generated by range-free techniques.

Using complementary hardware/software components could also improve the performance of the location determination system by providing supplementary information related for example to the mobile's motion. Having an internal datasource would also reduce the strain on the radio communication subsystem as the period of the precise location determination operation could then be increased.

Finally, range-based solutions implemented in a completely distributed fashion and propositions exploiting combined signals from various sources remain a field in which research still has improvements to bring.

REFERENCES

1. R. Want, A. Hopper, V. Falcão and J. Gibbons, "The Active Badge Location System," ACM Transactions on Information Systems (TOIS), Vol. 10, No. 1, 1992, pp. 91- 102. doi:10.1145/128756.128759
2. W. Fahs, "Diffusion d'Informations Partagees Entre Mobiles Cooperants Evoluant Sous une Meme Cellule d'un Reseau Sans fil Avec Infrastructure," Ph.D. Dissertation, Université de Clermont-Ferrand, 2009.
3. T. S. Rappaport, "Wireless Communications, Principles and Practice," 2nd Edition, Prentice Hall, Upper Saddle River, 1996.
4. D. Lieckfeldt, J. You and D. Timmermann, "Characterizing the Influence of Human Presence on Bistatic Passive Rfid-System," Wireless and Mobile Computing, Networking and Communications, Marrakech, November 2009, pp. 338-343.
5. Specific Requirements Part 15.4: Wireless Medium Access Control (MAC) and Physical Layer (PHY) Specifications for Low-Rate Wireless Personal Area Networks (WPANs) Amendment 1: Add Alternate PHYs, IEEE Standard for Information Technology Telecommunications and Information Exchange between Systems, August 2007.
6. F. Evennou, "Techniques et Technologies de Localisation Avancées Pour Terminaux Mobiles dans Les Environnements Indoor," Ph.D. Dissertation, Université Joseph Fourier, Grenoble, 2007.
7. J. Rabaey, F. Burghardt, D. Steingart, M. Seeman and P. Wright, "Energy Harvesting—A systems Perspective," International Electron Devices Meeting, Washington, December 2007, pp. 363-366.
8. http://en.wikipedia.org/wiki/Energy_harvesting
9. M. Ghavami, L. B. Michael and R. Kohno, "Ultra Wideband Signals and Systems in Communication Engineering," 2nd Edition, John Wiley & Sons Ltd., New York, 2007. doi:10.1002/9780470060490

10. T. K. K. Tsang and M. N. El-Gamal, "Ultra-Wideband (UWB) Communications System: An Overview," 3rd International IEEE-NEWCAS Conference, August 2005, pp. 381-386.
11. L. De Nardis and M.-G. Di Benedetto, "Overview of the IEEE 802.15.4/4a Standards for Low Data Rate Wireless Personal Data Networks," Workshop on Positioning, Navigation and Communication, Hannover, April 2007, pp. 285-289.
12. P. Bahl and V. N. Padmanabhan, "RADAR: An In-Building RF-Based User Location and Tracking System," Nineteenth Annual Joint Conference of the IEEE Computer and Communications Societies, Vol. 2, 2000, pp. 775- 784.
13. F. Dong, Y. Chen, K. Li and D. Guo, "A Calibration-Free Localization Solution for Handling Signal Strength Variance," Mobile Entity Localization and Tracking (MELT), Vol. 5801, 2009, pp. 79-90.
14. H. Liu, H. Darabi, P. Banerjee and J. Liu, "Survey of Wireless Indoor Positioning Techniques and Systems," IEEE Transactions on Systems, Man and Cybernetics, Part C: Applications, and Reviews, Vol. 37, No. 6, 2007, pp. 1067-1080.
15. S. Lanzisera, D. T. Lin and K. S. Pister, "RF Time of Flight Ranging for Wireless Sensor Network Localization," Workshop on Intelligent Solutions in Embedded Systems (WISES), Vienna, June 2006, pp. 1-12. doi:10.1109/WISES.2006.329127
16. S. Schwarzer,M. Vossiek, M. Pichler and A. Stelzer, "Precise Distance Measurement with IEEE 802.15.4 (ZigBee) Devices," IEEE Radio and Wireless Symposium, Orlando, March 2008, pp. 779-782.
17. Z. Liang, I. Barakos and S. Poslad, "Indoor Location and Orientation Determination for Wireless Personal Area Networks," International Workshop on Mobile Entity localization MELT, Vol. 5801, 2009, pp. 91-105.
18. T. He, C. Huang, B. M. Blum, J. A. Stankovic and T. Abdelhazer, "Range-Free Localization Schemes for Large Scale Sensor Networks," Proceedings of the 9th Annual International Conference on Mobile Computing And Networking, September 2003, pp. 81-95.
19. D. Niculescu and B. Nath, "Ad Hoc Positioning System (APS)," Global Telecommunications Conference, Vol. 5, November 2001, pp. 2926-2931.
20. W.-W. Ji and Z. Liu, "An Improvement of DV-HOP Algorithm in Wireless Sensor Networks," International Conference on Wireless Communications, Networking and Mobile Computing, Wuhan, September 2006, pp. 1-4.
21. L. Gui, A. Wei and T. Val, "A Two-Level Range-Free Localization Algorithm for Wireless Sensor Networks," 2010 6th International

Conference on Wireless Communications Networking and Mobile Computing (WiCOM), Chengdu, September 2010, pp. 1-4.
22. L. Doherty, K. S. J. Pister and L. El Ghaoui, "Convex Position Estimation in Wireless Sensor Networks," Twentieth Annual Joint Conference of the IEEE Computer and Communications Societies, Vol. 3, 2001, pp. 1655-1663.
23. N. Bulusu, J. Heidemann and D. Estrin, "GPS-Less LowCost Outdoor Localization for Very Small Devices," IEEE Personal Communications, Vol. 7, No. 5, 2000, pp. 28-34. doi:10.1109/98.878533
24. F. Reichbach and D. Timmermann, "Indoor Localization with Low Complexity in Wireless Sensor Networks," IEEE International Conference on Industrial Informatics, Singapore, August 2006, pp. 1018-1023. doi:10.1109/INDIN.2006.275737
25. H. S. AbdelSalam and S. Olariu, "HexNet: HexagonBased Localization Technique for wireless Sensor Networks," IEEE International Conference on Pervasive Computing, Galveston, March 2009, pp. 1-6. doi:10.1109/PERCOM.2009.4912871
26. J. Yi, J. Koo and H. Cha, "A Localization Technique for Mobile Sensor Networks Using Archived Anchor Information," Sensor, Mesh and Ad Hoc Communications and Networks, San Francisco, June 2008, pp. 64-72.
27. http://www.decawave.com/

4
CHAPTER

A NEW ADAPTIVE EDCA APPROACH TO QOS OF WIRELESS COMMUNICATIONS

Huibin Wang, Yang Hu, Lili Zhang, Wei Xia

ABSTRACT

Under the background of intelligent transportation application, QoS for various services is different in wireless communication. Based on the MAC layer protocol, this paper analyzes the QoS in IEEE 802.11 MAC protocol framework, and proposes a new design of a Differentiation Enhanced Adaptive EDCA (enhanced distribution channel access) approach. The proposed approach adjusts the window zooming dynamically according to the collision rate in sending data frames, makes random offset, and further distinguishes the competition parameters of the data frames that have the same priority, so as to reduce the conflict among the data frames, and improve the channel utilization. Experiments with different service cases were conducted. The simulation results show that: comparing with the conventional EDCA method, the proposed approach can ensure that high priority services are sent with priority, and the overall QoS is highly improved.

Keywords: Intelligent Transportation Systems, Wireless Communi-Cation, Mac Layer Protocol, Enhanced Distribution Channel Access

1. INTRODUCTION

In the intelligent transportation systems (ITS), there are two wireless communication modes: car to car and car to road. They are the foundation of providing information exchange and service. Wireless communication improves the service active safety technology further. In early days, the main research topics of the intelligent transportation wireless communication technology are vehicles location and distance measurement between vehicles etc [1,2] in collaborative driving application. In recent years, more and more researches focus on transmission of service information and mass data flows such as audio and video and so on. In researches and applications, the modern wireless communication technologies are adopted. For example, in Fleetnet-road network project (see http://www. et2.tu-harburg.de/fleetnet, Fleetnet) and CarTalk 2000 project (see http://www.cartalk2000.net, Cartalk 2000 web site), wireless LAN and cellular networks are employed.

In different intelligent transportation application scenarios, each service has stringent requirements for QoS of wireless communications. For example, in the active safety application scenarios, the delay of sending the se curity message requires to reach ms level. While in emergency voice communication application, packet loss rate of voice data stream can't exceed 1%. QoS can be described in different aspects in the communication, such as the network throughput and delays. At present, the method of improving QoS generally uses separate-layer or crosslayer. Cross-layer program has the advantage of improving the efficiency of the protocol stack. But its network protocol is complex and is difficult to maintain. Therefore, in the application background, the main issue is separate-layer QoS program in recent days.

The MAC layer protocol controls node accessing to wireless channel and message transmission occupying wireless medium, so as to provide reliable data transfer for upper layer, and guarantee the overall performance of the network. Therefore, MAC layer is the focus of communication protocol in intelligent transportation application. For example, in the EDCA (enhanced distribution channel access) mechanism of the IEEE 802.11p [3] standard, the service data accesses to channel by a competitive way. And further distinction of priority-based QoS through distributing competitive parameters is achieved. CarTalk 2000 Project proposed the mechanism of Ad Hoc MAC [4]. It adopts dynamic TDMA mechanism, and each service accesses to the channel by making an appointment. The article [5] proposed the D-MAC mechanism, which reduces transmission conflict by using the directional antenna. Kaichi F. et al. proposed the VRCP mechanism [6], which can work in Central control or ad hoc network. And it can dynamically switch according to whether the node is in the coverage area of the central controller.

However, the existing programs for intelligent transportation cannot fully meet the requirements in real-world applications. For network throughput and delay, this paper lays special stress on analyzing the MAC layer, and emphatically studies the performance of QoS of EDCA mechanism in intelligent transportation and designs a new mechanism to improve it. Then simulate the algorithm and analyze the experiment results. This paper is structured as follows: Section 2 introduces the principle of EDCA mechanism and the support to QoS. Section 3 designs and describes a differentiation enhanced adaptive EDCA mechanism. Section 4 simulates on the OPNET platform and analyzes the results. Section 5 is the summary.

2. RELATED WORK

The MAC layer provides three mechanisms and protocols in the intelligent transportation with wireless communication, which are described as follows:

1) CSMA/CD-based Protocol, such as the 802.11p. CSMA/CD uses the mechanism of listen-before-talk (LBT). Each node in the system senses the channel before sending the packet. If the channel is idle, send data. If the channel is busy, continue to sense until it becomes idle, and then use binary backoff mechanism to reduce the probability of collision.

2) TDMA-based protocol, such as RR-ALOHA. Protocol, a channel is divided into several time slots (called BASIC CHANNEL, BC). Besides the payload of each channel, FI (Frame Information) is also transmitted. FI includes the idle state of each slot. Any terminal can ask for the establishment of BC in idle time slot. And the channel is remained by the terminal that occupies it before in busy time slot.

3) Directional antenna-based mechanism, such as DMAC protocol. It is developed from IEEE 802.11. The protocol requires each node to know the location of themselves and their neighbors. Source node first launches a RTS/CTS/ACK handshake mechanism before sending data packets. RTS or Omni are sent directionally according to the situation of the transmission around. The directional antenna will be blocked when receiving the RTS or CTS, and the source nodes delay sending packets according to the received information. The conflict of transmission can be reduced by using directional antennas and channel reusing rate can be increased. But the

directional antenna systems are expensive, and it's hard to maintain in the actual implementation.

TDMA-based MAC protocols require strict time synchronization and its slot is fixed. When a node sends no data, its slots cannot be occupied by other nodes that need to send data. So it can't use channel resources effectively. For example, in Japan ALOHA is mainly used in MAC layer protocol of Dedicated Short Range Communication (DSRC) [7]. It is designed for the electronic toll collection and other applications, and it can't meet the demands in many service application scenarios and services.

CSMA-based protocol does not require time synchronization. It uses carrier sense method to judge whether the channel is idle before sending data. If the channel is busy, it adopts backoff mechanism to make the node return to sense, and tries again. So the conflict probability is small and the channel utilization is effective. CSMAbased protocol is more suitable for the rapid movement of vehicles. Therefore, this paper focuses on the MAC protocol based on CSMA/CD mechanism.

3. CONVENTIONAL EDCA MECHANISMS AND SUPPORT OF QOS

QoS refers to a series of performance indexes that need to be satisfied when the network transmits some data flows. QoS guarantee reflects the capabilities that network devices or protocols ensure data transmission and meet specific service. It is described by the throughput, delay, delay jitter and packet loss rate and other parameters. In order to support the application of the Intelligent Transportation System, IEEE 802.11 Working Group established wireless access standard for the vehicle environment, named IEEE 802.11p. In this standard, the MAC layer adopted the EDCA mechanism based on CSMA/CD.

The working principle of EDCA mechanism is shown in **Figure 1**.

EDCA provides the priority-based QoS function. This mechanism defines the access category (AC). Reference to 802.1D standard, it casts 8 priorities into 4 ACs, each of which has a separate backoff parameter. Hence, there are 4 separate competition entities in each node. The distinction between priorities is obtained by setting the values of different parameters such as Arbitration Inter Frame Space (AIFS), minimum contention window (CWmin) and maximum contention window (CWmax) are used for different ACs. For each AC[i] (i = 1,2,3), the backoff mechanisms are the same as DCF.

To achieve the distinction between priorities, for $0 \le i < j \le 3$, demand $CW_{min}[i] \ge CW_{min}[j]$, $CW_{max}[i] \ge CW_{max}[j]$ and $AIFS[i] \ge AIFS[j]$. As can be seen, higher-priorities have smaller CWmin, CWmax, AIFS parameters. They can enter the backoff process and finish it firstly in the competition. Thus they have a top priority to access to channel. If several entities finish backoff at the same time, the higher-priority entity will be sent as a priority.

Furthermore, the EDCA mechanism introduces the concept of transmission opportunity (TXOP). After AC competing for channels and having sent a frame successfully, if there are still data frames need to be sent, the channel can be used all the time in the TXOP time limit. When we reach the TXOP time limit or there are no data frames to send, TXOP is over and starts a new competition.

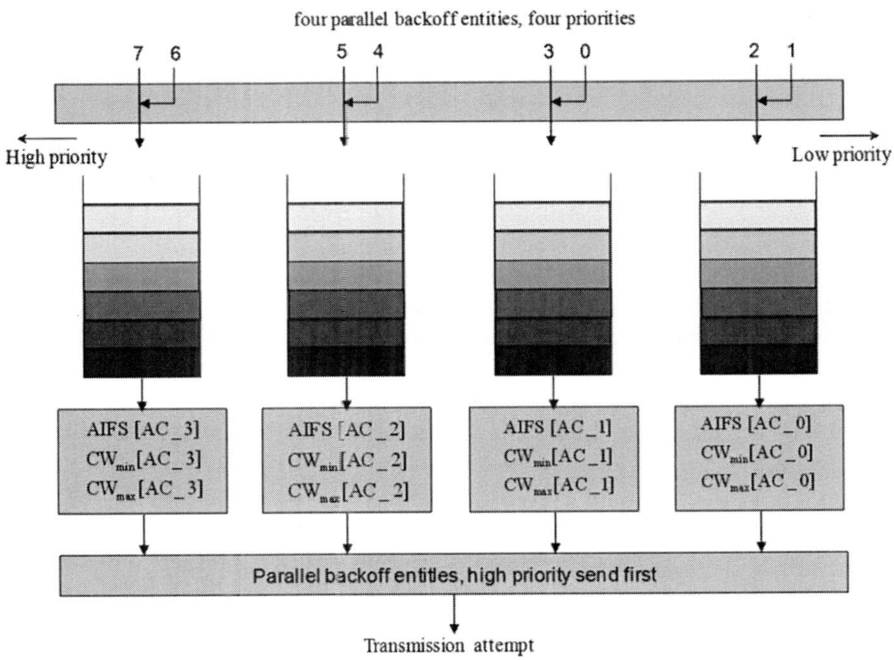

Figure 1. The working principle of EDCA mechanism

The EDCA mechanism ensures the higher-priority service to be sent as a priority by the means of distinguishing priorities. However, in high load, for middle and low-priority service flows, the performance will become rather poor. For EDCA mechanism, parameters configuration is the key of effectively using channel resources. And the reasonable parameter adjustment algorithm plays an important role in expressing the system performance.

4. NEW DESIGN OF DIFFERENTIATION ENHANCED ADAPTIVE EDCA

On one hand, the EDCA mechanism supports to send the high priority service firstly, but backoff parameters can not be adjusted dynamically with the change of network situation. Thus the network performance can't be utilized effectively. On the other hand, EDCA can distinguish competition parameters from different ACs that belong to the same node. It leads to data frames from the same AC of multiple sites choosing the same competition parameters, which tries to send at the same time, especially for the high priority AC with the same AIFS, its scope of the CW is small, thus the probability of choosing the same parameters is very large. This phenomenon is even more evident with the nodes increasing. For the above two aspects, this paper will design a Differentiation Enhanced Adaptive EDCA (DE-AEDCA). The mechanism can adapt to different network congestions by calculating the collision rate of the sending-data frames, dynamically adjusting the window zooming, maintaining an appropriate contention window range. While the random offset achieves the further distinction of data frames competition parameters between the same priority, and reduces their conflict probability. Consequently, it cuts down the numbers of idle time slots caused by conflict, and improves the channel utilization.

4.1 Selection of Determining the Network Status Parameters

In the successive, same interval of time period T_n $n = (1,2,3,\cdots)$, we record the collision rate of different $AC[i]$ $(i = 1,2,3)$ access types periodic. We assume the number of data frame conflicts to be $Num_n^{collision}[i]$ in one T_n, using parameter $P_n[i]$ to show the conflicts in the current time period T_n. The range of the $P_n[i]$ value is 0 to 1. The formula is shown in (1).

$$P_n[i] = Num_n^{collision}[i] / Num_n^{total}[i] \quad (n = 1, 2, 3, \cdots) \tag{1}$$

Collision rate changes in each time period T_n dynamically, which T_n is an integer multiple of slot duration. If T_n is too large, it cannot reflect real-time network status. On the contrary, if T_n is too small, the computation will increase. When OFDM technology is adopted in the physical layer, one time slot length is $9\mu s$, data interval is $0.01s$ in this article. Then we take the time data sources to produce 9 data frames as a statistical cycle, which means $T_n = 90000\mu s$. In order to suppress judging errors of network state due to the burst of data frame, we deal with the current $P_n[i]$ with weighted average method of multiple time periods collision rate. Weights after treatment is $P_n^{weight}[i]$, which is computed as (2):

$$\begin{cases} P_n^{weight}[i] = k \times P_n[i] + (1-k) \times P_{n-1}^{weight}[i], & (n \geq 2) \\ P_n^{weight}[i] = P_n[i], & (n = 1) \end{cases} \tag{2}$$

where $k(0 \leq k \leq 1)$ is the smoothness factor. It is changed by the emergent state of network, for instance, estimated by comparing $Num_n^{total}[i]$ with $Num_{n-1}^{total}[i]$. If the nodes generate stable value of data frames in each time period, k can take a higher value. In this paper, it is more stable to produce data of source node. According to experimental results, k takes 0.8 in the simulation of this paper.

4.2 Conflict Process of Data Frame

The algorithm for conflict process of data frame is given as:

$$\begin{cases} CW_{new}[i] = \min\{CW_{\max}[i], CW_{old}[i] \times (1 + 2^{\varphi})\}, \\ \varphi = P_n^{weight}[i], \end{cases}$$
$$(i = 0, 1, 2, 3) \tag{3}$$

where φ is the current weighted value of collision rate, $\in \varphi$[0,1]. The scope of $1+2^{\varphi}$ is [2,3], thus the competition window of $AC[i]$ can progressively increase in different multiples depending on collision rate of its data frame. The larger the value of $P_n^{weight}[i]$ is, the greater the progressively increasing multiples is, and vice versa, thus competition window can adaptively adjust depending on network status indication.

In this algorithm, each access category $P_n^{weight}[i]$ is statistical respectively, because the high-priority data has higher right to access to channel than that of the lowpriority data, so $P_n^{weight}[i] \le P_n^{weight}[j], (0 \le i \le j \le 3)$, it means that the higher the priority is, the smaller collision rate is. Thus the progressively increasing multiples of their competition window are smaller. Therefore, this algorithm still strictly ensure the priority relations among all categories of data, it is critical for the application of intelligent transportation system, because it must ensure that high-priority data (such as emergency short message data) to send first.

4.3 Judgement of Sending the Data Frames Successfully

The equations are an exception to the prescribed specifications of this template. You will need to determine whether or not your equation should be typed using either the Times New Roman or the Symbol font (please no other font). To create multileveled equations, it may be necessary to treat the equation as a graphic and insert it into the text after your paper is styled.

First, we define a variable γ_i whose value shows the current consecutively sending number of data frame of node $AC[i]$. Its value adds 1 when it successfully sends a data frame, and the value is cleared when a conflict occurred.

Secondly, we subtract the value $CW_{\min}[i] \times (1 - P_n^{weight}[i])$ of the contention window when $\gamma_i = 1$. Subtract $2 \times CW_{\min}[i] \times (1 - P_n^{weight}[i])$ of the contention window when $\gamma_i = 2$, and so on. That means the decreasing value increases when γ_i increases every time. Every time, the increasing value is Δ:

$$\Delta = CW_{\min}[i] \times (1 - P_n^{weight}[i]) \tag{6}$$

Finally, when the value of contention window reduced to half of the initial value, it achieves the congestion avoidance phase, in order to avoid excessive competition which results from contention window decreasing too fast. Hereafter, when γ_i keeps on increasing, it resumes taking $CW_{\min}[i] \times (1 - P_n^{weight}[i])$ as original decreasing value and beginning to accelerate declining as beforementioned steps until declining to $CW_{\min}[i]$.

The basic idea of this algorithm is similar to slow-start decreasing algorithm SSDS (Slow-start Decrease Scheme. But it increases the adaptive capacity of the decreasing window, and it adds congestion control to avoid further competition which results from contention window decreasing too fast. The range of $(1 - P_n^{weight}[i])$ is [0,1] in the algorithm, and the range of Δ is $[0, CW_{\min}[i]]$. The larger $P_n^{weight}[i]$ is, the more severe the current network congestion is, and the smaller Δ is, the slower the declining pace of contention window is, and vice versa. The algorithm can judge the current network state and adjust the contention window size dynamically, and maintain a good state of congestion avoidance.

The algorithm can be described as (7):

$$\begin{cases} CW_{\gamma_i}[i] = \max\{CW_{\min}[i], CW_{\gamma_i - 1}[i] - \Delta \times \gamma_i\}, & (\gamma_i \le m) \\ CW_{\gamma_i}[i] = \max\{CW_{\min}[i], CW_{\gamma_i - 1}[i] - \Delta \times (\gamma_i - m)\}, & (\gamma_i > m) \end{cases} \tag{7}$$

where γ_i is the number of sending data frame successfully after the congestion avoidance phase, and $CW_{\gamma_i}[i]$ is the value of the contention window which is calculated by the algorithm. For example, $CW_0[i]$ shows the initial contention window before the first time successfully sending data frame. m means the number of sending data frames continuous successful before the congestion avoidance phase.

$$\begin{cases} CW_m[i] \geq \frac{1}{2} CW_0[i], \\ CW_{m+1}[i] < \frac{1}{2} CW_0[i], \end{cases} \quad m \in \gamma_i \tag{8}$$

4.4 Discrimination of Same Access Categories

Another improvement in this paper is the differentiation enhancement. It is designed to resolve the problem that the same AC data frames of different nodes compete with each other leads to the drastic decline of the whole network's performance when EDCA is under the high load.

Usually, the waiting time before trying to transfer each AC data frame equals to the sum of the avoidance time and the delay time. The avoidance time is random generated between 0 and $CW[i]$. The delay time which is decided by $AIFS[i]$ is the time necessary to wait before avoidance. So the basic idea of this algorithm is to increase the probability of the same AC choosing different competition parameter value by designing an added random offset to $AIFS[i]$ and $CW[i]$, on this condition, it's easier to distinguishing the competition parameters of the same AC data frames of different nodes. The concrete processes of the algorithm are as follows:

- The CW-based offset The biggest offset $CW_{offset}^{\max}[i]$ is computed by

$$\begin{cases} CW_{offset}^{\max}[i] = [(1 + CW_{\min}[i])^{\varphi}], \\ \varphi = P_n^{weight}[i], \end{cases} \quad (i = 0,1,2,3) \tag{9}$$

Then a random value between 0 and $CW_{offset}^{\max}[i]$ is selected as the real offset by

$$CW_{offset}^{\max}[i] = Random(0, CW_{offset}^{\max}[i]), \quad (i = 0,1,2,3) \tag{10}$$

Finally, according to (11), the real avoidance time can be computed as

$$T = [random(0, CW[i]) + CW_{offset}[i] \times aSlotTime], \quad (i = 0,1,2,3) \tag{11}$$

Equations (10) and (11) show that the random offset changes with the state of the whole network. The larger the value of network congestion condition is, the larger $CW_{offset}^{max}[i]$ is, it means the larger the choice scope of the offset is, the smaller the probability of choosing the same offset is.

- AIFS-based offset Above all, this algorithm can further distinguish the same AC by AIFS offset. First of all, assign an offset scope $[N_i, M_i]$ for each AC by its priority. Random select an integer as the number of slot time of $AIFS[i]$.

$$AIFS_{offset}[i] = Random[N_i, M_i] \times aSlotTime \tag{12}$$

Secondly, we put it as offset and add it to $AIFS[i]$, then computing the $AIFS_{new}[i]$.

$$AIFS_{new}[i] = AIFS[i] + AIFS_{offset}[i] \tag{13}$$

Though the introduction of the offset, we decline the probability of the same AC data frame choosing the same competition parameter and reduce the collision rate between them, which is in favor of improving the performance of the whole network.

The flow chart of the algorithm is shown in figure 2.

5. SIMULATION STUDIES AND ANALYSIS

The simulation experiments are conducted on OPNET platform. The OPNET uses process domain model, node domain model and network domain model to react the relevant features of the networks comprehensively.

Finitestate machines, OPNET core function, and standard C and C++ can make up the process model to describe bottom layer algorithm and the pipe stage model. OPNET Modeler uses discrete-time driven simulation mechanism that can get very detailed simulation results. It greatly improves the efficiency of the simulation.

5.1 Design of Simulation Cases

The simulation in this paper only considers the impact of MAC layer protocol on QoS requirements of multiple intelligent transportation service. Meanwhile, to cast the impact of routing layer on the wireless communication performance, this paper only considers one-leap communication without considering the impact on wireless communication of barrier and relative speed. The results in [11] shows that in one-hop range, the impact on throughput and delay of distance and speed of the cars is not obvious, but it has much relation with the number of the vehicles communicating at the same time. This paper mainly simulates different cases by setting different numbers of vehicles. In the simulation, the service scenario can be divided into high density case, middle density case and low density case by the number of vehicles communicating at the same time.

5.2 Parameter Settings

The services in Intelligent transportation systems can be divided into four categories: active safety, such as warning of dangerous road, collision warning; public service, such as emergency vehicles (fire engine, ambulance) which can pass prior on congested road, emergency speech communications; driving assistance service, such as roads na vigation, video monitoring; business and entertainment service, such as Internet access, electronic map download.

In view of the above service categories, application-layer sets four services in the simulation: emergency short message, speech, video, background, and the priority are from high to low. The service parameters are in table 1. This paper only considers the MAC layer's effects for QoS, neglect routing layer and the above ones. The high layer is simulated by sending source modules.

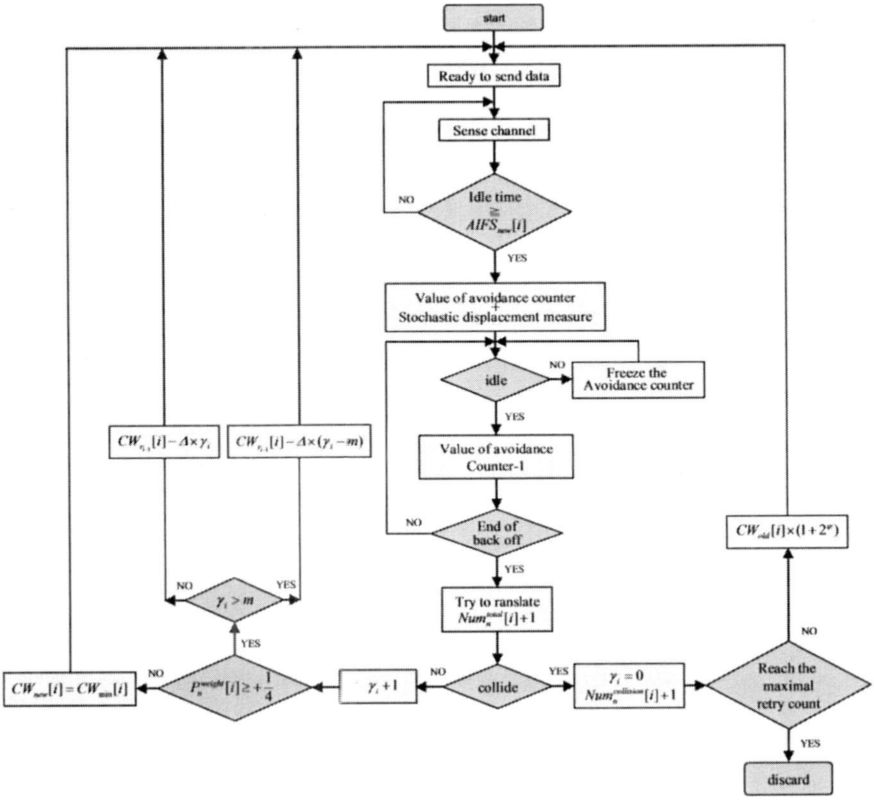

Figure 2. The flow chart of the differentiation enhanced adaptive EDCA

Physical layer uses OFDM modulation technique, its bandwidth is 20 MHz, transmission rate is 24 Mbits/s, SIFS is 16 us, DIFS is 34 us, and a SlotTime is 9 us. It distinguishes the different cases by setting different numbers of vehicle nodes. 32 nodes are set in high-density vehicle case; 20 nodes are set in middle-density vehicle case; and 12 nodes are set in low-density vehicle case. In each case every vehicle node carries only one service, and in different cases the numbers of vehicle nodes which carry different services are same. The parameters of MAC layer are showed in table 2, AC[3], AC[2], AC[1], AC[0] are emergency short message, speech, video, background services.

The AIFSN offset range of the mechanism proposed in this paper is in table 3.

Table 1. Parameter list

service	Inter-packet Gap(s)	Packet size	Transmission rate (Kbits/s)
emergence	0.01	12	9.6
speech	0.01	80	64
video	0.01	1250	1000
background	0.01	1500	1200

5.3 Simulation Results

In the simulation, the performance of network throughput and delay are analyzed. Network throughput refers to the summation of data packets received and successfully transmitted to application layer by MAC layer in all sites. Delay means different packets start when sending node in the application layer generates and end when destination node receives those packets, including media access delay and transmission delay (ms).

Figures 3, 4 and 5 are the simulation results with high density, medium density, and low density, respectively. The corresponding values of the four data streams in EDCA mechanism are described as the curves with round, left triangle, upper triangle and square symbols, according to their descending priority. The corresponding values of the four data streams in DE-AEDCA mechanism are described as the curves with the plus sign, lower triangle, right triangle and diamond symbols, by their priority from high to low. In figures, the x-axis represents the simulation time (sec.), the y-axis represents the delay time (sec.) and throughput (bit/s) separately. **Table 4** is the statistic results for the three types of cases in simulation.

The simulation results with high-density of vehicles, show that, in comparison to conventional EDCA, the delays of various services in the proposed DE-AEDCA mechanism are decreasing. Thus, it can meet the demands of security-related emergency short messages and QoS requirements of speech service. Meanwhile the throughput of video service increases from 6100 kbits/s up to 8000 kbits/s, and the packet loss rate drops to 6.25%. In addition, the throughput of background service rises from

Table 2. Mac layer parameter

	CW_{min}	CW_{max}	AIFS(us)
AC[3]	7	15	34
AC[2]	15	31	34
AC[1]	31	1023	52
AC[0]	31	1023	79

Table 3. Aifsn offset range of different Ac

	Ni	Mi
AC[3]	0	6
AC[2]	0	6
AC[1]	3	9
AC[0]	3	9

1850 kbits/s to 3250 kbits/s. Thus throughput of the video service and the background service respectively increases by 31.1% and 75.7%. Therefore the total throughput improves 38.6%. To sum up, the overall performances are improved.

For the middle-density case, the delay of various services has a certain decrease; the bandwidth needs of background service are met; no packet loss occurs; and the throughput performance is improved. In comparison to the EDCA mechanism, its overall performance is improved to some extent.

For the low-density vehicles, the delay is smaller than that in EDCA mechanism. Network packet loss does not occur any more. So the DE-AEDCA mechanism can meet the needs of various service applications properly.

As can be seen from the simulation results in the three situations of high-density and high network load, middledensity and high network load, as well as low-density and low network load, DE-AEDCA mechanism has both a higher network throughput and a smaller delay compared with EDCA mechanism. The reason is DE-AEDCA

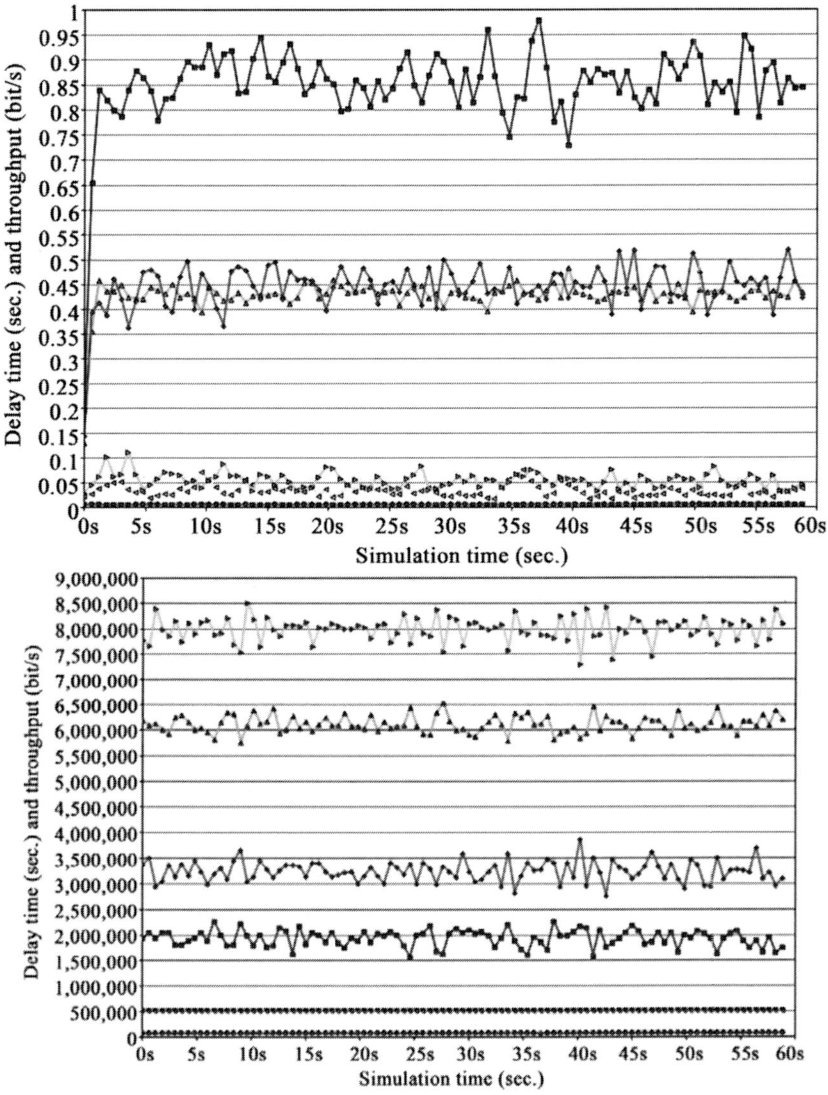

Figure 3. The simulate results at high density. (a) medium access delay; (b) network throughput

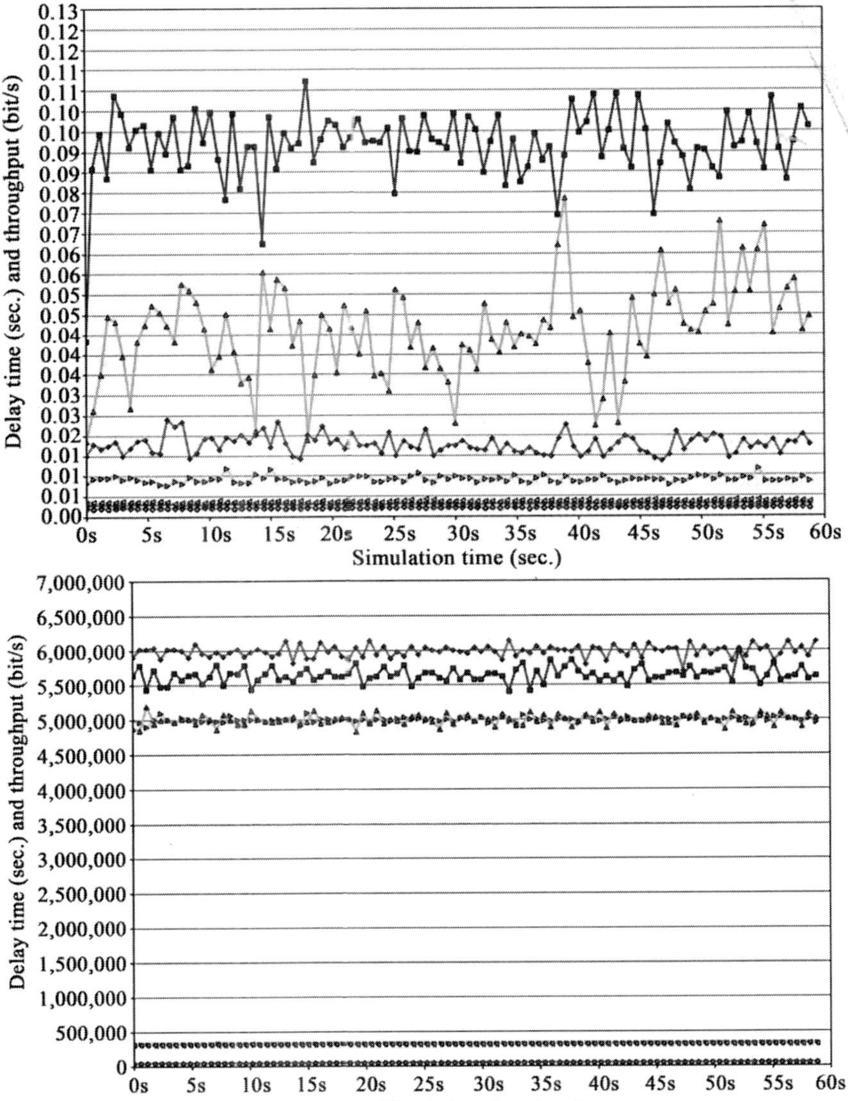

Figure 4. The simulate results at middle density. (a) medium access delay; (b) network throughput

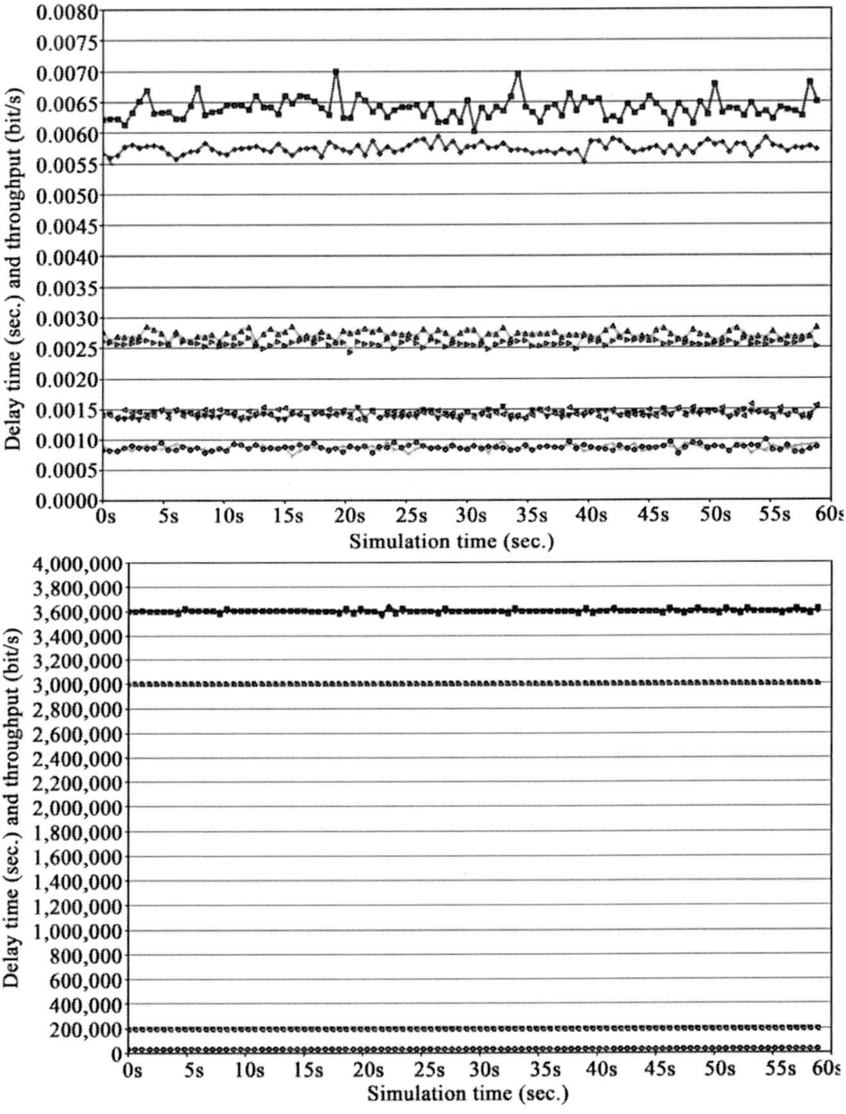

Figure 5. The simulate results at low density. (a) medium access delay; (b) network throughput

Table 4. The statistical results of simulations

		EDCA				DE-AEDCA			
		AC[3]	AC[2]	AC[1]	AC[0]	AC[3]	AC[2]	AC[1]	AC[0]
Delay/ average (ms)	high	6	34	430	850	2.5	4.6	55	450
	middle	2.1	3.7	43	91	1.9	2.8	9	18
	low	1.1	1.7	2.8	6.6	0.9	1.4	2.6	5.7
Throughput/ average (Kbits/s)	high	76.8	512	6100	1850	76.8	512	8000	3250
	middle	48	320	5000	5600	48	320	5000	6000
	low	28.8	192	3000	3600	28.8	192	3000	3600

mechanism considers the adjustment of competition parameters in conditions of a low network load and a high network load, so whenever in high-density and middledensity, or in the low-density vehicle cases, DE-AEDCA mechanism can have better QoS performance than EDCA mechanism.

In addition, among the data flows of same priority, DE-AEDCA mechanism increases the range of the parameters and reduces the possibility of choosing the same parameter by choosing a random offset, thus it can reduce conflict of the same priority data in network and improve the channel utilization. Experiments show that the DE-AEDCA mechanism always has better performance than EDCA mechanism in all kinds of intelligent service case.

6. CONCLUSIONS

This paper considers QoS demands of various services in intelligence transportation, and analyzes the QoS based on the EDCA mechanism in IEEE 802.11p standard framework. A Differentiation Enhanced Adaptive EDCA is designed, which improves the QOS of EDCA by dynamically adjusting the size of competition window and further distinguishing the competition parameters of the data frames that have the same priority. The simulation results show that in different situations of network congestion, the mechanism can discriminate the service priority, and make the high-priority service with low delay and high throughput. Meanwhile the overall performance of the network is improved significantly. Its performance is better than EDCA.

7. ACKNOWLEDGEMENTS

This work is supported by the National High Technology Research and Development Program of China (863 Program) (No. 2007AA11Z227).

REFERENCES

1. T. Yashiro, T. Kondo, H. Yagome, M. Higuchi and Y. Matsushita, "A Network Based on Inter-Vehicle Communication," Proceedings of Intelligent Vehicles Symposium, 14-16 July 1993, pp. 345-350.
2. H. Fujii, O. Hayashi and N. Nakagata, "Experimental Research on Inter-Vehicle Communication Using Infrared Rays," Proceedings of IEEE Intelligent Vehicles Symposium, Tokyo, 19-20 September 1996, pp. 266-271.
3. IEEE P802.11pTM/D6.0 Draft Standard for Information Technology Telecommunications and information exchange between Systems Local and metropolitan area networks-Specific requirements, Part 11: Wireless LAN Medium Access Control and Physical Layer specifications, November 2008.
4. F. Borgonovo, A. Capone, M. Cesana and L. Fratta, "ADHOC MAC: New MAC Architecture for Ad Hoc Networks Providing Efficient and Reliable Point-to-Point and Broadcast Services," Wireless Networks, Vol. 10, No. 4, July 2004, pp. 359-366.
5. Y. Ko, V. Shankarkumar and N. Vaidya, "Medium Access Control Protocols Using Directional Antennas in Ad Hoc Networks," Proceedings of IEEE INFOCOM 2000, Tel Aviv, vol. 1, March 2000, pp. 13-21.
6. K. Fujimura and T. Hasegawa, "A Collaborative MAC Protocol for Inter-Vehicle and Road to Vehicle Communications," Proceddings of IEEE Intelligent Transportation Systems Conference 2004, Washington, D.C., October 2004, pp. 816-822.
7. "Standard Specification for Telecomunications and Information Exchange between Roadside and Vehicle Systems — 5GHz Band Dedicated Short Range Communications (DSRC) Medium Access Control (MAC) and Physical Layer (PHY),"ASTM E2213-03, September 2003.
8. J. Wang, S. Makfile and J. Li, "A Random Adaptive Method to Adjust MAC Parameters in IEEE802.11e WLAN," Journal of Central South University of Technology, Vol. 16, pp. 629-634, 2009.

5

CHAPTER

A NOVEL OUTAGE CAPACITY OBJECTIVE FUNCTION FOR OPTIMAL PERFORMANCE MONITORING AND PREDICTIVE FAULT DETECTION IN HYBRID FREE-SPACE OPTICAL AND RF WIRELESS NETWORKS

Barnet M. Schmidt

ABSTRACT

This study develops an optimal performance monitoring metric for a hybrid free space optical and radio wireless network, the Outage Capacity Objective Function. The objective function—the dependence of hybrid channel outage capacity upon the error rate, jointly quantifies the effects of atmospheric optical impairments on the performance of the free space optical segment as well as the effect of RF channel impairments on the radio frequency segment. The objective function is developed from the basic information-theoretic capacity of the optical and radio channels using the gamma-gamma model for optical fading and Ricean statistics for the radio channel fading. A simulation is performed by using the hybrid network. The objective function is shown to provide significantly improved sensitivity to degrading performance trends and supports of proactive link failure prediction and mitigation when compared to current thresholding techniques for signal quality metrics.

1. INTRODUCTION

Current techniques of free-space optical (FSO) and radio frequency (RF) link performance measuring and monitoring are traditionally applied at the physical and lower link layer, at the RF and FSO signal in space (SiS) and at the bit/packet levels. Performance metrics commonly used include received signal level, optical signal-to-noise ratio (OSNR) for the optical links, quality (Q) factor, eye diagram, signal-to-noise ratio (SNR) or signal-to-interference ratio (SIR) for the RF links, and bit error rate (BER). Traditional methods of link health analysis include observation of these parameters in attempt to determine their trends. In good atmospheric conditions the received optical signal level and OSNR will be high and will fluctuate within some previously-defined normal range as determined by the optical channel fading characteristics. In good RF propagation conditions, the received signal level will be correspondingly high as the SNR/SIR, and in turn it will also fluctuate within a bounded acceptable range. Degrading optical conditions of the atmosphere will in turn degrade the received optical signal level and OSNR. Degrading RF propagation conditions, such as fading and multipath will have similar effects upon the received signal level, SNR, and/or SIR.

The error correcting codes (ECCs) used in most digital communications systems provide excellent performance until a specific threshold of SNR is exceeded at which point the error rate increases rapidly. In most cases, by the time the performance monitor (PM) component of the network management system (NMS) has determined that the SNR and other performance metrics have crossed the acceptable threshold, user data are lost. Current methods of thresholding by themselves are therefore inadequate to predict link failure.

In poor transmission conditions the NMS institutes corrective actions consisting of reducing the transmitted data rate, increasing the signal power, instituting more frequent retransmissions (usually managed independently by the link layer protocol used in the network), or combinations of these techniques. Traffic offloading and rerouting may also be used to reassign traffic and better performing links in lieu of those with degraded performance temporarily until transmission conditions improve and the PM observes the performance metrics to once again cross their acceptable thresholds.

The optimal performance monitoring function developed in this study is the Outage Capacity Objective Function ("Objective Function"). The objective function is defined as the outage capacity of the hybrid FSO/RF channel as a function of BER. It is demonstrated that the objective function provides an extremely sensitive measure of the network performance, identifying impending failure well in advance of signal quality thresholding alone.

2. NETWORK TOPOLOGY AND CHARACTERISTICS

This hybrid FSO and radio wireless network considered in this study has the topology shown in**Figure 1**. The network characteristics are described below:

1) Wavelength, Frequency, and Structure: The FSO channels operate at the "long" mid-infrared (MIR) wavelength of 10,000 nm. Studies in this range of wavelengths predict improvements in performance over shorter wavelengths (see Related Work, below). The frequency of the RF channels is 1.9 GHz, a widely used band in cellular radio systems.

2) Propagation: Line-of-sight (LOS) propagation is assumed for both the FSO and RF links.

3) Cellular Topology: The RF segment utilizes a cellular topology with 12 cells and K users per cell. Handoff is used as in commercial cellular systems.

4) MIMO: The FSO channels are analyzed for the single input-single output (SISO) case using only one transmitter and receiver, and the multiple input-multiple output (MIMO) case. The objective function is produced for both SISO and several MIMO levels.

Each RF transceiver is capable of communicating with the co-located FSO transceiver using a baseband digital link. This allows the optical and RF transceivers to balance user traffic between the FSO and RF links (for simplicity these cross-links are not shown on the network diagram).

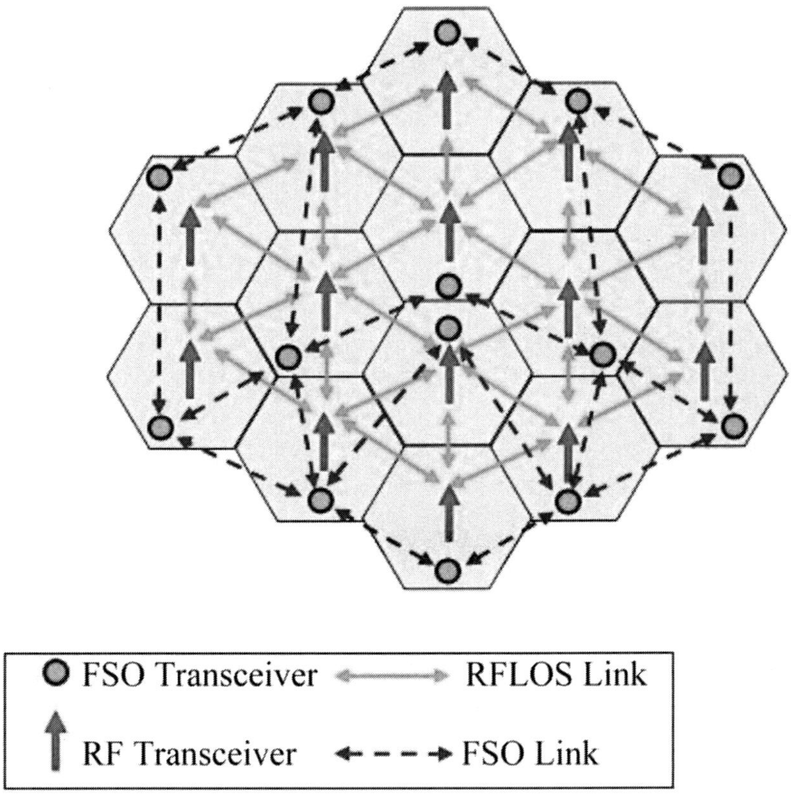

Figure 1. Hybrid network topology.

3. OUTAGE CAPACITY

The outage capacity, C, for both the optical and RF channels is defined as the maximum throughput (load) that the hybrid channel can support at a given BER:

$$C \equiv f(\mathrm{BER}) \tag{1}$$

The outage capacity of the optical and RF channels is determined from the application of information theory to the fading statistics of the channels.

The optical channel capacity is determined by the instantaneous mutual information of the channel and its fading distribution. Information theory shows that the capacity of the channel is the mutual information of the channel maximized over all possible distributions of inputs. This assertion is valid in the absence of fading. In the presence of fading, the instantaneous capacity of the channel is the product of the instantaneous mutual information conditioned upon the channel information and the instantaneous value of the signal intensity (fading) is determined by the instantaneous mutual information, T(p,x), where x is the instantaneous channel realization given the instantaneous channel state, i, conditioned upon the probability distribution function (PDF) of the irradiance (the fading distribution), $f_i(x)$, and p, the power distribution of the N MIMO transmitters:

$$p \equiv (p_1, p_2, \cdots, p_N) \tag{2}$$

Rewriting the outage capacity in terms of the mutual information and the fading distribution:

$$C = T(p, x|i) f_i(x) \tag{3}$$

3.1. Related Work

MIR wavelengths have been shown to improve FSO performance due to lower Mie scattering losses in [1] and [2]. Studies [2,3], and [4] have demonstrated that atmospheric FSO channel fading is accurately characterized by the Gamma-Gamma distribution. In the gamma-gamma statistics two independent fading models that characterize the fading effect of strong and weak turbulence, including scattering and scintillation are used. Each of the distributions is a gamma distribution and is an extension of prior studies that establish a doubly-stochastic model as the optical fading channel. Influence of diversity methods on the RF channels has been studied in [5]. Performance improvements in FSO at MIR wavelengths are considered in [6]. Effects cf atmospheric turbulence, as well as further support of the Gamma-Gamma fading distribution on FSO channel capacity have been studied in [7] and [8]. The effect of pointing errors and misalignment on the outage capacity of purely FSO channels has been investigated in [9] and [10].

3.2. Instantaneous Mutual Information

The instantaneous mutual information of the channel is a measure of the dependence between the random variables that characterize the channel: p, i, and x (4):

$$T(p,x|i) = \sum_{i_n \in i} \sum_{x_n \in x} \sum_{p_n \in p} p_{p,x,i}(p_n, x_n, i_n) \log \frac{p_i(i_n)\, p_{p,x,i}(p_n, x_n, i_n)}{p_{p,i}(p_n, i_n)\, p_{x,i}(x_n, i_n)} \quad (4)$$

In the information-theoretic sense, the mutual information quantifies the degree to which knowing any of the random variables p, x, and/or i reduces the uncertainty about the others. In many analyses of channel capacity (including this one) and/or error performance knowledge of the channel state i increases the mutual information. When channel state is unknown or unreliable, estimation methods may be used to obtain these values.

The channel capacity is the instantaneous mutual information of the channel maximized over all possible distributions of the transmitted information:

$$C_m = \max(p) \int_0^{\infty} \left(T(p,x|i) \right) f_i(x)\, \mathrm{d}x \quad (5)$$

Expanding this equation produces (6):

$$C_m = \max \int_0^{\infty} \sum_{i_n \in i} \sum_{x_n \in x} \sum_{p_n \in p} p_{p,x,i}(p_n, x_n, i_n) \cdot \log \frac{p_i(i_n)\, p_{p,x,i}(p_n, x_n, i_n)}{p_{p,i}(p_n, i_n)\, p_{x,i}(x_n, i_n)} f_i(x)\, \mathrm{d}x \quad (6)$$

and is used in developing the outage capacity for the gamma-gamma channel. The outage capacity is the maximum throughput of the channel at a given specified error probability (which, in turn, implies a specific BER at that rate).

3.3. Outage Probability

An analytical derivation of the exact outage probability for the gamma-gamma model is an intractable problem in closed form [4]. Instead, the asymptotic behavior of the outage probability as a function of OSNR and the lower bound is investigated.

The error probability with respect to (conditioned upon) the instantaneous mutual information, instantaneous channel realization, and the data rate R can be stated as:

$$P_e(\mathrm{OSNR}, R) = \Pr\left(T(p,x) < R\right) \tag{7}$$

The probability of an error at a given OSNR and data rate is simply the probability of the instantaneous mutual information of the channel being less than the data rate. This is intuitive since the instantaneous mutual information of the channel averaged over the inputs represents the information capacity of the channel under the given impairment conditions (that is, OSNR) while the data rate represents the transmission rate required by the user. The Shannon-Hartley theorem imposes an upper bound on the capacity of any channel as a function of the bandwidth and the SNR. The summations in the mutual information Equation (4) may be expressed in terms of the Shannon capacity and (2) rewritten as (8):

$$T(p,x) = \frac{1}{L}\sum_{l=1}^{L}\log_2 Q - E\left[\log_2\left[1+\sum_{q=2}^{Q}\exp\left(-\left(p_q x_q^2\right)+\sqrt{p_q x_q^2}\left(z_q - z_1\right)\right)\right]\right] \tag{8}$$

where: L is the length of the transmitted block in codewords, z_q are additive white Gaussian noise (AWGN) samples with zero mean and variance (normalized power) 1, x_q is the channel realization when codeword q is received, and Q is the number of codewords.

The inner term of the summation in (8) is the mutual information of an AWGN channel with an OSNR of $p_q x_q^2$. The asymptotic behavior of the outage probability can be determined by introducing the OSNR exponent, defined as:

$$d(\mathrm{OSNR}) = -\lim_{\mathrm{OSNR}\to\infty} \frac{\log \Pr\left(\mathrm{OSNR}, R\right)}{\log\left(\mathrm{OSNR}\right)} \tag{9}$$

where P is the total transmitted power of the M transmitters.

Evaluating (9) using the Shannon capacity while substituting the Shannon-Hartley equation for the rate R yields:

$$d\left(\mathrm{OSNR}\right) = \frac{MN}{2}\left[1+\left(B\left(1-R_c\right)\right)\right] \tag{10}$$

where B is the system bandwidth, N is the number of receivers in the MIMO system, and R_c is the binary code rate, which is:

$$R_c = \frac{R}{\log_2 Q} \tag{11}$$

A strategy to maximize the outage capacity in the gamma-gamma channel is to allocate the transmitted power among the N transmitters so as to satisfy the condition of (7) being minimized as much of the time as possible.

In this analysis, the transmit power is equally allocated to each of the M transmitters (instantaneous power at each transmitter is 1/M). Allocating the transmitted power to the transmitters is subject to both the short and long term power constraints of the system. In the short term, more power than the total available system power can be allocated to a subset of the transmitters if that optimizes the instantaneous mutual information rate under at that particular OSNR. The long term power, which is the expected value of the short term power, places a limit on how long such short-term allocation can be performed. The SNR exponent corresponding to the long term power constraint is the same as the short-term power constraint SNR exponent when the value of the SNR exponent is less than 1. If the value of the (short-term power constraint) SNR exponent is greater than 1, the long-term SNR exponent is infinite. In this case the outage probability curve slope is infinite and there is an instantaneous increase in the outage probability under these conditions.

The optimal OSNR exponent and thus the outage probability is a strong function of the number of transmitters and receivers in the MIMO system, specifically their product MN. As will be shown in the results, as the number of transmitters and receivers are increased (hence, the product MN) the ultimate floor of the outage probability drops at a sufficiently high OSNR. This demonstrates the value of spatial diversity which is one of the reasons for using a MIMO system. However, the slope of the outage probability curve increases with increasing MN. This results in a tradeoff of the minimum outage probability vs slope or tolerance to OSNR. In systems having low OSNR (which is the usual case for FSO systems) it is advantageous to maximize the number of transmitters and receivers. In this case, the high slope of the outage probability as a function of OSNR can be tolerated since the high OSNR/low outage probability mode will be the usual operating condition. If the system operates over a wide range of OSNR, particularly in low OSNR regime, it is advantageous to have a smaller number of transmitters and receivers. This is borne out by the fact that at low OSNR, the outage probability increases rapidly as the OSNR decreases. In this case, the ultimate minimum outage probability is traded off. If MN were to be large in this case, there would be frequent cases where the outage probability rapidly increases to 1 over a narrow range of (decreasing) OSNR.

3.4. Outage Capacity of the Optical Channel

In considering the effect of the gamma-gamma channel (or any other optical fading channel) it is realized that the received signal irradiance in any fading channel is the equal to the received signal in an AWGN channel scaled by the instantaneous value of the fading. The distribution of the received signal irradiance as previously described is simply the distribution of the signal received from an AWGN channel scaled by the distribution of the fading. This is simply a linear model of the channel, which is an AWGN channel followed by a spatial filter whose optical transfer function replicates the fading. We use this result to determine the instantaneous mutual information of the gamma-gamma channel.

The fading probability distribution function of an AWGN channel is a Gaussian distribution. The instantaneous mutual information of the fading channel is bounded by that of a Gaussian (AWGN) channel. Information theory proves that the Gaussian distribution contains the maximum entropy and therefore also represents the maximum mutual information. As a function of the optical signal irradiance variance σ_p^2 (which is the normalized signal power) and the two-sided noise power spectral density (PSD) N_0, the instantaneous mutual information becomes:

$$T(p,x|i) \leq \frac{1}{2}\log_2\left[1+\frac{2\sigma_p^2}{N_o}\right] \tag{12}$$

The upper bound of channel capacity is the outage capacity when conditioned upon the probability of failure. The capacity of the gamma-gamma channel is the upper bound of the mutual information of the Gaussian channel scaled by the filtering of the Gaussian-Gaussian channel. Since the ergodic capacity and the Gamma-Gamma distributions are independent, the upper bound is obtained from the product of the Gaussian mutual information upper bound and Gamma-Gamma distributions and then averaging over the irradiances. In this relationship P is the total transmitted power of the MIMO array, i is the received signal irradiance, ψ is a variable dependent upon the distance between the transmitted symbols in the complex signal space (the square of the incremental distance between the points in the transmitted On-Off Keying (OOK) constellation), and ξ is the normalized power per bit (p/R). The maximum capacity is (13):

$$C,\max \leq \frac{1}{2}\int_0^{\infty}\log_2\left[1+\left(\frac{1}{P}-1\right)\psi\zeta p^2 i^2\right] \cdot\left[\frac{2^{2(\alpha+\beta-2)}}{2\pi\log(2)\Gamma(\alpha)\Gamma(\beta)}\right]di \tag{13}$$

In this equation the final product term of the integrand is the gamma-gamma fading distribution. The gamma functions Γ(α) and G(α) quantify the fading contributions from weak and strong turbulence, respectively. The values of α and β are the number of eddy currents in the turbulent atmospheric flow, where a is the number associated with the small-scale refractive index changes and scattering, and β being the number of large scale eddy currents associated with strong turbulence [5]. These parameters depend upon the physical dimensions of the FSO system, the wavelength, transmission distance, and altitude:

$$\alpha = \exp\left[\frac{0.49\chi^2}{\left(1+0.18d^2+0.56\chi^{2.4}\right)^{1.167}}-1\right]^{-1} \tag{14}$$

$$\beta = \exp\left[\frac{0.51\chi^2\left(1+0.69\chi^{2.4}\right)^{-0.833}}{\left(1+0.9d^2+0.62d^2\chi^{2.4}\right)^{0.833}} - 1\right]^{-1}$$ (15)

where: $\chi^2 = 0.5C_n^2 k^{1.167} L^{1.833}$, $d = \sqrt{\dfrac{kD^2}{4L}}$, and the wave number, $k = \dfrac{2\pi}{\lambda}$.

D: Diameter of the receiver lens aperture (0.01 m).

L is the transmission distance (1500 m).

C_n^2 is the refractive index structure parameter (1.7×10^{-14}, altitude-dependent. The altitude is 100 m. in this study).

The integration in (13) is carried out to produce a closed form solution, resulting in (16):

$$C,\max \le \left[2^{(\alpha+\beta-2)} \frac{2^{(\alpha+\beta-2)}}{2\pi\log(2)\Gamma(\alpha)\Gamma(\beta)}\right] \cdot G_{6,2}^{1,6}\left[\left[\frac{\left(16\left(\frac{1}{P}-1\right)\psi\xi\right)P^2}{\alpha^2\beta^2}\middle| \begin{matrix} 1,1,\frac{1-\alpha}{2},\frac{2-\alpha}{2},\frac{1-\beta}{2},\frac{2-\beta}{2} \\ 1,0 \end{matrix}\right]\right]$$ (16)

where: $G_{p,q}^{m,n}\left(\begin{matrix} a_1,\cdots,a_p \\ b_1,\cdots,b_q \end{matrix}\right)$ is Meijer's G-Function [11].

3.5. Error Rate Performance

The error rate performance of the gamma-gamma channel is determined below. In the MIMO receiver some form of combining must be used to coherently add the received signal energy from each of the receivers in the array. The most frequently used combining method is equal gain combining (EGC). In EGC the outputs of the receivers are equally weighted and summed. While optical beamforming methods have been investigated (using complex weighting of the individual receivers' outputs in an attempt to maximize the resultant SNR) EGC remains the dominant method. The probability density function (pdf) of the irradiance (gamma-gamma distribution) is a component of the capacity Equation (13). To determine the error probability with combining at the receiver array we need to modify the irradiance pdf considering the combining gain introduced by the receiver array.

The moment generating function resulting from the introduction of diversity combining results a new pdf or conditioned probability that is a function of the gammagamma fading and the diversity combining (the function f(S) incorporates the subchannel state S which is discussed below) (17):

$$f_{ECG}(I) = \sum_{k=0}^{MN} \binom{MN}{k}$$

$$\cdot \sum_{j=0}^{\infty} \left(\frac{c_j(MN-k,k)}{\Gamma(j+(MN-k)\beta+k\alpha)} f(S)^{j+(MN-k)\beta+k\alpha-1} \right) \tag{17}$$

The error probability conditioned upon the irradiance as a function of the transmitted symbol X, the received symbol $\hat{X}$ and the PSD is:

$$P(X,\hat{X}|I) = Q\sqrt{\frac{\varepsilon(X,\hat{X})}{2N_o}} \tag{18}$$

where Q(.) is Marcum's Q function.

Each channel of the MIMO array has its own channel state that is specific to the mth transmitter and the nth receiver which is the subchannel state, S_{mn}. The state of the aggregate MIMO channel S is a linear combination of all the subchannel states for the system consisting of M total transmitters and N total receivers:

$$S = \sum_{m=1}^{M} \left[\sum_{n=1}^{N} S_{MN} \right] \tag{19}$$

With EGC, all the receiver outputs, r_n, are equally weighted. The combined receiver output is therefore:

$$R = M \sum_{n=1}^{N} r_n + \varepsilon \tag{20}$$

where ε is a sum of Gaussian random variables, x_n:

$$\varepsilon = M \sum_{n=1}^{N} x_n \tag{21}$$

Having a variance of:

$$\sigma^2 = \frac{M^2}{N\sigma^2(\varepsilon_n)} \tag{22}$$

Under the assumption that the channel state information is perfectly known to the receiver the combined receiver output is:

$$R = \sqrt{\frac{E_S}{N_o}} I_S \tag{23}$$

where E_S is the energy and I_S the irradiance per symbol, respectively. Substituting the combined receiver output from (20) into the OSNR in the expression for the conditioned error probability yields:

$$P(X, \hat{X} | S) = Q\sqrt{\frac{E_S S^2}{4N_o\sigma^2}} \tag{24}$$

Conditioning with the gamma-gamma fading pdf and averaging over the fading (that is, the entire gammagamma pdf conditioned upon the channel state S) results in the error probability as:

$$P\left(X,\hat{X}\middle|S\right)=\int_0^\infty Q\sqrt{\frac{E_S S^2}{4N_o\sigma^2}}f\left(S\right)\mathrm{d}S \tag{25}$$

Substituting the fading pdf, which is the last term of the gamma-gamma capacity Equation (13) into the conditioned error probability (17) results in the bit error probability for the gamma-gamma channel with scattering, where γ is the Mie scattering coefficient (see Equation (26)).

4. OUTAGE CAPACITY OF THE RADIO CHANNEL

The radio frequency (RF) segment of the hybrid network is designed using a cellular topology, similar to that of commercial cellular radio systems. In this study, the cellular RF network consists of twelve cell sites, each with N channels. The allocated bandwidth per user is 30 MHz. As with the optical channel, we apply the fundamental capacity limits of information theory to determine the outage capacity given the total allocated bandwidth and SNR.

The RF segment outage capacity is determined for AWGN and Ricean fading channels. Ricean fading is often encountered in urban environments containing buildings and other structures that give rise to shadowing and multipath. The outage capacity is determined using two forms of spatial diversity: transmit diversity and beamforming. The transmit diversity scenarios utilizes EGC while the beamforming scenario utilizes equal weighting of each beam.

4.1. AWGN

Outage capacity was discussed as a measure of link capacity in the optical domain earlier in this paper. As in the optical domain we define the outage capacity as the rate of the channel conditioned upon outage probability. It is the greatest possible data rate with probability 1 – Pr, where Pr is the outage probability:

$$C_O\left(p\right)=\mathrm{Rate}_{\mathrm{max}}\mid P=\left(1-\mathrm{Pr}\right) \tag{27}$$

We can now calculate the pdf of the transmit diversity channel capacity is determined using the methods of [5]:

$$f(C) = \chi_{2M}^{2} \frac{\dfrac{\left|\sum_{m=1}^{M} w_m a_m\right|}{\mathrm{d}C}}{\mathrm{d}\sum_{m=1}^{M} w_m a_m} \tag{28}$$

Then:

$$\frac{\mathrm{d}C}{\mathrm{d}\left|\sum_{m=1}^{M} w_m a_m\right|} = \frac{\dfrac{Z}{2M}}{1 + \dfrac{\left|\sum_{m=1}^{M} w_m a_m\right|}{2M} z} \log 2 \tag{29}$$

where χ_{2M}^{2} is a chi-square distribution of the M transmitters, Z is the total SNR, z is the SNR at an individual receiver, and w_m and a_m are the complex weight and amplitude of the m_{th} received signal, respectively.

The pdf of the outage capacity can then be expressed as:

$$f(C) = \frac{2^{C+1} M \log 2}{z} \chi_{2M}^{2} \left(2^{C} - 1\right) \frac{2M}{z} \tag{30}$$

For the beamforming case, the pdf of the outage probability as:

$$f(C) = \frac{2^{C+1} \log 2}{Mz} \chi_{2}^{2} \left(2^{C} - 1\right) \frac{2}{Mz} \tag{31}$$

Using (31), the outage probability is:

$$\begin{aligned} \Pr(O) &= \int_{-\infty}^{C(p)} f(c)\,\mathrm{d}C \\ &= \frac{-2^{C+1} \left(\log(2)\right) \chi_{2}^{2} \left(2^{C} - 1\right)}{M^{2} z} \end{aligned} \tag{32}$$

The outage probability differs between the beamforming and the transmit diversity system in the number of degrees of freedom of the chi-square distribution.

For transmit diversity the number of degrees of freedom contains the factor of M whereas for beamforming it is fixed at 2.

We consider here two measures of channel capacity in the forward link. The maximum normalized channel capacity is the Shannon capacity, which is equal to $\log_2(1+z)$. Substituting (30) into the Shannon capacity equation we obtain the capacity with transmit diversity as:

$$C_T = \int_0^M \log_2\left(1+\frac{Z}{2M}\right)\left|\sum_{m=1}^{M} w_m a_m\right| \mathrm{d}m \tag{33}$$

Note that the integration in the above two equations yields the expected value of the Shannon capacity with respect to the variation of signal-to-noise ratio with the diversity level M. For beamforming, the Shannon capacity is similar to that for receive diversity, with this factor added for the beamforming gain, M:

$$C_{BF} = \int_0^M \log_2\left(1+\frac{MZ}{2}\right)\left|\sum_{m=1}^{M} w_m a_m\right| \mathrm{d}m \tag{34}$$

The channel capacity C_T for the 12-cell system is the sum of the individual cell capacities C_i across all users:

$$C_T = \sum_{i=1}^{12K} C_i \tag{35}$$

Applying the summation of (35) to (33) and (45) for transmit diversity and beamforming respectively, yields the total system capacity as:

$$C_{T,TD} = B_{ds}\sum_{i=1}^{12K}\int_0^M \log_2\left(1+\frac{\Gamma_{i,DS/FH}}{2M}\right)\left|\sum_{m=1}^{M} w_m a_m\right| \mathrm{d}m \tag{36}$$

$$C_{T,BF} = B_{ds} \sum_{i=1}^{12K} \int_0^M \log_2\left(1 + \frac{M\Gamma_{i,DS/BF}}{2}\right) \left| \sum_{m=1}^{M} w_m a_m \right| \mathrm{d}m$$

(37)

where B_{ds} is the total allocated RF bandwidth (30 MHz), M is the total number of rays, and $\Gamma_{i,DS/FH}$ And $\Gamma_{i,DS/BF}$ are the signal-to-interference ratios (SIR) at the ith user for transmit diversity and beamforming, respectively.

4.2. Transmit Diversity in the Ricean Channel

To calculate the capacity with transmit diversity of the RF segment with frequency selective (Ricean) fading, we calculate the distribution of SNR imposed by the Ricean pdf. This yields a Ricean-distributed sequence of ray amplitudes and phases, whose individual capacities we then integrate over the total number of rays, m, to find the total system capacity. The signal envelope of the Ricean fading channel is defined by the pdf [5]:

$$v_T(t) = \left(\frac{1}{X}\right) \exp\left[-\frac{2t + s^2}{2X}\right] J_0\left(\frac{\sqrt{2ts}}{X}\right)$$

(38)

where X is the contribution to the mean signal power of the diffused signal component, t is time, and s is 2ry – $v_t(0)$ (where y is average interference-to-noise ratio).

The pdf of the interference-to-noise ratio is the sum of identical interfering signal powers. The pdf of the total interference-to-noise ratio for Ricean fading is thus (39):

$$h_T(r) = \int_0^{\infty} \frac{y}{X} \exp\left[-\frac{2ry + s^2}{2X}\right] J_0\left(\frac{\sqrt{2rys}}{X}\right)$$

$$\frac{y^{I-1}}{Y(I-1)!} \exp\left[-\frac{y}{Y}\right] \mathrm{d}y$$

where J_0 is the zeroth order Bessel function, Y is mean power of ith interferer, and I is the total number of interferers. The channel capacity for transmit diversity with Ricean fading is similar to AWGN case, using the pdf of the SNR in the Ricean channel and the topology of the cellular RF network [5] (40):

$$C_{Ri,T} = B_{ds}\sum_{i=1}^{12K}\int_0^M \log_2\left(1+\frac{M\Gamma_i}{2\left(G_p M^3+2.31KM\Gamma_i\right)}\right)\left|\sum_{m=1}^{M} w_m a_m\right| h_T(r)\,dm$$

4.3. Beamforming in the Ricean Channel

To calculate the capacity of beamforming in the RF segment with frequency selective (Ricean) fading, we calculate the distribution of signal-to-noise ratio resulting from the application of the Ricean fading. This yields a Ricean-distributed sequence of ray amplitudes and phases, whose individual capacities we then integrate over the total number of rays, m, to find the total system capacity.

The Shannon capacity of the RF segment with beamforming is similarly determined to that for the AWGN case, using the pdf of the SNR in the Ricean channel:

$$C_{Ri,BF} = B_{ds}\sum_{i=1}^{12K}\int_0^M \log_2\left(1+\frac{M\Gamma_{i,DS/FH}}{2}\right)\left|\sum_{m=1}^{M} w_m a_m\right| h_T(r)\,dm \tag{41}$$

5. OBJECTIVE FUNCTION

The objective function for the hybrid channel is produced by summing the outage capacity of the optical and RF channels with respect to (w/r/t) to the joint BER. The joint BER is the joint error rate pdf of the optical and RF channels. Since the FSO and RF fading pdfs are independent, the joint BER is the product of the two pdfs.

Figure 2 illustrates the objective function for the hybrid FSO/AWGN RF channel with beamforming. The outage capacity of the hybrid channel, in Mbps, is the sum of (16) and (37). The BER is the product of (26) and the AWGN error probability [7],

$$P_e = Q\sqrt{2Z} : \quad :(42)$$

Figure 3 provides the objective function for the Ricean

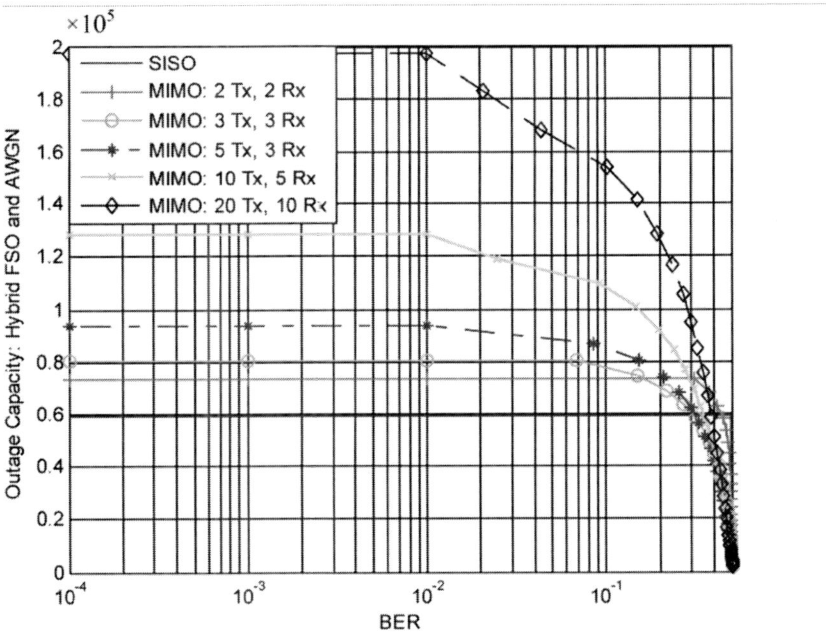

Figure 2. Objective function—FSO/AWGN beamforming.

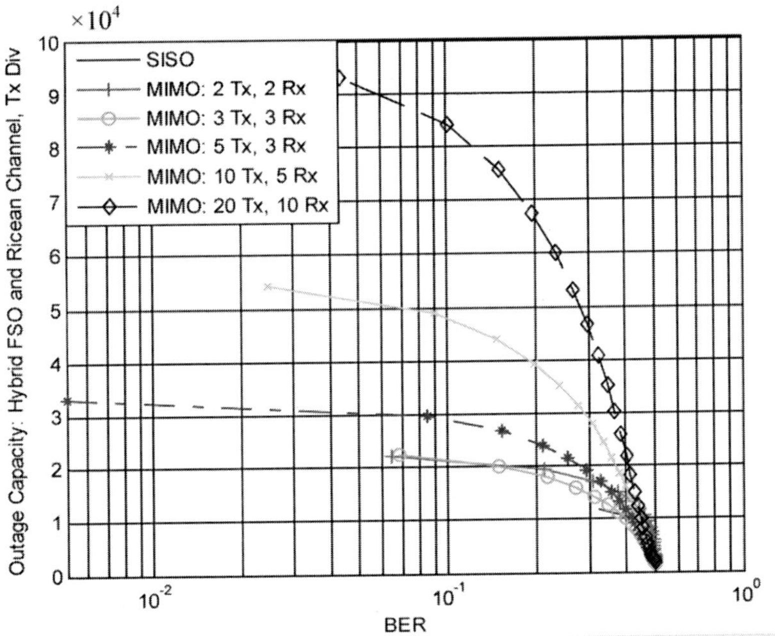

Figure 3. Objective function—FSO/Ricean transmit diversity.

RF channel with transmit diversity, The outage capacity of the hybrid channel is the sum of (16) and (40). The BER is the product of (26) and the Ricean error probability with transmit diversity [7],

$$P_{e,T} = \left[2 + \frac{\delta^2 U_{AVG}}{1+\delta^2}\right]^{-1} e^{-\frac{U_{AVG}}{2(1+\delta^2)+\delta^2 U_{AVG}}} \tag{43}$$

where U is the SIR, U_{AVG} is the average value of U, and δ is the ratio of the power of the diffuse signal components to the direct component.

Figure 4 incorporates the Ricean channel with beamforming. The outage capacity of the hybrid channel is the sum of (16) and (41). The BER is the product of (26) and the Ricean error probability with transmit diversity [7],

$$P_{e,B} = \left[\frac{\delta^2 U_{AVG}}{N_b \left(1+\delta\right)^2} \right]^{-1} e^{-\frac{U_{AVG}}{2\left(1+\delta^2\right)+\delta^2 U_{AVG}}} \tag{44}$$

where N_b is the number of beams.

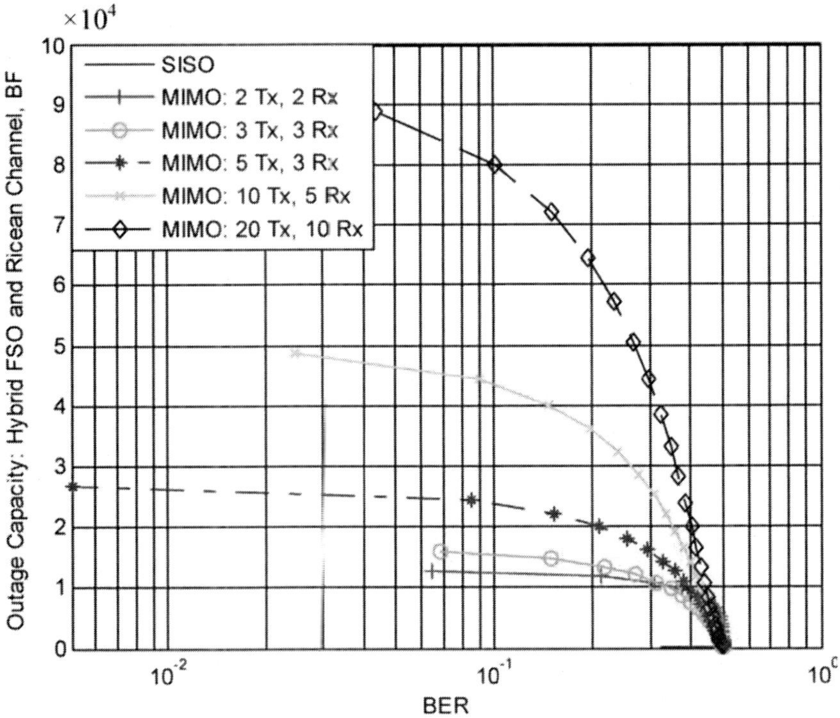

Figure 4. Objective function—FSO/Ricean beamforming.

6. SIMULATION

The objective function for the FSO and AWGN RF hybrid channel is simulated using the network described in Section 1. The objective function is characterized for the AWGN and Ricean cases by loading the hybrid channel and observing the BER until failure occurs while evaluating the corresponding outage capacities over the gamut of BER in the simulation model.

The optical and RF channel parameters used in the simulation are provided in **Table 1**represent moderately turbulent conditions. and $\beta = 2.1$. The optical channels are operated as MIMO using 10 transmitters and receivers. A Gaussian optical pulse shape is used for the FSO channels, while for the RF channels a sinc(x) chip shape is used. The optical transmit power is 200 mW, which is typical of contemporary laser diodes. RF transmit power is 200 mW, an average value of microcell base stations. The transmission distance for both the FSO and RF links is 1500 m. (1.5 km). This distance is sufficient to operate the gamut of OSNR required and is at the upper limit of generally reliable FSO operation with low-altitude (100 m.) turbulence and other atmospheric impairments. The values of $\alpha = 3.0$

Due to the high available bandwidth, the FSO channels use OOK modulation (1 bit per symbol) whereas due to limited bandwidth the RF channels use quadrature amplitude modulation (QAM) with two bits per symbol.

Table 1. Hybrid channel parameters

Table 1. Hybrid channel parameters.

FSO	RF
$\lambda = 10{,}000$ nm	Frequency = 1.90 GHz
$\alpha = 3.0, \beta = 2.1$	Bandwidth = 30 MHz
Tx Power = 200 mW	Tx Power = 300 mW
No. of Transmitters: 10	No. of Transmitters: 10
No. of Receivers: 10	No. of Receivers: 10
Pulse Shape: Gaussian	Chip Shape: sinc(x)
Distance = 1500 m.	Distance = 1500 m.
Modulation: OOK	Modulation: QAM
FEC: None	FEC: None
Altitude: 100 m.	Altitude: 100 m.
$G_p = 1$ (0 dB)	$G_p = 6$ (7.8 dB)

A total RF bandwidth of 30 MHz is used. The RF channels use quadrature amplitude modulation (QAM) and direct-sequence spread spectrum (DSSS) with a processing gain (ratio of spread to unspread bandwidth) of $G_p = 6$. The FSO channels do not use DSSS, therefore $G_p = 1$.

Figure 5 demonstrates a scenario in which the hybrid network experiences degrading optical propagation conditions:

.6.1. Initial Condition: Still Air, No Rain

With no turbulence or rain attenuation, BER is less than 5×10^{-4} and outage capacity is greater than 5.01×10^4 Mbps. The channel is assigned a load of 5.01×10^4 Mbps. No loss of data or users occurs under this condition.

6.2. Onset of Rain

BER transitions from the low to the high regime, increasing to 0.20. Outage capacity is now 3.6×10^4 Mbps which is lower than the offered load of 5.01×10^4. The channel experiences an outage since the offered load is greater than the outage capacity under this BER condition (note that the BER scale has been expanded).

6.3. Increasing Turbulence

The onset of turbulence causes BER increase to 0.24. Outage capacity is reduced to 3.3×10^4 Mbps which is still lower than the offered load of 5.01×10^4. The channel outage continues. Applying load shedding as commanded by the optimal FM service results in the following behavior:

6.4. Load Shedding and Correction

BER has increased to 0.24. The PM service observes the increasing BER trend. Prior to allowing the outage capacity to drop below the offered load, the PM service reduces the offered load (by removing users from the channel) from 5.01×10^4 to 2.6×10^4 Mbps corresponding to a maximum BER of 0.29. Since the offered load is now less than the outage capacity at the actual BER (0.24), which is 3.3 × 10⁴ Mbps service of the hybrid channel is restored.

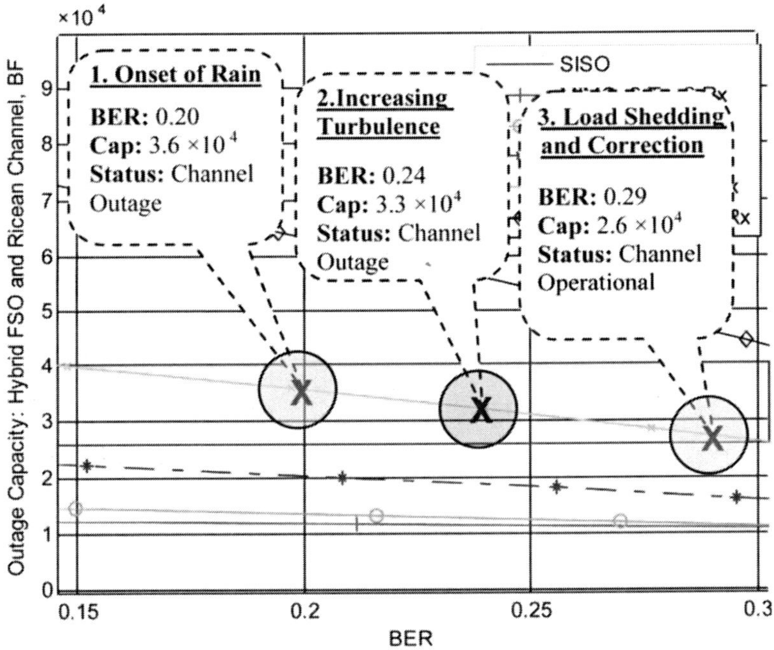

Figure 5. Outage capacity objective function in the hybrid FSO/Ricean channel onset of turbulence.

The feedback dynamics of the traffic balancing and load-shedding algorithms are designed to effect the required adjustment in traffic while maintaining the actual load at an acceptable margin above the outage capacity at the current BER.

Figures 6 and 7 compare the sensitivity of the objective function with that of a measurement based purely on error rate vs. OSNR in the high BER regime. The derivative of both functions is determined as a measure of the sensitivity of the parameter with respect to the independent variable (in the case of the objective function, the derivative of the outage capacity with respect to the BER, and for the BER/OSNR measurement, the derivative of the BER with respect to the OSNR).

The high BER example here is the same as that in the previous section demonstrating the use of the objective function in a degrading environment. To compare sensitivities in a typical example, we calculate the derivative of the objective function (outage capacity) with respect to BER, and the derivative of the BER with respect to OSNR for the hybrid channel with Ricean RF (beamforming).

The first observation made in the high BER regime sensitivity comparison is that the objective function's sensitivity is greater than that of BER vs. OSNR (corresponding to this high BER range) measurement by a factor of 1300 - 1700 or 31 - 62 dB. The BER vs OSNR slope is essentially constant in this BER regime while the objective function's slope exhibits an increase with BER. Thresholding based on OSNR/SIR alone often yields less than desirable results since the error rate exhibits a large slope with respect to SNR near the knee of the error rate performance curve. The control system utilizing the objective function is provided with this gain improvement and can thus effect a significantly faster adjustment of the offered load than one using a BER vs OSNR measurement. A significant observation is that the slope of the BER vs OSNR measurement in the high BER regime is very small, starting with an optical MIMO system of 10 transmitters. For any size system above 10, the response of the BER measurement to changes in OSNR is very low compared to that of the outage capacity.

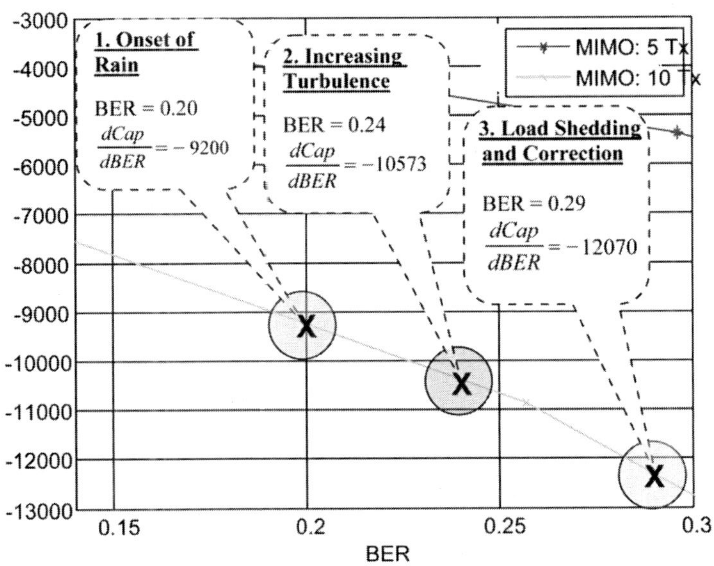

Figure 6. Derivative of objective function, hybrid FSO and ricean channel with beamforming.

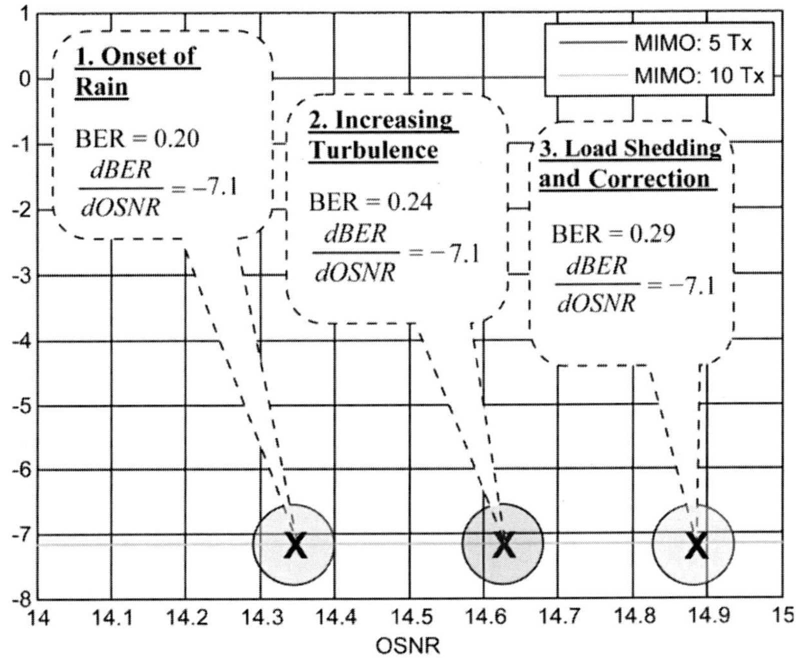

Figure 7. Derivative of BER w/r/t OSNR, hybrid FSO/ Ricean channel.

The objective function shows that outage capacity of the network may be traded off for BER. In applications in which the maximum capacity is the most important factor, a higher error rate may be tolerated in favor of high capacity. The outage capacity objective function allows equally reliable and optimal performance monitoring in both types of applications, high capacity/high BER and reduced capacity/low BER. In many applications BERs in the high BER regime are unacceptably high. The previous example applies to cases in which a high BER may be tolerated.

The low BER regime below 0.02 experiences similar behavior. In the low BER regime it is noted that the objective function's sensitivity is once again greater than that of BER vs OSNR. The BER vs OSNR slope exhibits a small variation of −2.1 to −2.4 in this range. In this regime the control system is provided with an average 46 dB of gain improvement. The speed of adjustment of the offered load using the objective function is thus improved using the objective function in low BER regime over that of measuring BER vs OSNR alone. As in the high BER case the slope of the BER vs OSNR measurement in the high BER regime is low, starting with an optical MIMO system of 10 transmitters.

7. CONCLUSIONS

The sensitivity of the objective function to channel performance degradation is significantly higher than that of the BER measurement as demonstrated by the previous comparisons. The derivative of the BER vs OSNR measurement in the high BER regime shows no change between 0.2 and 0.29 BER, and for much of the high BER regime starting with an optical MIMO system of 10 transmitters. For any size system above 10, there is very little response of the BER measurement to changes in OSNR. The control system that is adjusting the offered load based on a BER measurement can effect very little correction in this case, due to low or zero gain, in this regime for large MIMO arrays. As a result of the low sensitivity of BER upon the SIR/SNR and OSNR, the network manager, using these performance metrics, is unable to offload the traffic from degrading channels prior to failure. In contrast, the objective function provides high sensitivity. The NMS will be able to effect an early correction by offloading the degrading channel to maintain an acceptable margin below the outage capacity before any outage is experienced by the user.

In addition to high sensitivity, the objective function has several major advantages as a performance monitoring parameter over traditional figures of merit including simplicity—it is a function of a single independent variable), ease of measurement—error rate measurement is provided by most optical and RF transmission equipment, cross-domain (FSO and RF) characterization, and intrinsic correction—the objective function continuously provides the adjustment in traffic load required to restore channel operation.

8. FURTHER RESEARCH

The objective function has been characterized for other RF fading environments such as Rayleigh and Nakagami-m, Nakagami-n, and Nakagami-c. Research is also conducted to characterize the objective function with the Hata-Nakamura transmission model that is widely used in characterizing performance in urban environments. In the case of Rayleigh and Nakagami environments, the fading distributions associated with the

Rayleigh and Nakagami PDFs are analytically determined and the capacity model conditioned upon them—using the method developed for the Ricean case in this study. The HataNakamura model, for which much numerical performance data exists, is currently being incorporated into the network simulation. The simulation, parameterized with the Hata-Nakamura propagation data for the RF channels, will then be operated over the OSNR/SIR gamut to empirically derive the objective function.

9. ACKNOWLEDGEMENTS

The author thanks Profs. R. Martini, W. Carr, and K. Stamnes for their valuable suggestions in the course of this work.

REFERENCES

1. J. Li and M. Uysal, "Optical Wireless Communications: System Model, Capacity, and Coding," Proceedings of the IEEE Vehicular Technology Conference, Orlando, 6-9 October 2003, pp. 168-172.

2. E. Bayaki, R. Schober and R. K. Mallik, "Performance Analysis of Free-Space Optical Systems in GammaGamma Fading," IEEE Global Telecommunications Conference, (GLOBECOM 2008), New Orleans, 30 November-4 December 2008, pp. 1-6.

3. X. Zhu and J. Kahn, "Free-Space Optical Communication through Atmospheric Turbulence Channels," IEEE Transactions on Communications, Vol. 50, No. 8, 2002, pp. 1293-1300.

4. W. Gappmair and S. Muhammad, "Error Performance of PPM/Poisson Channels in Turbulent Atmosphere with Gamma-Gamma Distribution," Electronics Letters, Vol. 43, No. 16, 2007, pp. 880-882.

5. B. Friedlander and S. Scherzer, "Beamforming vs Transmit Diversity in the Downlink of a Cellular Communications System," IEEE Transactions on Vehicular Technology, Vol. 53, No. 4, 2004, pp. 1023-1034

6. R. Martini and E. Whittaker, "Quantum Cascade LaserBased Free Space Optical Communications," Journal of Optical and Fiber Communications Reports, Vol. 2, No. 4, 2005, pp. 279-292.

7. B. Schmidt, "Optimal Performance Monitoring for a Hybrid Long-Wavelength FSO and Radio Frequency Wireless Network in Fading Channels," Ph.D. Dissertation, Department of Physics, Stevens Institute of Technology, Hoboken, 2013.

8. A. Garcia-Zambrana, C. Castillo-Vazquez and B. Castillo-Vazquez, "On the Capacity of FSO Links over Gamma-Gamma Atmospheric Turbulence Channels Using OOK Signaling," European Society for Information Processing (EURASIP) Journal on Wireless Communications and Networking, Vol. 2010, 2010, Article ID: 127657.

9. A. Farid and S. Hranilovic, "Outage Capacity Optimization for Free-Space Optical Links with Pointing Errors," IEEE Journal of Lightwave Technology, Vol. 25, No. 7, 2010, pp. 1702-1710.

10. A. Farid and S. Hranilovic, "Diversity Gain and Outage Capacity Probability for MIMO Free-Space Optical Links with Misalignment," IEEE Transactions on Communications, Vol. 60, No. 2, 2012, pp. 1702-1710

11. L. Andrews, "Special Functions for Engineers and Applied Mathematicians," MacMillan, New York, 1985.

6
CHAPTER

METIS RESEARCH ADVANCES TOWARDS THE 5G MOBILE AND WIRELESS SYSTEM DEFINITION

Jose F Monserrat[1*], Genevieve Mange[2], Volker Braun[2], Hugo Tullberg[3], Gerd Zimmermann4 and Ömer Bulakci[5]

ABSTRACT

The Mobile and wireless communications Enablers for the Twenty-twenty Information Society (METIS) project is laying the foundations of Fifth Generation (5G) mobile and wireless communication system putting together the point of view of vendors, operators, vertical players, and academia. METIS envisions a 5G system concept that efficiently integrates new applications developed in the METIS horizontal topics and evolved versions of existing services and systems. This article provides a first view on the METIS system concept, highlights the main features including architecture, and addresses the challenges while discussing perspectives for the further research work.

Keywords: 5GBeyond 2020METISSystem concept

1 INTRODUCTION

In the decade beyond 2020, it will be necessary to support 1,000 times a higher mobile data volume per area [1] together with new wireless broadband communication services coming from a plethora of different market segments. These requirements will go beyond the natural evolution of IMT-Advanced technologies, which show the need for a new mobile generation, with certain disruptive features with respect to legacy technologies. Although there is no unanimous agreement so far, this seems to be the birth of the Fifth Generation (5G) technologies.

Mobile and wireless communications Enablers for the Twenty-twenty Information Society (METIS) is an integrated research project partly funded by the European Commission under the Framework Programme 7 (FP7) research framework [2]. METIS aims at laying the foundation for the beyond 2020 wireless communication systems by providing the technical enablers needed to address the very challenging requirements foreseen for this time frame. Such a system has to (1) be significantly more efficient in terms of energy, cost, and resource utilization than today's systems, (2) more versatile to support a significant diversity of requirements (e.g., payload size, availability, mobility, and Quality-of-Service (QoS)) and new scenario use cases, and (3) provide better scalability in terms of number of connected devices, densely deployed access points, spectrum usage, energy, and cost. The technical goals derived from these main objectives for METIS [3] are:

- 1,000 times higher mobile data volume per area,
- 10 to 100 times higher typical user data rate,
- 10 to 100 times higher number of connected devices,
- 10 times longer battery life for low-power devices,
- 5 times reduced end-to-end (E2E) latency, reaching a target of 5 ms for road safety applications.

The key challenge is to meet these goals at a similar cost and energy consumption as today's networks.

From the point of view of METIS, a complete redesign of the Internet is discarded. Since the current Internet has become so large, the implementation of new architectural principles is impractical due to the commercial and operational difficulties it poses. Some concepts such as information-centric networking (ICN) [4] are considered unrealistic, although some of its fundamentals, as the universal caching, have to be taken into account in order to efficiently distribute the traffic load in the network. If 5G includes a revolution, it will come from the radio interface design, where some new paradigms under discussion represent a radical change in the current view we have on mobile networks in addition to some fundamental changes in the mobile core functionalities, e.g., driven by new concepts of mobility.

In this challenging task, the METIS project has defined a set of five essential Horizontal Topics (HTs) to be consolidated into a single 5G system concept. These HTs are direct device-to-device communications (D2D), massive machine communications (MMC), moving networks (MN), ultra-dense networks (UDN), and ultra-reliable communications (URC) [5].

The present paper discusses a number of design choices and features that change the way we understand mobile networks. It is worth noting that this article represents the common vision of the set of network operators, industrial, and academic partners that work together in the METIS project. It does not provide the final solution, but highlights the most promising research lines, which combine efforts in common directions in the project.

In the next section, we present the METIS view on the key functional and technical features of the next generation mobile system, detailing the network architecture in Section 3. Section 4 highlights the technical enablers and research challenges that address the technical goals of the project for 5G. We finally draw the main conclusions of the paper in Section 5.

2 MAIN FEATURES OF THE 5G SYSTEM

All mobile generation changes so far have been based on a new radio link concept and have provided an increase of the peak data rate of approximately two orders of magnitude. The 5G system must meet the requirements of increased rate and capacity needed beyond 2020 and requirements on reduced latency. However, the integration of new services and application areas is as important as increasing rate and reducing latency. The 5G system will be the wireless enabler of the Internet of things and, in addition to human users, must cater to different machine-type communications with widely different requirements. All together, the range of requirements will increase as compared with current mobile broadband (MBB) technologies. For instance, the data rates will range from very low for sensor data to very high rates for high-definition video. Latencies will range from extremely low for safety-critical applications to applications where latency is not really a constraint. Packet sizes will vary from small for, e.g., smartphone applications to large for, e.g., file transfers.

By using a system architecture that supports also D2D communication and UDN deployments, METIS foresees a multi-Radio Access Technologies (RAT) system that efficiently integrates fundamental building blocks as:

- *Evolved mobile broadband (eMBB)* will provide high data rates and low latency communications improving quality of experience (QoE) for the users. As depicted in Figure 1, eMBB will go much beyond current MBB solutions.

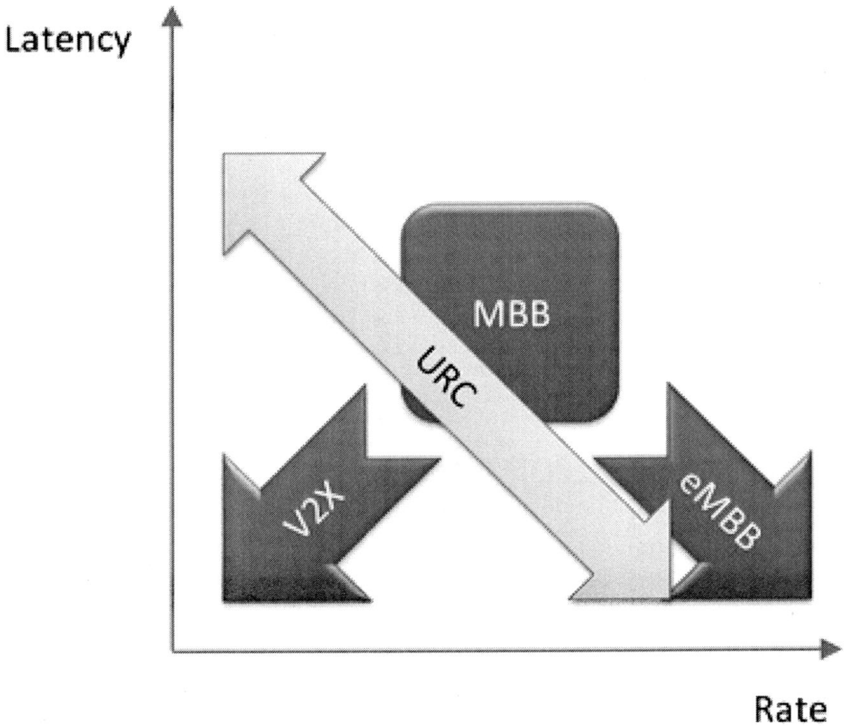

Figure. 1.extension of requirements as compared with current mbb technologies.

- *Massive machine communications (MMC)* will provide up- and down-scalable connectivity solutions for tens of billions of network-enabled devices, where scalable connectivity is vital to the future mobile and wireless communications systems.

- *Vehicle to Vehicle, Device and Infrastructure (V2X) and driver assistance services* require cooperation between vehicles and between vehicles and their environment (e.g., between vehicles and vulnerable road users over smartphones) in order to improve road safety and traffic efficiency in the future. Such V2X services for MNs require reliable communication links that enable the transmission of data packets with guaranteed maximum latencies even at high vehicle speeds.

- *Ultra-reliable communications (URC)* will enable high degrees of availability. It is required to provide scalable and cost-efficient solutions for networks supporting services with extreme requirements on availability and reliability. Reliable service decomposition provides mechanisms for graceful degradation of rate and increase of latency, instead of dropped connections, as the number of users increases.

To address these challenges, new flexible air interfaces, new possible waveforms, and new multiple access schemes, medium access control (MAC), and radio resource management (RRM) solutions (e.g., [6,7]), and signaling protocols must be investigated to discard the idea that physical layer improvements are already close to their upper limit. However, the foreseen steps towards the ubiquitous high bit-rate services pass mainly by managing multi-cell and multi-user MIMO together with new paradigms of network deployment with multiple RATs and multilayer networks.

The key supporting enablers include:

- *Dense and dynamic RAN* providing a new generation of dynamic radio access networks (RANs). The term RAN 2.0 could be also used referring to this flexibility of the RAN.

- *The spectrum toolbox* contains a set of enablers (tools) to allow 5G systems to operate under different regulatory and spectrum sharing scenarios.

- *Flexible air interface* incorporating several radio interface technologies operating as a function of the user needs.

- *Massive MIMO*, in which the number of transmit and receive antennas increase over an order of magnitude.

- *New lean signaling/control information* is necessary to guarantee latency and reliability, support spectrum flexibility, allow separation of data and control information, support large variety of devices with very different capabilities, and ensure energy efficiency.

- *Localized contents/traffic flows* allow offloading, aggregation, and distribution of real-time and cached content.

2.1 Dense and dynamic RAN

There is a consensus that future traffic demand will require a high density of a variety of access nodes [8,9]. UDN refers to this new paradigm of wireless communication network, which includes network cooperation and ultra-dense availability of access points. Nomadic and moving nodes, mounted on a car, bus, or train, can provide connectivity to users in their proximity and increase data rates by reducing the radio distance to the nearest access node. D2D communications will guarantee the ubiquity of high-quality services and offload the infrastructure transport network [10].

Network densification has been used for wireless evolution during last years and will continue this trend in the next decade [11]. Even some authors have speculated about a 10 year-out situation in which the number of access points could surpass that of active users [12]. UDNs present serious challenges in terms of mobility support, interference management, and operation (cost, maintenance, and backhaul).

In the context of mobility in UDNs, the classical approach of user-initiated mobility management based on measurement reports poses a severe problem, provided the increased signaling and device measurement overhead. Therefore, current trends of self-configuration of handover parameters [13] must be revisited to adapt to the 5G scenarios with random layout and a variety of cell ranges. Moreover, mobility prediction will be fundamental to initiate the handover with enough time in advance.

On the other hand, from a radio planning point of view, the disparity in transmission powers in heterogeneous deployments invalidates traditional DL-based cell association mechanisms. Cell range expansion (RE) seems to be the solution to overcome this problem. An additional benefit is a more effective offloading from the macro layer to the small cells [14]. Whereas RE partially compensates UL interference issues, DL interference is increased at the new small-cell edge due to the DL transmissions from the macro cell to other users. For this reason, 3GPP has already introduced improvements in the inter-cell interference coordination mechanisms of LTE-A [15,16]. To sum up, higher densification is nowadays being pursued by applying patches to the old ideas, and this has reached its limit with the current densification values. Given this, it is clear that new system designs are needed for 5G networks to tackle the root of the problem, that is, the huge imbalance between the nature of mobile devices and cell infrastructure. This leads to the idea of asymmetric user association and the fact that UL and DL should be treated as independent networks. This way, the unity of the cell is broken, and UL and DL are totally decoupled and served by different access points [17].

This decoupling philosophy is also being proposed in the 3GPP for the splitting of control and user planes (C/U-plane) [18]. Given this, it is possible to transmit the C-plane, and consequently HO messages, from a stable macro cells whereas the U-plane is transmitted from the closest small cell (see Figure 2). The recent work by Li Yan et al. argues about the suitability of this approach for future 5G networks and provides interesting insights on practical aspects [19].

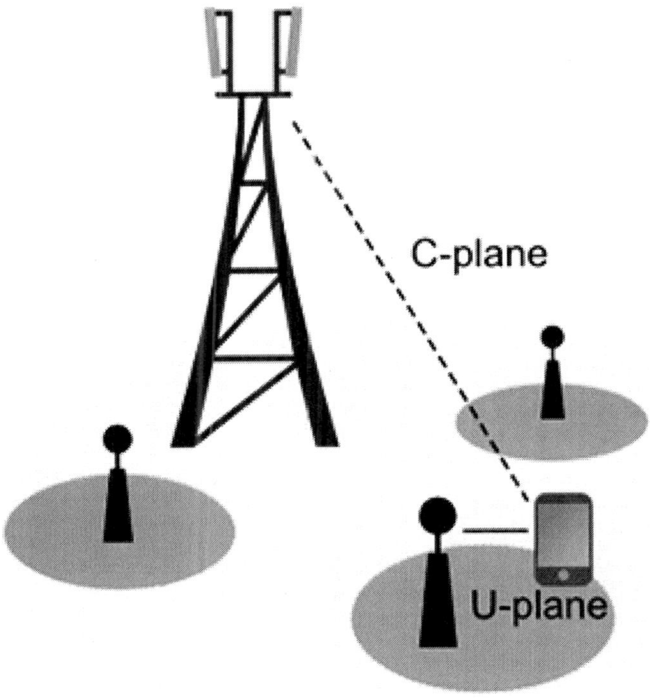

Figure.2.decoupling of control and user plane in udn.

Another important problem is the provision of high-quality services within vehicles on the move. One possible solution to address the problem in the future will be the deployment of a relay within the vehicle. The relay can be connected to the vehicular rooftop antenna, in order to establish a robust (relay/backhaul) link to the cellular network. On the other hand, relays allow for optimum coverage conditions inside the vehicle and a certain control of the consumed resources through the usage of closed subscriber groups. Since the use of a relay eliminates the vehicle penetration loss (VPL) and provides additional antenna gains due to better characteristics of the rooftop antenna, this solution improves significantly the link budget [20] and allows for a more efficient resource utilization of the cellular network. However, in current LTE networks, an additional overhead burden due to signaling exists (around 10%), and therefore, the benefit from the use of those relays is to be demonstrated.

Concerning the research on mobile relaying, most studies focus on network architecture aspects [21-23], mostly on the basis of its usage in LTE/LTE-Advanced (see Figure 3). In this case, the mobile Relay Node (RN) is connected as a regular user to a base station, known as donor e-Node B (DeNB). Moreover, regarding the system performance assessment, the expected gain derived from the usage of moving relays is investigated, for instance in [23]. Other studies evaluate the impact of deploying mobile femtocells from the point of view of spectral efficiency and signaling overhead considering opportunistic use of spectrum and scheduling [24,25]. Current research on the integration of moving networks into 5G focuses on improved procedures, channel prediction, reduced signaling, and optimum backhauling [26].

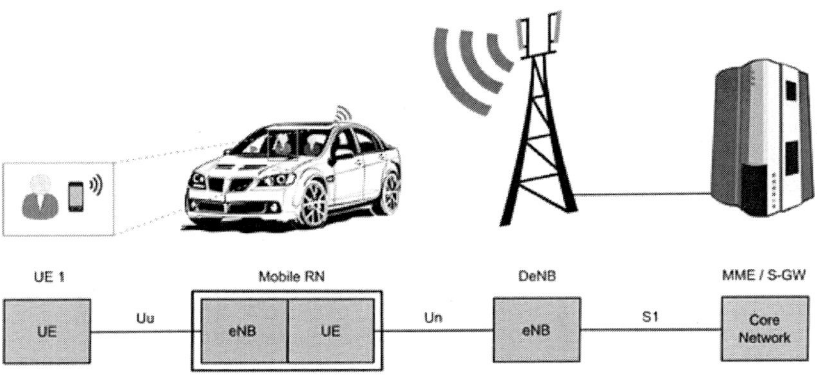

Figure. 3.current lte-a architecture for moving networks.

In order to reduce the link distance, the 3GPP initiated in Release 12 the study item on the use of device-to-device (D2D) communications under the name of proximity services (ProSe) [27]. The 3GPP defines D2D communication as a type of communication between two nearby UEs in which there is not any routing through the core network. ProSe in LTE-

Advanced is also aimed at critical situations and must be functional when cellular networks are not operational.

In 3GPP, ProSe studies focus on two aspects, namely D2D discovery, also known as proximity detection, and D2D communication. D2D discovery aims at identifying other neighbor UEs for possible D2D communication, whereas D2D communication is the actual direct link for data transfer. For both operations, it is of paramount importance to design and efficient control plane to synchronize terminals, ensure radio link security, and keep track on the mobility during the communication phase. Figure 4 shows the current approach for this functional split in 3GPP. More details on this control plane design for D2D communications can be found in [28] and references therein.

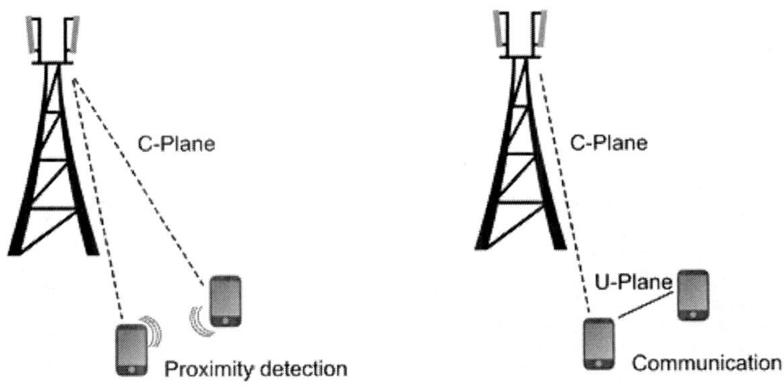

Figure. 4.D2D APPROACH IN 3GPP.

D2D discovery is usually localized in static resources while D2D communications should use resources in an opportunistic manner, only when necessary. However, this communication mode may introduce severe interferences into the overlaying cellular network, which results in the need for dynamic and network-controlled radio resource allocation to the D2D link [29]. Several methods have been proposed to deal with this coexistence [30]. Feng et al. [31] proposed a resource allocation scheme for D2D communications operating in opportunistic manner focusing on the guarantee of QoS requirements in the heterogeneous network. On the other hand, Xu et al. in [32] designed an iterative auction for resource allocation whereas Lingyang et al. proposed a game for this resource allocation [33].

Beyond these concepts, METIS foresees a new generation of dynamic RAN, referred to as RAN 2.0. Following an analogy with the Web 2.0 concept, in

RAN 2.0, the user devices can become active entities of the RAN, in addition to existing network nodes (see Figure 5). This user-centric network (UCN) is a new paradigm of wireless networking in which user devices will temporarily take over the role of access nodes for other users, e.g., to guarantee the ubiquity of high-quality services. This multi-hop cellular solution opens the door for an interconnection of devices that collaborates within a group of users they trust searching for mutual benefit, similar to a social network. RAN 2.0 implies a flat architecture from service point of view resulting in low latency. It is also accompanied by an agile infrastructure support, since *ad hoc* and smart coordinated setup of networks is expected under this service/user-centric model. Finally, RAN 2.0 also entails a different distribution of intrinsic functions that can be executed in any node depending on its hardware and software capabilities.

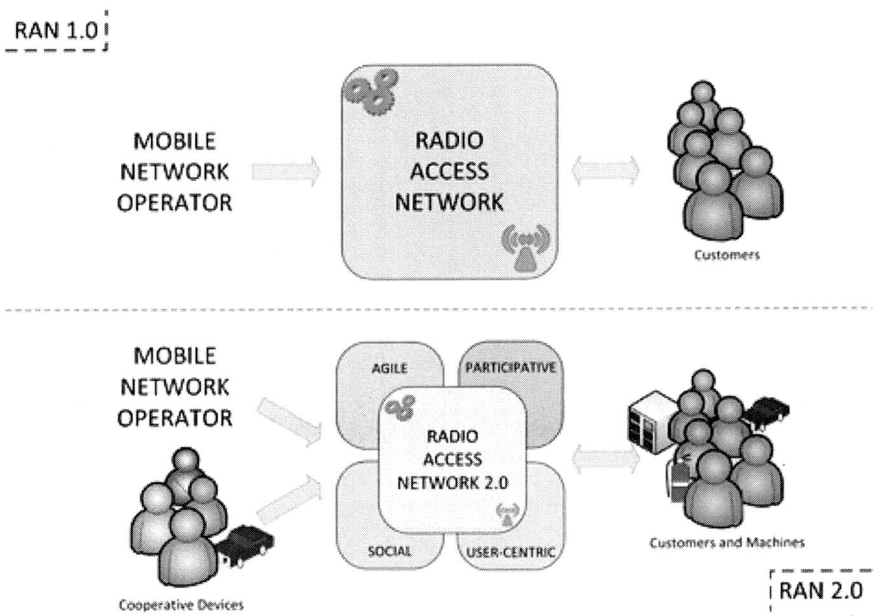

Figure. 5. RAN 2.0 CONCEPT.

This concept has been also discussed under the name of Spontaneous Smartphone Network (SSN) [34], since it depends heavily on the network functionalities provided by the smartphone operating system. All current operating systems have tethering capabilities to route packets from the cellular link to another short-range wireless interface like Wi-Fi, but operating at IP layer. However, this tethering characteristic is not working transparently to the end user, lacks adequate security control procedures for the pairing and ciphering of the information, cannot make use of cooperation and does not support self-organizing principles to manage the links, and ensures minimum interference and dynamic configuration to face unexpected changes in the channel conditions. Current research is focusing mainly on the dynamic configuration of the network [34] and on security [35], likely the main showstopper of this 5G enabler.

2.2 The spectrum toolbox

Of course, new spectrum bands must be identified in the International Telecommunication Union-Radio-communication (ITU-R) Radio Regulations to support the increase of traffic demand. In this direction, the worldwide allocation of a number of bands to International Mobile Telecommunications (IMT) technologies is on the agenda for the next World Radio Conference 2015 (WRC-15). Figure 6 shows the possible candidate bands for IMT under WRC-15 agenda [36]. Despite the fact that the current practice of predominantly using dedicated licensed spectrum will remain the main stream, new regulatory tools and approaches of sharing the spectrum and optimizing its use must be devised [37]. On the other hand, the coverage of higher frequency bands may be limited, for example, to hot spots or dense areas. MMC and low latency on the other hand are required in wide area coverage.

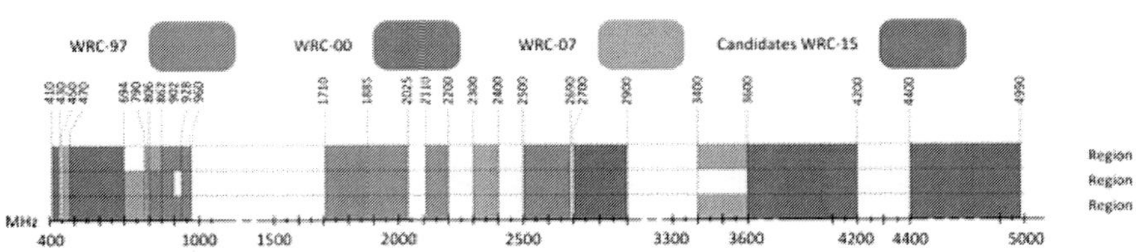

Figure. 6.POSSIBLE CANDIDATE BANDS FOR IMT UNDER WRC-15 AGENDA ITEM 1.1.

For its special relevance, we need to mention here the use of millimeter wave (mmW) communications (30 to 300 GHz). Although conventionally beyond 6 GHz spectrum has been discarded due to its high attenuation, recent research points towards its potential use for backhaul, D2D and even cellular communications [38]. Moreover, semiconductor technologies are getting mature enough to reduce cost and power consumption of wireless communications in these new bands [9]. On the other hand, propagation problems can be tackled thanks to the use of massive MIMO and the resulting huge antenna array gains. Recent measurements at 28 and 73 GHz in New York University [39] have revealed that, with an accurate beam-selection, signals can be detected at a distance between 100 to 200 m even in the absence of line of sight (LoS). This result has motivated an increased interest of the wireless community on cellular mmW communications. This proposal still requires intense research to overcome its main challenges, that is, the support of mobility for the optimum beam selection and the design of transmission techniques that alleviate the need for feedback on channel state information (CSI) to the transmitter [40].

2.3 Flexible air interface

The variety of spectrum options leads to a system concept incorporating a number of air interfaces, each optimized for a specific set of use cases. The air interfaces including their protocol stacks will be configurable to a great extent in order to meet the diverging traffic demands of different applications. They will be integrated in smart and dynamic wireless software-defined networks (SDNs), [41], which means a generalization of the software-defined radio (SDR) concept [42]. In contrast to existing cellular systems, sophisticated technologies such as ultra-dense access node deployments and multi-node coordination will be supported to achieve optimum performance [43]. Frequency division duplex (FDD) and time division duplex (TDD) systems are expected to further coexist, while TDD-only operation is expected to become more widely used in higher frequency bands. Full duplex is under investigation, but its use will probably be restricted to low-power radio nodes, e.g., for indoor and outdoor small cell applications (including in-band wireless backhaul). Evolved versions of existing communication systems need to be efficiently integrated.

2.4 Massive MIMO

In massive MIMO, we see a very large antenna array at each base station and in general an order of magnitude more antenna elements as compared with conventional systems. Massive MIMO can be used for a more efficient backhaul wireless link or even for the access link, in which a large number of users are served simultaneously. Here, massive MIMO is interpreted as multiuser MIMO with lots of base station antennas. Although massive MIMO has been historically associated with the use of beamforming, new trends are arising towards the use of spatial multiplexing techniques in conjunction with beamforming to make the most of the richness of the channel [44]. With the incorporation of nomadic and moving nodes, sophisticated MIMO solutions can be implemented in a vehicle node, making advanced multi-user MIMO feasible.

Massive MIMO can increase system capacity by a factor of 10 and, at the same time, increase the energy efficiency on the order of 100 times [45]. This boost in capacity comes from the use of aggressive schemes of spatial multiplexing whereas the improvement in energy consumption is due to the concentration of power in very concrete space regions.

The main concern when the concept of massive MIMO comes to real deployments relies on the factor form of an antenna of hundreds of element. In this sense, the concept of massive MIMO is intrinsically related to its use in high frequencies, e.g., in the centimeter wave (cmW) and mmW bands, which will reduce the size of each of these elements [46]. Many different configurations for the real antenna arrays can be used, as depicted in Figure 7, ranging from distributed antenna systems, to planar and cylindrical arrays.

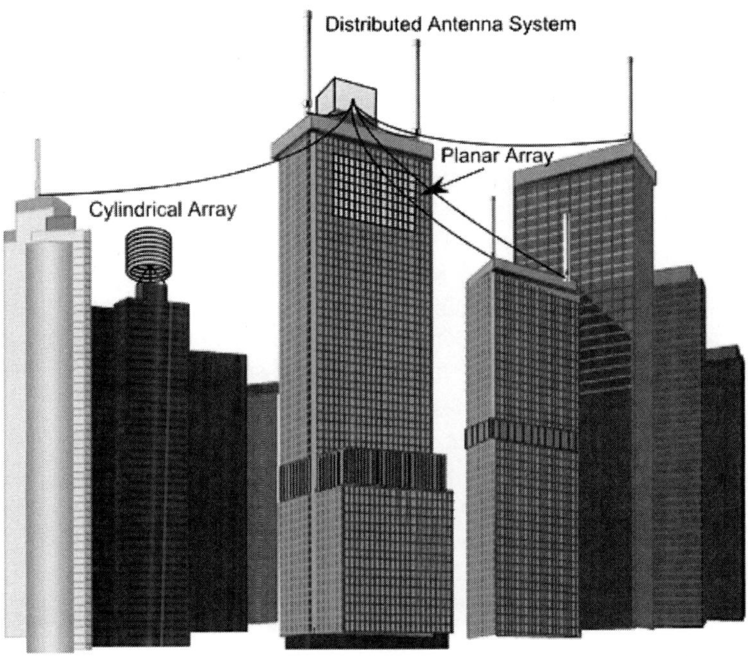

Figure. 7. Different alternatives for the deployment of massive mimo, including distributed antennas, planar arrays and cylindrical arrays.

Extensive use of beamforming will be an essential tool to improve link budgets. In addition, spatial multiplexing using massive MIMO techniques at mmW along with small cell geometries seems to be a symbiotic convergence for throughput boost [47]. The success of these techniques might be threatened by impairments in the transmission and reception chain, such as hardware limitations and lack of appropriate channel knowledge. This will be a matter of exhaustive research in the coming years. Another important problem with massive MIMO is pilot contamination. The number of orthogonal uplink pilot sequences is limited and depends on the coherence interval, which is very small at higher frequency bands. When reusing the same pilot from two transmitters, the negative consequences derived from interferences are referred to as pilot contamination. A beamforming scheme based on the channel estimates obtained from contaminated pilots results in interferences directed to those transmitters using the same pilot [48]. Therefore, new transmission schemes based on blind channel estimation are to be designed for the next generation mobile [49].

2.5 New signaling/control information

The control information/signaling needs to be fundamentally readdressed in 5G systems to accommodate the different needs of different services. In order to facilitate the spectrum-flexible multi-layer connectivity for eMBB services, a separation of control and data plane can be used. One example is mmW communication, where the control channel can be established at lower frequencies. Another example is network-controlled communication of content via D2D connection, offloading the cellular network. More generally, the user centricity of 5G systems shown in Figure 5 induces separation of control and data planes.

MMC, on the other hand, benefits from a closer coupling between the control and data plane, even integration of the control and data planes. MMC also requires optimized sleep mode solutions for battery-operated devices and mobility procedures with a minimum of signaling and measurements.

URC requires guaranteed latency and reliability. Here, it should be noted that the successful reception of control information is a prerequisite for communication of the data part. For very reliable, low-latency V2X connection, the device discovery can be assisted by the wide-area network. The increasing number of network nodes requires lean signaling for energy performance boost.

Another aspect that requires further attention is security. In the sensible case of D2D communications, the wide-area system shall provide security parameters to both communications links based on the internal credentials of the users. This network-assisted security establishment is a challenging issue in the development of D2D communications.

2.6 Localized contents and traffic flows

One of the main goals of METIS is to reduce current E2E latency performance to less than 5 ms. However, the delay budget analysis of legacy technologies reveals that most of the delay comes from the Internet and the core network parts of the E2E connection. Therefore, localized traffic flows, including data traffic offloading, aggregation, caching, and local routing, contribute to meeting this target.

The network operator improves the user experience by providing, e.g., authentication and security features while reducing the load on the data transport. In this framework, context information and network assistance for D2D discovery are of paramount importance to enable such direct communication.

In MMC, the use of concentrators acting as local gateways could allow direct communication among sensors located in capillary networks without the need to reach the core network gateway. For MMC, the localized traffic flows allow low-power access to the network. The network edge nodes can provide aggregation and information fusion of sensor data reducing the transport load and provide local added information value. Further, the necessary context information for MMC operations can be stored locally.

For delay-sensitive services, e.g., road safety applications, it is necessary to turnaround the traffic flow and perform critical computations close to the user to meet the latency constraints of less than 5 ms.

Moreover, the concept of caching could be shifted to the network edges, reaching access nodes or even the own devices that could act as proxies in case of having the requested content in the memory [50].

3 FLEXIBLE NETWORK ARCHITECTURE FOR 5G

The 5G architecture development is driven by three key aspects, namely, flexibility, scalability, and service-oriented management. All of these three aspects are complementary to each other such that the diverse set of 5G technologies which aim to fulfill broad range of service requirements can be efficiently supported. Flexibility is the core aspect to enable dynamic configuration of the necessary network functionalities for the efficient realization of a target service and use case. Furthermore, the future-proof architecture shall be flexible enough to handle diverse requirements of the use cases and services that we foresee today and the ones that may emerge in the future and cannot be anticipated now. On this basis, scalability will be assisted by flexibility to fulfill the requirements of extremely contradicting services, e.g., MMC verses multi-user ultra high definition (UHD) tele-presence.

As addressed above, the new generation of RAN networks envisaged for 5G needs to efficiently handle multiple layers and a variety of air interfaces in the access and the backhaul domains. They have to control and cope with the dynamics of traffic, user behavior, and active nodes involved and need to be able to differentiate a larger variety of QoS characteristics, such as ultra-low latency traffic, ultra-reliable communications, and broadcast traffic. SDN, Network Function Virtualization (NFV) [51], and advanced self-organizing network (SON) technologies will play an important role in the implementation and control of the network functionalities, in particular, to improve scalability and reliability as shown later in this paper with respect to achievement of METIS goals.

The generic METIS 5G network architecture, as sketched in Figure 8 from the topological point of view, must accommodate a number of technical enablers and communication paradigms while taking into account existing and emerging evolutionary and revolutionary architectural trends. The network topology will comprise various flavors of Cloud-RAN (C-RAN) [52], traditional access nodes as well as new virtual access nodes where the fixed-cell concept disappears in favor of device-centric communications. Traditional access nodes are the ones that we have today with the hardware and software components existing together, e.g., dedicated hardware implementations of base stations. The virtual access node is a new notion anticipated for 5G mobile and wireless communications networks, where the functionalities at an access node will be run by various virtual machines that can be modified or extended based on a need basis of the services and/or use cases to be supported.

Moreover, 5G network architecture will also be scenario and use-case specific, e.g., it may be different in areas with low user density compared to deployments in ultra-dense areas, such as Mega-Cities.

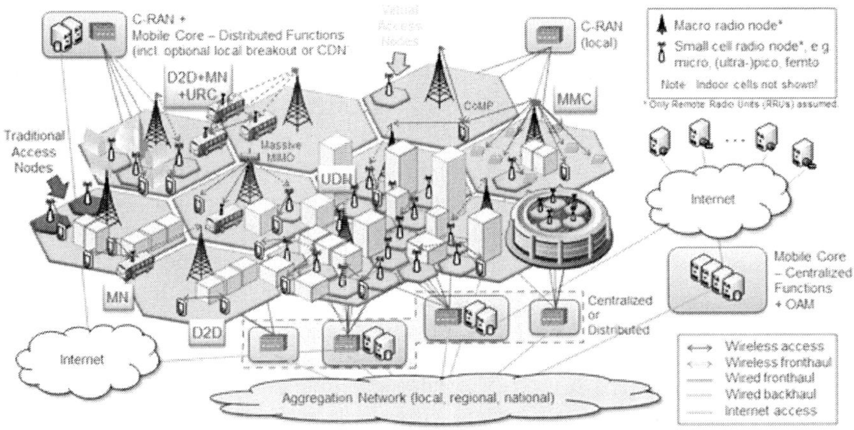

Figure. 8. Generic 5g network architecture (high-level topological view).

Implementation of radio network and service functions in C-RAN environments will simplify the mapping of SDN and NFV features (known from core function virtualization in data centers) to the RAN. It also increases flexibility with respect to integration of decentralized core functions in C-RAN processing units like local mobility management, local breakouts, and content delivery networks (CDNs) with caching capabilities [50]. Due to centralized processing and minimum delay among baseband processing units (BBUs), C-RAN environments allow simplified clustering of cells for joint RRM and interference coordination (including coordinated multipoint transmission/reception (CoMP)). Dependent on network infrastructure availability of the mobile network operator and delay limitations set on back-/fronthaul links [b], e.g., by CoMP schemes, C-RANs can be deployed in a distributed or more centralized way, which differentiate especially in terms of the number of contained BBUs (realizable via virtual machines on general purpose processors, possibly with support of hardware accelerators for dedicated physical layer computing tasks). Nevertheless, the BBU number of a local C-RAN can be also high in case of UDN, e.g., for a stadium scenario.

A first consideration of the main functional building blocks to appear in the METIS 5G RAN architecture from the logical point of view is shown in Figure 9. Flexibility shall enable a dynamic configuration of the necessary network functionalities in order to fulfill the requirements of a target service. This can necessitate new or tailored functionalities that will be made available on-demand based on the capabilities, such as processing and memory, of the network elements. On this basis, the functions may be updated or enabled on different time scales and spatial dimension. For example, a machine may be updated by cluster-head functionality when there are at least a minimum number of other machines in the proximity, and the candidate cluster-head machine has the capabilities of running that functionality. The Service Flow Management, as depicted in the logical view of architecture, has the task of analyzing the services and outlining the requirements for data flows through network, which will be orchestrated by the Function Coordinator. We note that to realize such flexibility end-user devices, e.g., nomadic nodes, D2D relays, and MMC gateways, are to be integrated as network elements. Function Coordinator has an interface with Function Agents by which it can configure the Base-Band Elements (BBEs) and core network elements (CNEs) based on the service requirements, also known as service orchestration, and forms the service chain. For example, addition of new functions and updates of the existing functions at BBEs and CNEs are performed by the Function Agent considering the configurations received from the Function Coordinator. Here, service requirements are determined by the Service Flow Management. Wireless SDN (WSDN) Controller sets up the service chain on the physical network infrastructure taking into account the configurations orchestrated by the Function Coordinator. WSDN Controller then constructs the data plane processing for the data flow, i.e., it builds up the connections for the service chain of CNEs and BBEs in the physical network through which the data flow passes. Accordingly, WSDN Controller configures the routers/switches (i.e., has interfaces to routers/switchers) also by factoring in radio-specific functionalities like resource scheduling and mobility management. It is worth noting that conventionally C-RAN infrastructure comprises BBUs. This corresponds to BBEs in the envisioned logical architecture. Nevertheless, thanks to the flexibility, which is anticipated for 5G, some core functionalities can also be moved to C-RAN. Therefore, C-RAN can compromise multiple BBEs as well as CNEs as depicted in Figure 5.

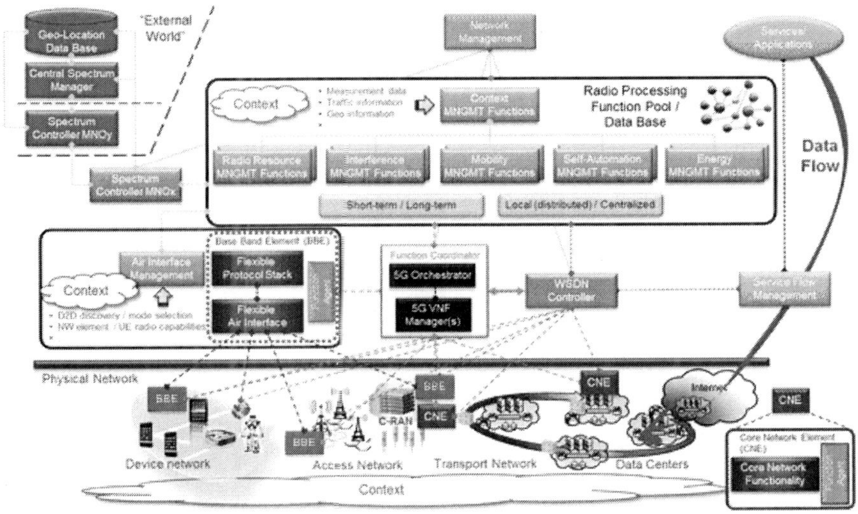

Figure. 9 Generic 5g Network Architecture (High-Level Functional/Logical View).

The envisioned architecture will provide the necessary flexibility to realize efficient integration and cooperation of functional blocks according to the individual service and network function needs of the HTs as well as functions for future evolution of existing cellular and wireless networks. The functions can be flexibly modified, tailored, and created by the Function Coordinator (consisting of 5G NFV Orchestrator and 5G virtual network function (VNF) Managers) according to the data flows and can be moved to the relevant network elements on-demand. In this regard, the context information, which is collected from the whole network, will be efficiently utilized by management functions. The architecture is based on WSDN approach to enable on-demand creation of customized virtual networks using shared resource pools and allowing effective service-adaptive decoupling of control and data plane in order to optimize routing and mobility management across the whole service transport chain. From the perspective of the end-to-end flow, even protocol stacks will be optimized and customized via software according to the required services and the network topology, i.e., the protocol stack becomes more and more a set of functional building blocks flexibly/dynamically combined on-demand to fulfill the very specific tasks for a certain service. On this basis, BBEs comprise flexible protocol stack, flexible air interface, and Function Agent.

Moreover, it is worth noting that WSDN Controller can also be virtualized running as a VNF with its Element Management System (EMS) & VNF Manager [53].

In contrast to present 4G, certain functionalities of the user equipments (UEs) may be partially controlled by the operator, e.g., for D2D relaying. Furthermore, not only do the UEs connect to the network, but also a wide range of devices, such as, sensors and robots, will co-exist, and the operator may have further degrees of freedom to coordinate the functionalities of these devices. It has to be noted that flexibility may be limited by capabilities of the network elements involved, such as sensors, which may not be updated with all new functionalities, e.g., due to hardware limitations.

Especially for operators with both fixed and mobile network infrastructure, cost reduction in future deployment stages is of great importance. For reuse of common network infrastructure on transport and access layer (Fixed-Mobile Convergence (FMC)), WSDN and NFV are, as well, seen as the key enablers [54] allowing multi-operator network infrastructure and resource sharing in a cost-optimized way. The integration of joint core functionalities belongs to the next steps towards FMC.

An exemplary use case is described in the following. The motivation is to enhance the user experience of an extremely high density of active users and machines in an area where normally the mobile access network nodes are sparsely deployed. The challenge is then to fulfill the requirements of such a massive connectivity of a mixture of eMBB and MMC devices for a certain time period by the network that is highly under-dimensioned for such a use case. Within the framework of the envisioned flexible architecture, service flow management determines that there is an over-flow of mixture of services to be provisioned to massive number of devices and triggers the Function Coordinator regarding this situation. The Function Coordinator (here 5G Orchestrator) in turn updates the service chain to handle these requirements. One solution to tackle the signaling congestion is to perform clustering mechanisms where a cluster head coordinates the scheduling of a group of devices. In this regard, Function Coordinator (here VNF Manager) commands the Function Agents of the selected cluster heads (selection can be performed by the Context Management Entity according to the collected context, e.g., device capabilities reported by Function Agents) in the target region to install the cluster-head functionality in the BBEs of those selected cluster heads. The functionality definition is taken from the Radio Processing Function Pool via Function Coordinator (here 5G Orchestrator) and Function Agents. It is worth noting that the cluster head functionality is only utilized for a given time period when the service is required. Due to limited machine capabilities, e.g., small memory, such cluster head functionality, may be removed depending on the service and energy efficiency requirements. Based on the input from Function Coordinator (here 5G Orchestrator), the WSDN Controller can now perform the re-routing of the data flows according to service chain in line with the updated network topology.

Although in the envisioned flexible architecture, there are additional logical entities introduced, which are needed to enable the anticipated flexibility, compared to today's network architecture, there is a general consensus that adopting these architectural trends has various advantages in terms of cost savings (see, e.g., [51]). For instance, a virtualized network enables faster service provisioning, i.e., increased speed of time to market, and extension of the "classical" service portfolio to new offerings for, e.g., 3rd party connectivity providers (XaaS incl. network sharing), which implies revenue increase for the operators as the impact on the hardware deployments is minimized. As instead of dedicated hardware implementations, commercial-off-the-shelf hardware can be utilized, a better capital efficiency can be reached. Along with the cost savings, better optimization of the network resources can contribute to the energy savings in the network, which is in turn coupled with the cost savings, as well. It is worth noting that this trend is already ongoing on the fixed network side [55], and this vision for 5G mobile network environment can create synergy among fixed and mobile networks.

4 MEETING THE METIS GOALS FOR 5G - RESEARCH CHALLENGES

In this section, we briefly discuss the primary methods envisaged for the main features of the 5G system concept that can contribute most beneficially to achieve the METIS goals stated at the beginning of this paper.

4.1 1,000 times higher mobile data volume per area

The primary methods to address this goal are to use more frequency spectrum, to use more network nodes, and to enhance network performance and spectral efficiency.

4.1.1 Use more frequency spectrum

Both higher data rates and higher data volume require access to more frequency spectrum for mobile communications. As stated before, in WRC-15, it is expected that clearly more spectrum will be released for mobile communications. This new spectrum lies in the frequency range between 300 MHz and 6 GHz. However, for the future 5G system, these new spectrum opportunities will not be sufficient. Additionally, the release of new spectrum in the mmW bands was postponed to WRC-18.

Moreover, new spectrum access schemes are being considered such as Licensed Shared Access (LSA) [56], which would allow using further spectrum for mobile communications especially in higher frequency bands in a more flexible way. Accordingly, research is conducted on further novel spectrum access schemes, e.g., for coordinated spectrum sharing between different UDNs, supporting network densification as addressed below.

Also, wireless local area network (WLAN)-type systems [57] operating in the unlicensed bands, such as the ISM and U-NII bands at 2.4 and 5 GHz, as well as the 60 GHz band, can be more tightly integrated.

4.1.2 Use more network nodes

Densification of networks by deploying more nodes is already applied today with femto or pico cells providing local connectivity for increased capacity in addition to the macro cells in the wide-area. This trend is expected to continue, where the small cells can make use of higher frequency bands, such as the 3.5 GHz bands or the mmW bands. The use of TDD with higher frequencies can become more prominent as compared with 3G/LTE today. The different network layers/cell hierarchies and RATs are to be integrated by means of smarter networks. Dedicated implementation solutions such as distributed RRUs can be developed for specific deployments such as for providing coverage of very crowded places, e.g., stadiums and events. To this end, intensive research is conducted to support MN and UDN, e.g., solutions for in-band backhaul, where the access and backhaul share the same spectrum, specific distributed RRM schemes, or dynamic nomadic node selection for backhaul optimization.

4.1.3 Enhance network performance and spectral efficiency

Enhanced network performance and spectral efficiency are to be achieved by applying more sophisticated protocol and air interface technologies as compared to LTE-A [58]. The primary air interface technologies in this regard are the use of multiple antennas, possibly involving massive antenna constellations, multi-stream transmission such as spatial multiplexing or non-orthogonal multiple access, and cooperation techniques such as coordinated multi-node transmission or solutions based on interference alignment.

4.2 10 times to 100 times higher typical user data rate

The primary methods to address this goal are to use spectrum aggregation and more dynamic spectrum access, usage of higher frequencies up to mmW frequencies and short-range communications.

4.2.1 Spectrum aggregation and more dynamic spectrum access

Spectrum aggregation is already defined by 3GPP LTE-A and enables peak data rates up to about 1 Gbps. This may evolve further to take into account the new bands released during WRC-15. Aggregation of multiple bands over a wide range of frequencies poses challenges concerning the cost and efficiency of radio frequency filter and transceiver solutions, in particular for the UEs. Global research must be conducted to propose new schemes for spectrum opportunity detection and assessment and spectrum management concept for LSA. Based on the need for spectrum, these technical enablers shall help to obtain spectrum by negotiating access to it with other entities and also assessing the use of spectrum.

4.2.2 Usage of higher frequencies

Usage of higher frequencies up to mmW frequencies will further increase the achievable data rates to 10 Gbps or beyond. To compensate for the path loss at such high frequencies, high antenna directivity is required if transmission range shall be beyond a few meters. Accordingly, flexible air interfaces for mmW frequencies must be developed. The identification of suitable frequency bands and the characterization of the propagation channel demand new investigations.

4.2.3 Short-range communications

Finally, reducing the transmission distance, approaching short-range communications, e.g., D2D or UDN, is also an efficient means to increase user data rate. Indeed, node densification and dynamic RAN (cf. RAN 2.0 described previously) can allow confining interference to very specific areas of use while reducing the signal attenuation. Both effects shall boost the signal to interference plus noise ratio (SINR) thus increasing the user data rate by a logarithmic law.

4.3 10 times to 100 times higher number of connected devices

The primary methods to address this goal are to reduce signaling overhead, to use dynamic profiles of users, and to implement traffic concentration, in particular for MMC applications.

4.3.1 Reduce signaling overhead

Signaling overhead can be reduced by designing more efficient protocols and by thinning out some protocol layers, e.g., for the simplification of end-to-end procedures. On the air interface, the signaling for random access, time adjustment, and/or resource assignment can be reduced, for example, by using contention-based access, waveforms that are more tolerant to timing mismatch as compared to orthogonal frequency division multiplexing (OFDM), and/or by resource reservation, making the air interface more flexible and suitable for small payload traffic. For instance, methods dedicated to contention-based access for massive number of devices without scheduled access and multi-user detection facilitated by compressive sensing are good candidates to reduce signaling.

4.3.2 Use dynamic profiles of users

NFV technologies can play a role to overcome current limitations of a rigid protocol stack by allowing flexible configuration of the network functionalities. The future air interface shall support this flexibility by using dynamic profiles of users, of which entries are configurable to pre-defined use cases. In this sense, some radio functions will be activated/deactivated on demand depending on the specific needs of the service and the status of the network. Again, the user-centric paradigm applies, since the network shall adapt its topology and operation mode according to the needs of each user.

4.3.3 Traffic concentration

Finally, traffic concentration can be achieved by means of gateways, for example, by linking a capillary MMC network to a wide area cellular network. In this framework, the RAN might configure one specific device to act as such concentrator (cluster head) by using D2D communications to collect all data from the surrounding machines. It is expected that this last hop link should work on a different radio interface to host a large number of mesh devices with severe energy constraints. For instance, techniques are developed to reduce the amount of traffic in gateway-based MMC solutions by context-based device grouping and signaling.

4.4 10 times longer battery life for low power MMC devices

The primary methods to address this goal are to improve air interface, procedures, and signaling and to reduce the distance between the MMC devices and the access node. Techniques to improve the air interface include the use of sleep modes, energy-efficient modulation, coding and multiple access schemes, for example, use of constrained modulation techniques such as Gaussian Minimum Shift Keying (GMSK), very robust coding and/or spreading allowing very low transmission powers, and simplified procedures such as those introduced for the previous goal.

The MMC devices can, in some scenarios, be brought closer to the access node, for example, by gateways connecting a capillary network to a wide area network or by denser deployment of small cells.

4.5 Times reduced End-to-End (E2E) latency

The primary methods to address this goal are to use more efficient network architectures, more efficient air interface designs, signaling and procedures, better QoS differentiation, and direct D2D communication.

4.5.1 Efficient network architectures

Low-latency network architectures are concerned with shortening the distances and number of hops between the user and the content, for example, by distributing some network functions that are centralized today or by applying local and universal caching as well as local breakouts to external networks like Internet, service provider, or enterprise networks.

4.5.2 Efficient air interface designs

The air interface may adapt the frame structures in order to reduce the Transmission Time Interval (TTI) and/or Hybrid Automatic Repeat reQuest (HARQ) round trip times. While this can efficiently be implemented at the higher frequencies where more bandwidth is available, this can become very inefficient for wide area coverage. Signaling and procedures can be thinned out, by simplifying their components.

4.5.3 Better QoS differentiation

Better QoS differentiation is important in this context. As the low-latency transmission may become inefficient, it should only be applied with the services that require such low latencies and should be avoided with delay-insensitive services.

4.5.4 D2D

D2D can be used between devices within transmission range of each other which saves a radio hop and the routing of traffic through the network. In some case, it can be applied also for local broadcasting, for example, where vehicles broadcast position/velocity information to enhance traffic safety.

5 CONCLUSIONS

This paper has presented the current view on the 5G system concept from the METIS project. Within the 5G system concept and its main features, a generic overview for the envisioned flexible 5G network architecture is provided. More details on this architecture proposal can be found in [59]. In contrast to current 3GPP work, D2D communications must be a cornerstone of the new 5G technology going towards the new RAN 2.0 paradigm. Flexible radio air interface can offer the best solutions targeted to different use cases. Moreover, the ultra density of access nodes, including vehicles, devices, and base stations, requires further research into the definition of more efficient mobility procedures and signaling simplification. On the other hand, a crucial difference of the future 5G networks in comparison to existing 4G networks is that the system must be from-start designed not only for human-centric, but also for machine-centric traffic in MMC with different service demands. Intensive research is required in the fields of overlaying multiple transmissions, new air interfaces, smart pre-allocation of resources, reduction of the synchronization requirements, and context/service-aware configuration of the radio access. Finally, special attention must be paid to the development of URC techniques, capable of supporting new services like tactile Internet or V2X communications for traffic safety.

Table. 1. Main enablers for the 5g design

Technique	Impact	Pros	Cons
D2D communications	System capacity can be increased by a factor of 2 using the same bandwidth when some RRM mechanisms are used for opportunistic access and there exists full cooperation among devices. Latency is also reduced to the order of the TTI length (e.g., 1–2 ms).	With properly selected safety distance, D2D communication can use the spectrum allocated to small cells (mostly uplink band) without affecting the small cell performance.	How to motivate cooperation of users is still one open issue. Battery consumption is something critical in current systems and D2D communication model is only accepted by the end user for him/her own benefit. Specification is far from being ready for D2D integration. Opposition from some mobile network operators is also an important barrier to overcome.
New air interfaces	Access time can be reduced down to 1.5 ms if new air interfaces are coupled with efficient access procedures (as shown by FBMC evaluations [60]). Due to the very good frequency localization, the transmit power can be concentrated on only very few subcarriers to eventually enhance significantly the expected coverage or to reduce battery consumption. This is of special relevance for machine-type communications, in which payload is very small.	The well-localized signal energy in frequency domain of the multi-carrier signal also allows for efficient access to fragmented spectrum and efficient spectrum sharing, as a minimum amount of guard bands are needed for the signal separation in frequency.	Changing the waveform impacts the signal structures, implying a revolutionary step towards a new radio generation. Backward compatibility thus cannot be guaranteed, however, for a new radio generation this should not be crucial requirement.
Ultra-dense network	Capacity is directly proportional with the number of nodes, provided centralized interference coordination. For indoor cases, the coefficient of proportionality could be as high as 0.73 with ISD of 10 m.	Together with increasing the bandwidth, the use of more nodes is an easy means to achieve desired levels of capacity.	Cost is the main issue of this approach, together with the need for interference coordination. In case of certain isolation between cells, this need for coordination is relaxed.
Traffic concentration	The use of accumulators or concentrators for machine-type communications improves range of coverage and sensors' throughput. The improvement factor is between two and three thanks to those relays. For the same throughput and coverage needs, traffic concentration reduces battery consumption.	Concentrators also reduce signaling overhead thanks to the coupling of parallel signaling flows.	Performance heavily depends upon the appropriate relay selection.
Moving networks	In vehicles equipped with two access points, one for outside transmission/reception and another for inside users, their best reception chain improves the link budget for the end user up to 9 dB, in cases where the user is outside the car, and up to 24 dB, when the user is within the car. This results in better coverage or higher user throughput, mostly in the cell-edge.	Battery is not a big issue for vehicles, which opens the door for more active collaboration between cars and end-users. Moreover, the number of antennas integrated in vehicles can be much higher than in the handheld devices. This allows for massive-MIMO solutions.	Mobility of access points increase management complexity mostly due to the higher dynamicity of the network.
Localized traffic flows	With a dedicated bandwidth of 80 MHz, end-to-end latency is reduced to 60%, as compared with current LTE-A system. Moreover, half of the traffic can be offloaded from the cellular system.	Use of localized traffic flows is simple to implement and can be easily integrated into current networks.	Universal caching requires storage resources as well as the design of specific signaling mechanisms.
Massive MIMO	Spectral efficiency can be increased by a factor of 20 when using 256 antennas in the transmitter and receiver side as compared with four antenna systems. For the same spatial multiplexing capability as legacy systems (8 streams), beamforming gain reaches 15 dB.	Simple way of increasing cell efficiency mostly for small cell deployments. Fits together with the use of higher frequencies above 6 GHz due to the reduced antenna size.	Pilot contamination is one of the main showstoppers of the use of massive MIMO. Moreover, TDD mode seems to be a must to reduce signaling overhead thanks to the use of channel reciprocity principle. Form factor forces the use of centimeter wave or mmW to compact massive antennas. Finally, performance is very sensitive to mobility and computation burden could make multi-user solutions unaffordable.
Spectrum above 6 GHz	Meeting the 5G user expectations will require much larger bandwidths as today, in the order of 2 GHz, which can only be available at frequencies above 27 GHz (e.g., 27.5 to 29.5 GHz; 40.5 to 42.5 GHz; 47.2 to 50.2 GHz; 57 to 76 GHz; or 81 to 86 GHz).	No restrictions on previous allocations of paired FDD spectrum allows for the use of TDD mode, which is much more efficient for channel estimation. Higher frequencies also reduce antenna size, thus permitting massive MIMO implementations.	Path losses are much higher at such high frequencies, which reduces the coverage to small distances or relies on the use of higher order beamforming to overcome such high attenuations.

Simulation results in METIS have proven the potential of some of the techniques discussed along this paper. Table 1 summarizes the impact of most significant techniques together with some comments concerning the pros and cons of their use in 5G systems. Note that the quantitative analysis provided in Table 1 is derived from several simulations in up to 12 different test cases. For the sake of simplicity, we are not going to include in this paper the details of all these test cases. Please refer to METIS deliverables for additional information [2,60].

Of significant relevance is the use of D2D communications, which could increase average system capacity by a factor of 2 (with high density of users and common interests in the consumed contents), and moving networks, in which cars can operate in relaying mode for other users, e.g., out-car users in case of nomadic node operation [61,62]. Thanks to relaying, vehicle penetration losses are avoided and the reception chain improves, due to the better characteristics of the antenna and the receptor in the vehicle. The mobile relay operates in a separate band in full duplex mode, that is, is totally transparent for the network point of view that treats the vehicle cell like a regular user. Moreover, the collaboration of cars allows for a feasible implementation of universal caching, by which cars forward cached content directly to interested user. Results have proven two positive effects of this universal caching. First, caching reduces end-to-end latencies of cached users. This reduction increases with smaller packets, that is, when the non-radio-transmission delay is more relevant over the whole end-to-end latency. The second positive effect of universal caching is the offloading of traffic. According to current shape of cellular network contents, one fourth of the users may be offloaded from the classical network. With the current traffic type distribution, about one half of the traffic load might be offloaded.

5.1 Endnotes

[a]In comparison with data traffic in 2010.

[b]Backhaul is denoted as the link between aggregation (alternatively core) network and BBU of the radio node. The fronthaul is denoted as the link between BBU and Radio Remote Unit (RRU). This denotation is valid also for the C-RAN case [52].

ACKNOWLEDGEMENTS

Part of this work has been performed in the framework of the FP7 project ICT-317669 METIS, which is partly funded by the European Commission. The authors would like to acknowledge the contributions of their colleagues in METIS with special thanks to Petar Popovski, Peter Fertl, David Gozalvez-Serrano, Andreas Höglund, Zexian Li, and Krystian Pawlak. Also thanks to Josef Eichinger and Malte Schellmann for the fruitful discussions during the revision of this article.

REFERENCES

1. Cisco, in Global Mobile Data Traffic Forecast Update, 2014–2019 White Paper, February 2015.
2. METIS, in Mobile and wireless communications Enablers for the Twenty-twenty Information Society, EU 7th Framework Programme project,
3. ICT-317669 METIS project, in Scenarios, requirements and KPIs for 5G mobile and wireless system, Deliverable D1.1, May 2013,
4. B Ahlgren, C Dannewitz, C Imbrenda, D Kutscher, B Ohlman, A survey of information-centric networking. IEEE Commun Mag **50**(7), 26–36 (2012)
5. A Osseiran, F Boccardi, V Braun, K Kusume, P Marsch, M Maternia, O Queseth, M Schellmann, H Schotten, H Taoka, H Tullberg, MA Uusitalo, B Timus, M Fallgren, Scenarios for the 5G mobile and wireless communications: the vision of the METIS project. IEEE Commun Mag **52**(5), 26–35 (2014)
6. D Gomez-Barquero, D Calabuig, JF Monserrat, N Garcia and J Perez-Romero, Hopfield neural network - based approach for joint dynamic resource allocation in heterogeneous wireless networks, in Proceedings 64th IEEE Vehicular Technology Conference (VTC), Montreal. 2006
7. JF Monserrat, P Sroka, G Auer, J Cabrejas, D Martin-Sacristan, A Mihovska, R Rossi, A. Saul, R. Schoenen, Advanced Radio Resource Management for IMT-Advanced in WINNER+ (II), in Proc. Future Network and Mobile Summit, pp.1-9, June 2010.
8. F Boccardi, RW Heath, A Lozano, TL Marzetta, P Popovski, Five disruptive technology directions for 5G. IEEE Commun Mag **52**(2), 74–80 (2014)

9. JG Andrews, S Buzzi, C Wan, SV Hanly, A Lozano, ACK Soong, JC Zhang, What will 5G be? IEEE J Sel Area Comm **32**(6), 1065–1082 (2014)
10. MN Tehrani, M Uysal, H Yanikomeroglu, Device-to-device communication in 5G cellular networks: challenges, solutions, and future directions. IEEE Commun Mag **52**(5), 86–92 (2014)
11. N Bhushan, L Junyi, D Malladi, R Gilmore, D Brenner, A Damnjanovic, R Sukhavasi, C Patel, S Geirhofer, Network densification: the dominant theme for wireless evolution into 5G. IEEE Commun Mag **52**(2), 82–89 (2014)
12. K Okino, T Nakayama, C Yamazaki, H Sato, Y Kusano, Pico Cell Range Expansion with Interference Mitigation toward LTE-Advanced Heterogeneous Networks, in Proc. of IEEE International Conference on Communications (ICC), 2011.
13. P Mugen, L Dong, W Yao, L Jian Li, C Hsiao-Hwa, Self-configuration and self-optimization in LTE-advanced heterogeneous networks. IEEE Commun Mag **51**(5), 36–45 (2013)
14. I Siomina, D Yuan, Load Balancing in Heterogeneous LTE: Range Optimization via Offset and Load-coupling Characterization, in Proc. of IEEE Int. Conference on Communications (ICC). June 2012.
15. KI Pedersen, Y Wang, B Soret, F Frederiksen, eICIC Functionality and Performance for LTE HetNet Co-Channel Deployments, in Proc. of IEEE Vehicular Technology Conf, Sep 2012
16. X Gu, X Deng, Q Li, L Zhang, W Li, Capacity Analysis and Optimization in Heterogeneous Network with Adaptive Cell Range Control, Int. J. Antennas. Propag. **2014**(215803), 10 (2014)
17. K Smiljkovikj, P Popovski, L Gavrilovska, Analysis of the Decoupled Access for DL and UL in Wireless Heterogeneous Networks, in IEEE Wireless Communications Letters, in press,
18. P Agyapong, M. Iwamura, D. Staehle, W. Kiess, A. Benjebbour, Design considerations for a 5G network architecture. IEEE Commun Mag **52**(11), 65–75 (2014)
19. L Yan, X Fang, Reliability Evaluation of 5G C/U-plane Decoupled Architecture for High-speed Railway. EURASIP J Wirel Commun Netw **2014**, 127 (2014)
20. B Zafar, S Gherekhloo, M Haardt, Analysis of multihop relaying networks: communication between range-limited and cooperative nodes. IEEE Veh Technol Mag **7**(3), 40–47 (2012)
21. Study on Mobile Relay for Evolved Universal Terrestrial Radio Access (E-UTRA), 3GPP TR 36.836, V2.0.2, July 2013.
22. A Krendzel, LTE-A Mobile Relay Handling: Architecture Aspects, in Proc. of the 19th European Wireless Conference (EW), Guildford, UK, pp. 1–6, 2013.

23. M Khanfouci, Y Sui, A Papadogiannis, and M Färber, Moving Relays and Mobility aspects. ARTIST4G project deliverable D3.5c-v2.0, 2012.
24. F Haider, M Dianati, and R Tafazolli, A Simulation Based Study of Mobile Femtocell Assisted LTE Networks, in Proc. Of the 7th International Wireless Communications and Mobile Computing Conference (IWCMC), Istanbul, Turkey, pp. 2198–2203, 2011
25. F Haider, W Haiming, H Haas, Y Dongfeng, W Haiming, G Xiqi, Y Xiao-Hu, E Hepsaycir, Spectral efficiency analysis of mobile Femtocell based cellular systems, in Proc. of the 13th International Conference on Communication Technology (ICCT), Jinan, pp. 347–351, September 2011.
26. ICT-317669 METIS project, Initial report on horizontal topics, first results and 5G system concept, Deliverable D6.2, April 2014,
27. Study on LTE Device to Device Proximity Services, 3GPP TR 36.843, 2014.
28. V Yazıcı, UC Kozat, M Oguz, Sunay, A new control plane for 5G network architecture with a case study on unified handoff, mobility, and routing management. IEEE Commun Mag **52**(11), 76–85 (2014)
29. F Malandrino, C Casetti, C-F Chiasserini, Toward D2D-enhanced heterogeneous networks. IEEE Commun Mag **52**(11), 94–100 (2014)
30. A Asadi, Q Wang, V Mancuso, A survey on device-to-Device communication in cellular networks. IEEE Commun Surv Tutor **16**(4), 1801–1819 (2014)
31. D Feng, L Lu, YY Wu, GY Li, G Feng, S Li, Device-to-device communications underlaying cellular networks. IEEE Trans Commun **61**(8), 3541–3551 (2013)
32. C Xu, L Song, Z Han, Q Zhao, X Wang, X Cheng, B Jiao, Efficiency resource allocation for device-to-device underlay communication systems: a reverse iterative combinatorial auction based approach. IEEE J Sel Area Comm **31**(9), 348–358 (2013)
33. S Lingyang, D Niyato, H Zhu, E Hossain, Game-theoretic resource allocation methods for device-to-device communication. IEEE Wireless Commun **21**(3), 136–144 (2014)
34. G Aloi, M Di Felice, V Loscrì, P Pace, G Ruggeri, Spontaneous smartphone networks as a user-centric solution for the future internet. IEEE Commun Mag **52**(12), 26–33 (2014)
35. PA Frangoudis, GC Polyzos, Security and performance challenges for user-centric wireless networking. IEEE Commun Mag **52**(12), 48–55 (2014)

36. ITU-R M.2079, in Technical and operational information for identifying Spectrum for the terrestrial component of future development of IMT-2000 and IMT-Advanced, 2006
37. AB MacKenzie, LA DaSilva, Application of signal processing to addressing wireless data demand [in the spotlight]. IEEE Signal Process Mag **29**(6), 168–166 (2012
38. X Cheng, Y Koucheryavy, Y Li, F Zhao, T Znati (ed.), Dynamic Spectrum Access for Throughput, Delay, and Fairness Enhancement In Cognitive Radio Networks, EURASIP J Wirel Commun Netw, November 2014
39. MR Akdeniz, Y Liu, MK Samimi, S Sun, S Rangan, TS Rappaport, E Erkip, Millimeter wave channel modeling and cellular capacity evaluation. IEEE J Sel Area Comm **32**(6), 1164–1179 (2014)
40. A Adhikary, E Al Safadi, M Samimi, R Wang, G Caire, TS Rappaport, AF Molisch, Joint spatial division and multiplexing for mm-wave channels. IEEE J Sel Area Comm **32**(6), 1239–1255 (2014)
41. K Pentikousis, Y Wang, W Hu, Mobileflow: toward software-defined mobile networks. IEEE Commun Mag **51**(7), 44–53 (2013)
42. E3 D2.4. Cognitive Function mapping to Networks Architectures, Standard Engineering and Software Technologies for Cognitive Radios, E3 Project Deliverable 2.4, December 2009.
43. R Wang, H Hu, X Yang, Potentials and challenges of C-RAN supporting Multi-RATs toward 5G mobile networks. IEEE. Access. **2(**1187), 1195 (2014)
44. V Jungnickel, K Manolakis, W Zirwas, B Panzner, V Braun, M Lossow, M Sternad, R Apelfrojd, T Svensson, The role of small cells, coordinated multipoint, and massive MIMO in 5G. IEEE Commun Mag **52**(5), 44–51 (2014)
45. E Larsson, O Edfors, F Tufvesson, T Marzetta, Massive MIMO for next generation wireless systems. IEEE Commun Mag **52**(2), 186–195 (2014)
46. W Roh, S Ji-Yun, P Jeongho, L Byunghwan, L Jaekon, K Yungsoo, C Jaeweon, C Kyungwhoon, F Aryanfar, Millimeter-wave Beamforming as an Enabling Technology for 5G Cellular Communications: Theoretical Feasibility and Prototype Results. IEEE Commun Mag **2**(2), 106–113 (2014)
47. AL Swindlehust, E Ayanoglu, P Heydari, F Capolino, Millimeter-wave massive MIMO: the next wireless revolution? IEEE Commun Mag **52**(9), 56–62 (2014)
48. L Lu, GY Li, AL Swindlehurst, A Ashikhmin, Z Rui, An overview of massive MIMO: benefits and challenges. IEEE J Sel Top Signal Process **8**(5), 742–758 (2014)

49. S Roger, D Calabuig, J Cabrejas, JF Monserrat, Multi-user non-coherent detection for downlink MIMO communication. IEEE Signal Process Lett **21**(10), 1225–1229 (2014)
50. X Wang, M Chen, T Taleb, A Ksentini, V Leung, Cache in the air: exploiting content caching and delivery techniques for 5G systems. IEEE Commun Mag **52**(2), 131–139 (2014)
51. ETSI ISG NFV (Operator Group), Network Functions Virtualisation – Network Operator Perspectives on Industry Progress, Updated White Paper, October 2013
52. NGMN Alliance, in Suggestions on potential solutions for C-RAN, White Paper, January 2013
53. ETSI ISG NFV, Network Functions Virtualisation (NFV); Virtual Network Functions Architecture, v1.1.1, Dec 2014.
54. A Tzanakaki, MP Anastasopoulos, GS Zervas, BR Rofoee, R Nejabati, D Simeonidou, Virtualization of heterogeneous wireless-optical network and IT infrastructures in support of cloud and mobile cloud services. IEEE Commun Mag **51**(8), 155–161 (2013)
55. A Manzalini, R Saracco, C Buyukkoc, P Chemuouil, S Kukliński, A Gladisch, M Fukui, W Shen, M Fujiwara, K Shimano, E Dekel, D Soldani, M Ulema, W Cerroni, F Callegati, G Schembra, V Riccobene, C Mas Machuca, A Galis, J Mueller, Software-Defined Networks for Future Networks and Services: Main Technical Challenges and Business Implications, IEEE Workshop SDN4FNS, 1–16, 2013
56. CEPT ECC, in Licensed Shared Access (LSA), ECC Report 205, February 2014
57. IEEE 802.11, in IEEE 802.11-2012 Part11: Wireless LAN Medium Access Control (MAC) and Physical Layer (PHY) Specifications, IEEE Standard, March 2012
58. D Martín-Sacristán, JF Monserrat, J Cabrejas-Peñuelas, D Calabuig, S Garrigas, N Cardona, On the way towards fourth-generation mobile: 3GPP LTE and LTE-Advanced. EURASIP J Wirel Commun Netw **2009**, 10 (2009)
59. ICT-317669 METIS, Final report on architecture, Deliverable D6.4, January 2015,
60. ICT-317669 METIS, Report on simulation results and evaluations, Deliverable D6.5, February 2015,
61. Ö Bulakci, Z Ren, C Zhou, J Eichinger, P Fertl, S Stanczak, Dynamic Nomadic Node Selection for Performance Enhancement in Composite Fading/Shadowing Environments, (IEEE VTC 2014-Spring, Seoul, South Korea)
62. ICT-317669 METIS, Final report on network-level solutions, Deliverable D4.3 Version 1, February 2015

7

CHAPTER

SURVEY ON WIRELESS SENSOR NETWORK TECHNOLOGIES FOR INDUSTRIAL AUTOMATION: THE SECURITY AND QUALITY OF SERVICE PERSPECTIVES

Delphine Christin [1,*], Parag S. Mogre [2] and Matthias Hollick [1]

ABSTRACT

Wireless Sensor Networks (WSNs) are gradually adopted in the industrial world due to their advantages over wired networks. In addition to saving cabling costs, WSNs widen the realm of environments feasible for monitoring. They thus add sensing and acting capabilities to objects in the physical world and allow for communication among these objects or with services in the future Internet. However, the acceptance of WSNs by the industrial automation community is impeded by open issues, such as security guarantees and provision of Quality of Service (QoS). To examine both of these perspectives, we select and survey relevant WSN technologies dedicated to industrial automation. We determine QoS requirements and carry out a threat analysis, which act as basis of our evaluation of the current state-of-the-art. According to the results of this evaluation, we identify and discuss open research issues.

Keywords: Wireless Sensor Networks; Industrial Automation; Quality of Service; Security; State-of-the-art; WirelessHART; ISA100.11a-2009; ZigBee; Wireless Interface for Sensors and Actuators; 802.15.4e Factory Automation MAC Layer

1. INTRODUCTION

Industrial automation has been successfully introduced in a countless amount of industries ranging from food to energy industries. Even if the products differ from one industry to another, the automated processes can be classified according to three main layers as proposed by [1]: the plant-floor automation layer, the manufacturing execution system layer and the enterprise resource planning layer. Internet technology can be considered as the link that interconnects all these layers and allows for information exchange. For example, it serves as a backbone to interconnect different production locations within one enterprise, to transfer production control data in near real-time to the headquarters or to integrate suppliers into a production workflow.

Within the scope of this survey, we focus on the plant-floor automation layer including sensors, switches, programmable controllers and motor starters (Fig. 1) that ensure the correct operation of machines and execution of processes, while the remaining layers are dedicated to the optimization of the production by managing resource allocation and operation scheduling for example. In addition to productivity gain and precision improvement, the automation of processes at the plant-floor automation layer allows the replacement of workers in harsh and hazardous environments or assigned to tedious tasks [2]. The WSNs are part of this layer and can be used for multiple purposes, such as monitoring synchronous or asynchronous events that require periodic data collection or detecting exceptional events, respectively [3]. For example, vibration, heat or thermal sensors can be deployed in proximity of machines to monitor their health. The analysis of the measured parameters can allow the detection of abnormal operating conditions and aids therefore in preventing potential machine failure. In addition to machine monitoring, WSNs can be deployed to measure basic physical quantities such as pressure, temperature, flow or more complex events such as process quality or automotive performance in industrial environments [4].

Although wired sensor networks can also be deployed for such monitoring scenarios, WSNs present additional advantages. In fact, their wireless capability allows deployments in hostile environments, where vibrations or moving parts may prevent the use of cables that would be damaged or even broken. In addition to reduce cabling costs, the WSNs provide network flexibility, as the sensor nodes may be relocated quickly without necessitating time-consuming cable installation and maintenance. However, the nature of the wireless medium opens up security and QoS issues. For example, potential attackers may easily eavesdrop or manipulate wireless communication in absence of security mechanisms and the wireless channel has to be efficiently allocated between the different devices to provide the required QoS.

Our contributions are as follows:

We first select relevant WSN technologies dedicated to industrial automation and we provide an exhaustive survey of their characteristics.

We then analyze industrial automation applications to determine QoS requirements and evaluate the selected standards according to each identified QoS requirement. Open issues are discussed based on the results of this evaluation.

We carry out a threat analysis to identify pertinent security requirements and we investigate if and how the selected standards fulfill the previously identified security requirements. Related open issues are finally highlighted and discussed.

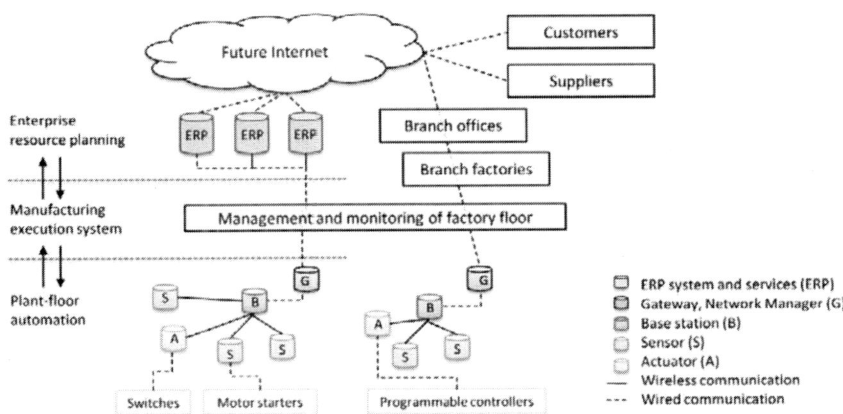

Figure 1. From the sensors to the customers.

The paper is structured as follows. Section 2 provides a detailed overview of the following WSN technologies: Wireless Interface for Sensor and Actuators (WISA), WirelessHART, ISA100.11a, ZigBee, ZigBee PRO, and 802.15.4e Factory Automation MAC Layer. Section 3 and Section 4 focus on QoS and security respectively. Both sections are composed of an analysis of the respective requirements and an evaluation of the selected specifications. Open issues are listed and discussed at the end of both sections. Section 5 concludes our work.

2. SELECTED WIRELESS SENSOR NETWORKS STANDARDS: STATE-OF-THE-ART

Wireless communication in industrial automation is mostly based on standardized technologies, such as the IEEE 802.11 [5] and IEEE 802.15 standard families [6], also designated as Wireless Local Area Networks (WLAN) and Wireless Personal Area Networks (WPAN). Both of these standard families were conceived for application purposes different than industrial automation. In fact, the IEEE 802.11-based standards offer high data rates in the order of tens of Mbit/s and ranges up to tens/hundreds of meters, while the IEEE 802.15-based standards only supports data rates of hundreds of kbit/s to several Mbits/s with ranges from a few meters up to hundreds of meters. However, to provide greater data rate and range, IEEE 802.11 technology consumes a greater energy budget that can limit the benefits obtained by wireless communications. Indeed, the sensor nodes are either powered by cables or batteries. In the former case, the advantages provided by wireless communication are partially negated, whereas in the latter case, the scarce energy resource has to be parsimoniously consumed in order to avoid frequent human interventions to recharge the batteries. Energy is thus a major concern in both previous cases and we therefore focus on the IEEE 802.15-based standards, and particularly on the IEEE 802.15.1 [7] and IEEE 802.15.4 [8] standard, within the scope of this work.

2.1. IEEE 802.15.1-based Standards

The IEEE 802.15.1 standard, also known as Bluetooth®, can be classified to fall between the IEEE 802.11 and IEEE 802.15.4 standards in terms of energy consumption and data rates. With medium data rates and lower energy consumption than the IEEE 802.11 standard, IEEE 802.15.1 offers an interesting compromise between energy consumption and data rate, and is therefore particularly suited for high-end applications requiring high data rates as well as applications with strong real-time requirements such as factory automation. The Wireless Interface for Sensor and Actuators (WISA) has been selected as a representative 802.15.1-based specification for further discussion.

Wireless Interface for Sensor and Actuators (WISA)

Released by ABB and presented in [9], the proprietary Wireless Interface for Sensors and Actuators (WISA) specification is based on the IEEE 802.15.1 physical layer and targets factory automation WSNs with packet error rate less than 1 −9 and cycle time of 2ms [10].

Network Elements

WISA networks [11] can be deployed in cellular topology with up to three cells (Fig. 2). Each cell uses a different transmission frequency, and is composed of a base station and up to 120 end devices including sensors and/or actuators organized in a star topology. The end devices communicate wirelessly via standard Bluetooth transceivers, while the base station is equipped with a specific transceiver, which is able to receive up to four channels in parallel [12]. Additionally, the base stations exchange information with the network manager via wired fieldbus such as DeviceNet [13] and Modbus [14].

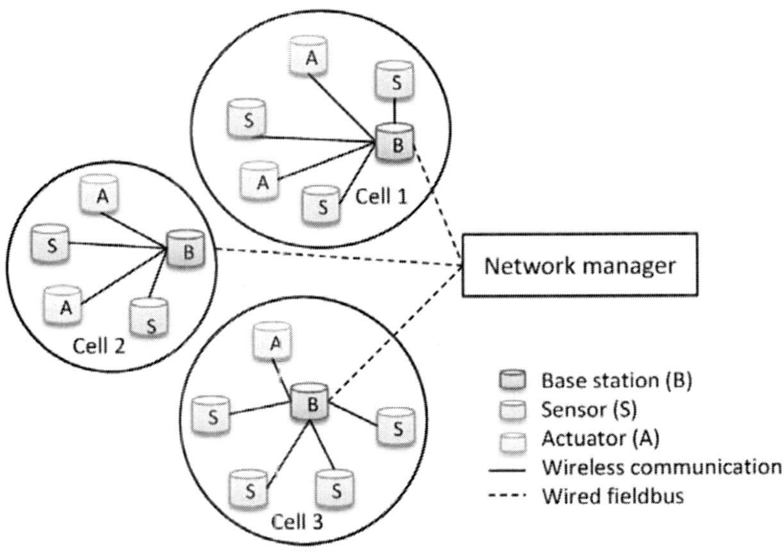

Figure 2. WISA network elements.

Architecture

The WISA architecture is limited to the physical and MAC layers, as sensors and actuators communicate exclusively with a central base station in a star topology within each cell. As mentioned previously, the WISA specification relies on the physical layer of the IEEE 802.15.1 standard operating in the 2.4 GHz frequency band at a data rate of 1 Mb/s. The WISA specification [15] is based on Time Division Multiple Access (TDMA) and Frequency Division Duplex (FDD), meaning that communication from base station to sensors, defined as downlink direction, and from sensors to base station, defined as uplink direction, occur at different frequencies. The downlink direction is exclusively reserved for the base station that remains continuously active and manages the TDMA scheme occurring in the uplink direction. The four uplink channels are divided into superframes of 2048 μs , which are composed of 30 timeslots able to support packets up to 64 bit length (Fig. 3). These time slots are allocated by the base station to each sensor willing to transmit data. In case of successful transmission of the sensor data, an acknowledgement is sent by the base station in the downlink. Otherwise the sensor retransmits the data in the next frame. To avoid interference and improve the reliability, frequency hopping is additionally applied after each superframe with a carrier spacing of 1 MHz.

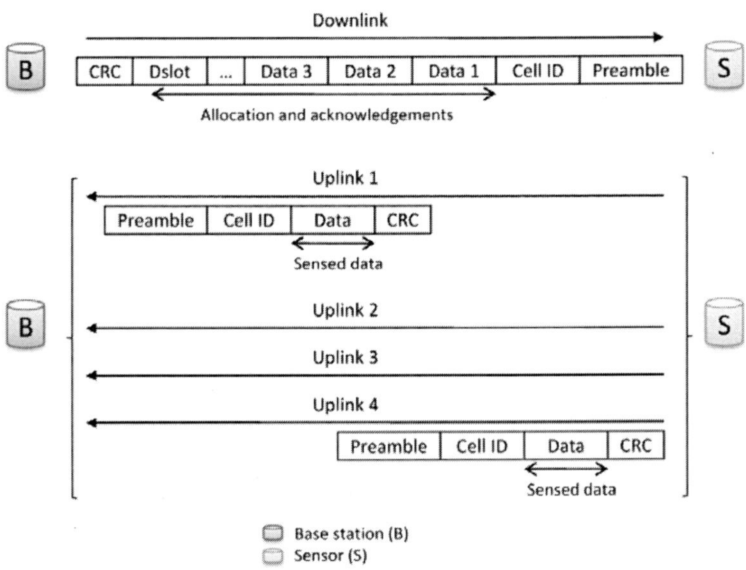

Figure 3. WISA superframe structure [16].

2.2. IEEE 802.15.4-based Standards

In comparison with Bluetooth®, the IEEE 802.15.4 standard presents lower data rates, but requires also lower energy budgets. According to [6], the standard is appropriate for infrequent exchanges of small packets, when power consumption is an important issue. The standard is therefore suited for process automation applications, where continuous production streams are monitored.

The IEEE 802.15.4 physical layer is common to all the following standards and operates in the 2.4 GHz frequency band as well as the 868 MHz and 915 MHz bands in Europe and North America, respectively. The 2.4 GHz frequency band is divided into 16 channels with a maximal data rate of 250 Kbits/s per channel and separated by a 5 MHz gap, while the 915 MHz band is divided into 10 channels with a maximal data rate of 40 Kbit/s each. The single channel in the 868 MHz frequency band presents a data rate of 20 Kbit/s. However, the effective data rates are smaller than the announced nominal values in reality, as mentioned in [6]. In addition to WISA, we have selected the WirelessHART, ISA100.11a, ZigBee, ZigBee PRO, and 802.15.4e Factory Automation MAC Layer technologies. These technologies address the complete protocol stack from the physical layer to the application layer (Table 1) and thus provide a complete system, except for the 802.15.4e FA MAC and the WISA technology that mainly focus on the data link layer.

Table 1. Scope of the selected technologies.

	WISA	WirelessHART	ISA100.11a	ZigBee	802.15.4e MAC
PHY/MAC layers	802.15.1-based	802.15.4-based	802.15.4-based	802.15.4-based	802.15.4-based
NET/TRANS APP layers	unspecified	specified	specified	specified	unspecified

Wireless HART

The HART Communication Foundation is an independent and not-for-profit organization that ensures the development of the HART Protocol. As technology owner and central authority, the foundation released the open WirelessHART™ standard in 2007, considered by [17] as the only released open wireless standard suitable for process measurement and control applications.

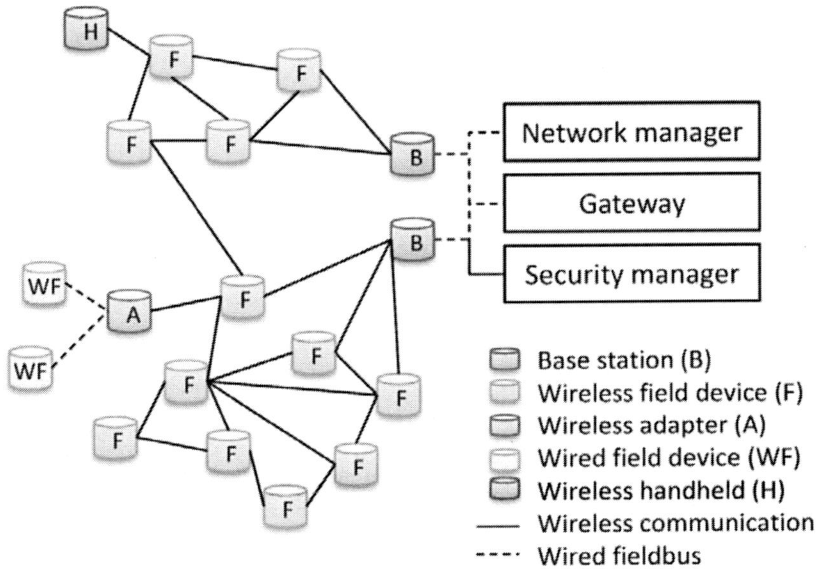

Figure 4. WirelessHART network elements [19].

Network Elements

WirelessHART networks are composed of different devices as illustrated in Fig. 4, including field devices, gateways, network and security managers. The field devices are organized in either star or mesh topology. However, the star topology is not recommended by [18]. The gateway is a bridge between the field device network and the host application. The gateway is configured by the network manager using HART commands and allows buffering large sensor data, event notifications, diagnostics, and command responses. In addition to the gateway configuration, the network manager configures the remaining devices and maintains the whole network. WirelessHART networks may include several network managers for redundancy reasons, however only one should be active at a time. The active network manager also schedules communication, manages routing tables, and monitors network health. In addition, the network manager can receive input from the host application and also queries the field devices about particular information via the gateway. The security manager collaborates with the network manager and prevents intrusion and attacks against the network by generating session keys, joint keys and network keys. Each security manager can collaborate with several network managers responsible for the key distribution to the concerned field devices. Additionally to these devices required by the standard, adapters and handheld devices may be added. The adapters connect HART devices to WirelessHART networks, whereas handheld devices configure and maintain WirelessHART compliant devices.

Architecture

The WirelessHART standard is presented by the HART Communication Foundation, as a reliable, secure, and robust standard. We consider successively the data link, network as well as transport and application layers that compose WirelessHART's protocol stack. At the data link layer, the WirelessHART standard coordinates and manages each device's transmission time by using TDMA with timeslots of 10 ms. Each time slot may be allocated to one source or may be shared between several sources using the Carrier Sense Multiple Access with Collision Avoidance (CSMA/CA) mechanism. In the former case, minimal latency can be reached, whereas the latter case supports efficient bandwidth utilization. Possible collisions due to multiple sources sending simultaneously are reduced by setting random back-off intervals for each source. During the timeslot assignment, the data to transmit are first prioritized by the network manager. The data originating from the network manager have the highest priority, followed by sensed data and event information. The timeslot allocation is then communicated by the network manager in the superframe to each device. At least one superframe (Fig. 5) is continuously repeated at fixed rate and further superframes can be added to support additional traffic [20]. The length of the superframes can be adapted to the needs of the application. However, the length is fixed once the superframe becomes active. In addition to the timeslot allocation, the network manager indicates the transmission channel in the superframe. This frequency hopping helps in reducing multi-path fading and interferences. Moreover, the faulty channels are eliminated by blacklisting.

The WirelessHART standard uses two possible routing protocols [22] at the network layer: graph and source routing. In graph routing, the network manager determines the different paths forming the graph. The paths are stored by each device and are then used to identify the next node to forward the packet. In case of source routing, the packet header contains the list of the devices from the source to the destination. In addition to the paths stored for the graph routing, each device maintains tables about its communication statistics and neighbor activities. Furthermore, the WirelessHART network layer offers broadcast, multicast and unicast transmissions.

At the transport layer, the WirelessHART standard supports connection-oriented as well as connectionless communication. The connection-oriented communications are set up for applications requiring a reliable transfer of data between the host application and the field device for example. The connection

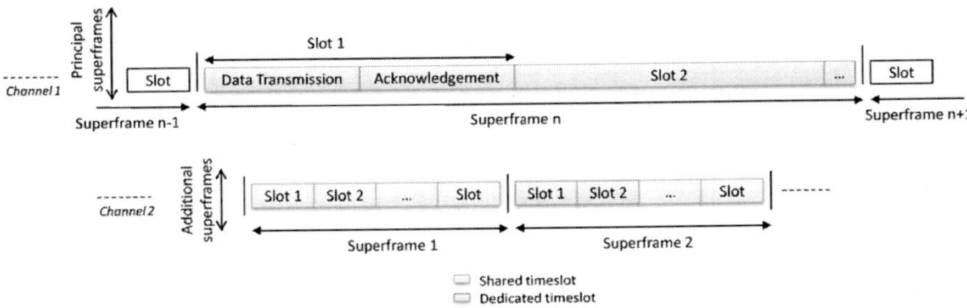

Figure 5. WirelessHART superframes [21].

set up starts by opening a dedicated port on the targeted field device with a specific HART command and the transmission data rate is then negotiated with the network manager, before the data transmission between the both entities can begin. Once all packets have been transmitted, the connection established between both entities is removed. The reliability of the data transfer is ensured by the acknowledgement and the order of the packets is maintained from the source to the destination. Nevertheless, these guarantees introduce additional overhead and a trade-off between reliability and overhead has to be found. For applications supporting out-of-order packet delivery, connectionless communication can be sufficient. Each datagram contains the full destination address and is routed independently from the others through the network [23]. Depending on the required level of reliability and the tolerance of packet loss, end-to-end acknowledgement can be introduced, but would also cause additional overhead. In connection-oriented and connectionless communication, the transport layer is also responsible for the segmentation of the data blocks at the source and their reassembling at the destination in order to allow transparent transfers for the upper layers. The WirelessHART application layer is based on the HART commands and extended by additional features allowing data publishing only when required for example.

ISA100.11a

The ISA100.11a-2009 standard [24] has been developed by the ISA100 standards committee, part of the non-profit International Society of Automation (ISA) organization, and approved by the ISA Standards and

Practices Board in September 2009. This first release focuses only on process applications that tolerate delays up to 100ms [12]. However, further releases addressing factory and building automation applications are expected in the next years. In parallel to the main ISA100.11a working group, different working groups address complementary issues including the compatibility of the ISA100.11a standard with existing wired and wireless standards.

Network Elements

ISA100.11a WSNs can be organized according to different topology schemes and are composed of field devices, gateway(s), and handheld device(s) as depicted in Fig. 6. Some of the field devices responsible for sensor data collection and actuator management can also provide routing functionalities. According to the standard, there is no limitation of the amount of subnets that form the network, and therefore the total amount of devices is not limited. However, the amount of devices per subnet is restricted by the addressing space to 30 000 devices. One or several gateways ensure the connection between the WSN and the user application. The gateways also support the interoperability with different standards such as WirelessHART by translating and tunneling information between the networks and could act as security and network managers. Moreover, handheld devices support device installation, configuration and maintenance.

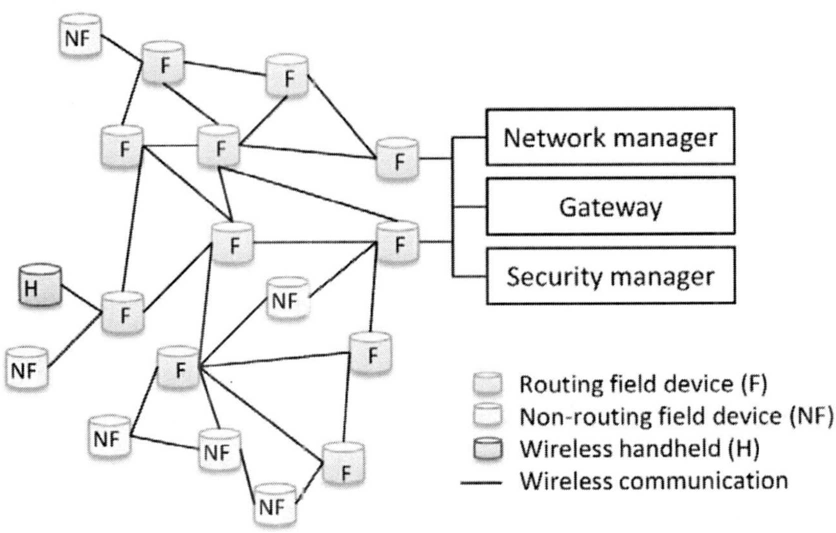

Figure 6. ISA100.11a network elements [25].

Architecture

The ISA100.11a standard addresses all OSI layers: physical, data link, network, transport and application layers. The physical layer is based on the IEEE 802.15.4 standard. However, additional requirements were defined. First of all, the ISA100.11a-2009 supports only the frequency band at 2.4 GHz; the lower sub-bands are not supported. In addition to the IEEE 802.15.4 standard, the ISA100.11a standard supports frequency hopping and also blacklisting that eliminates faulty frequency bands in order to improve the robustness against interferers. Additionally, the carrier sensing scheme can be disabled to reduce possible delay transmission.

The ISA100.11a data link layer is responsible for the management of the employed TDMA schemes by configuring the timeslot durations and managing the superframes. The configuration of timeslots can be done according to two different patterns: slotted channel hopping and slow channel hopping. The former scheme optimizes the bandwidth utilization and is adapted to energy-constrained routers, while the latter smooths the time synchronization requirements between neighbors by powering their receivers continuously during well-defined periods. The last scheme is therefore only adapted to routers offering unlimited energy budget. Both patterns can however be combined in a hybrid fashion by mixing superframes of both types. In addition to manage TDMA, the network manager assigns paths and links between the devices composing the WSNs. Each link is associated to one or multiple timeslots of a superframe and its type can be transmit and/or receive (Fig. 7). Information about the neighbors, the channel offset from the superframe hopping scheme as well as possible alternatives for the transmission and reception are also included. The ISA100.11a data link layer supports the previously described graph routing as well as source routing.

The network layer provides schemes for routing and QoS. It allows energy and bandwidth savings by mapping and translating the 128-bit addresses of the application endpoints to 16-bit short addresses used within subnets and vice versa. These savings can be increased by adapting the packet formats in function of the desired addressing, routing or QoS. Moreover, the network layer headers are compatible with the headers specifications conceived by the IETF 6LoWPAN working group and described in [26] in order to support future compatibility. Packet fragmentation and reassembly are also ensured at this layer. The packets can be routed at the backbone and the mesh levels, as defined in the standard. The first routing level is ensured by the data link layer, while the second is performed at the network layer by the end devices with routing capabilities. QoS is addressed in detail in the next section.

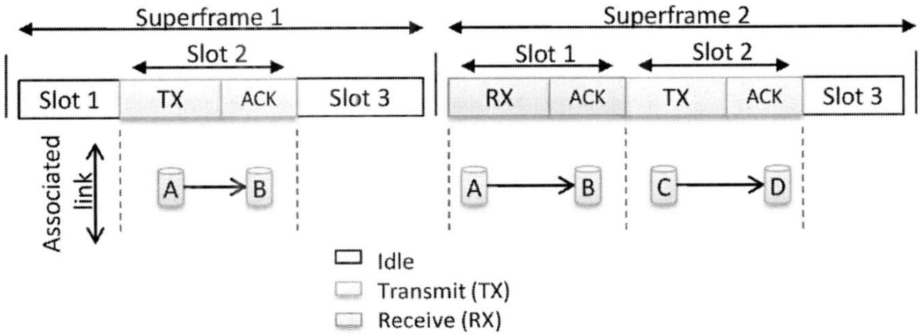

Figure 7. ISA100.11a superframes [24].

Depending on the level of reliability required by the application, the ISA100.11a transport layer can support end-to-end acknowledgements as well as unacknowledged communication. Additionally, flow control, segmentation and reassembly as well as security (see Section 4) are supported at transport layer, whereas the application layer ensures standard interoperability by using tunneling and native protocols at the gateways. The former carry protocols used in existing standards such as HART or FOUNDATION Fieldbus, while the latter provide efficient bandwidth utilization and therefore increase the battery lifetime.

ZigBee and ZigBee PRO

The ZigBee®standard, described in [27], was developed by the ZigBee Alliance and was originally designed for home automation. A new ZigBee PRO variant was released in 2007 to fulfill the industrial requirements. The ZigBee PRO standard is still based on the IEEE 802.15.4 physical and MAC layers and provides network and application layers with enhanced security features. However, the ZigBee PRO standard supports only frequency agility that consists of scanning available channels to determine the channel with the least interference, which is then selected and used by all ZigBee devices. Within the scope of this survey, we refer to both ZigBee and ZigBee PRO variants as ZigBee, except for the explicitly mentioned specificities.

Network Elements

ZigBee networks support hundreds of devices and should thus be suitable even for large deployments. They can be organized into star, tree or mesh topologies. The ZigBee standard is based on the two defined IEEE 802.15.4 device classes including Full-Function Device (FFD) and Reduced-Function Device (RFD) and proposes three different types of devices: ZigBee coordinator, ZigBee router and ZigBee end devices (Fig. 8). A unique FFD

ZigBee coordinator manages the network by supervising the network formation as well as information storage, and bridges it with others ZigBee networks. The ZigBee routers are complementary to the network manager and also FFD devices with additional routing capabilities, responsible for linking group of devices and supporting multi-hop communications. ZigBee end devices are either RFD or FFD. They transmit the collected sensor or actuator data to a unique FFD including router or coordinator functionality. Consequently, a FFD becomes the master of RFDs organized according to a star topology. Furthermore, the ZigBee specifications introduce a trust center to manage the keys and the end-to-end configuration. Only one center trusted by all devices should be active and be associated with all network devices.

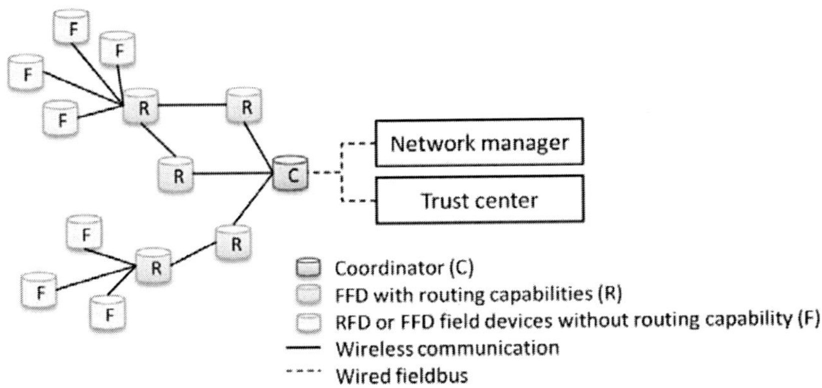

Figure 8. ZigBee network elements [28].

Architecture

The ZigBee stack is composed of the IEEE 802.15.4 physical and MAC layers as lower layers, and of the network and application layers specified by the ZigBee standard. After having set the selected common frequency for all devices, data transfers between ZigBee devices are possible. Two data transmission mechanisms are possible in ZigBee networks: with or without beacon. In the mode with beacon, the FFD sends a first beacon to synchronize all RFD sleeping phases and announces the superframe structure to manage the communication from end devices to the FFD. The first part of the superframe is slotted and CSMA/CA is used as channel access mechanism, while the second is composed of slots reserved for particular nodes by the network coordinator (Fig. 9). The FFD announces first the data transfer in the beacon to transfer data from the FFD to the RFD. Then, the concerned RFD must send a data request to the FFD to begin the data transmission. In case of FFD to FFD communication, the mechanism is similar, as one FFD acts as end device and is synchronized by the beacon originating from the second FFD. In the mode without beacon, no beacon and superframe are transmitted . The channel access is based on unslotted CSMA/CA. Each FFD coordinator remains continuously active to receive data coming from end devices during their limited active phase. RFDs send data requests to the FFD to receive data from the FFD. FFDs are permanently active and can thus communicate easily. In addition to transmission management, the MAC layer partially supports the admission of new devices in the network. The admission process starts by the scan procedure, during which the RFDs listen for beacon requests sent by a FFD. Request and acceptance notification are then exchanged at the MAC layer to complete the admission process. However, the decision to accept or reject a device is left to the security mechanisms supported by the upper layers and in case of acceptance, a 16-bit short address is assigned to the new device.

The network layer is specified by the ZigBee standard and is responsible for network formation, address assignment as well as routing over the ZigBee network. The network layer is complementary to the MAC layer and takes part in the join procedure by initiating a network discovery mechanism to detect surrounding ZigBee networks. After the selection of the network by the application layer, the network layer chooses a parent to attach the joining device and requests the MAC layer to begin an association procedure, where the network layer assigns the 16-bit address to the joining device. The ZigBee network layer employs the Ad hoc On Demand Distance Vector routing algorithm (AODV) as route discovery mechanism to manage routing in mesh networks.

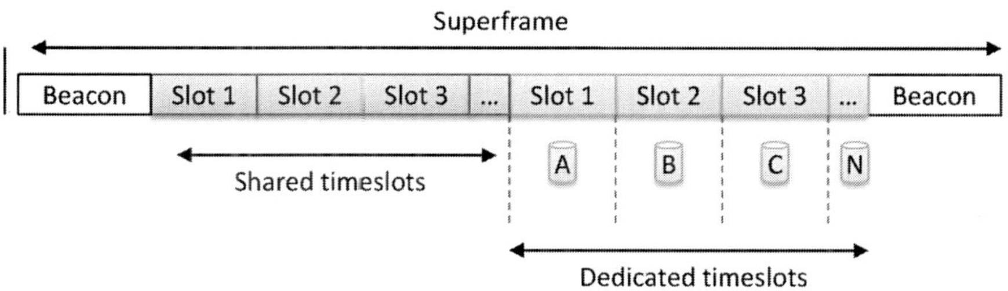

Figure 9. ZigBee superframe.

The ZigBee application layer proposes a framework for distributed application development and communication [27]. This application framework is composed of up to 240 Application Objects (APO). They consist of software units controlling dedicated device hardware and are disseminated over network devices. Each APO manages a set of variables and offers the possibility to set and read its values as well as report value changes. These functions are accessible by using the APO local number, which extends the device address. Additionally, the Application Sub Layer (APS) provides an interface to ensure security and data services between APO and ZigBee Device Objects (ZDO), which manage APO discovery services. Finally, application profiles described in the ZigBee specifications define formats and protocols for intra APO communication allowing the interoperability of ZigBee devices with the same application profile.

802.15.4e Factory Automation MAC Layer

The IEEE 802.15 Task Group 4e is currently developing a MAC layer [29] dedicated to factory automation and based on the IEEE 802.15.4 standard. The 802.15.4e Factory Automation MAC layer defines a deterministic TDMA communication scheme to fulfill the real-time requirement.

Network Elements

The network is composed of sensors and actuators organized in star topology around a gateway (Fig. 10). The network manager configures each end device via the gateway and allocates the dedicated time slots. After the configuration phase, sensor to gateway communication is unidirectional, whereas actuator/gateway communication is bidirectional.

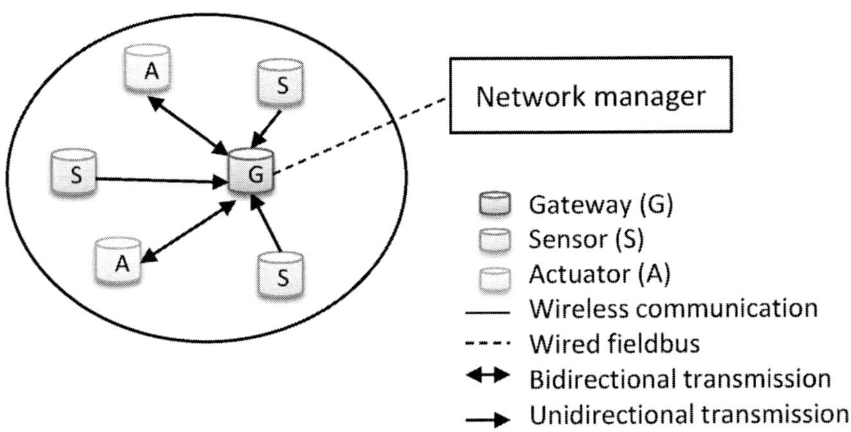

Figure 10. 802.15.4e Factory Automation MAC Layer network elements.

Architecture

The 802.15.4e Factory Automation MAC layer is based on the IEEE 802.15.4 physical layer and develops particular superframe formats as well as transmission modes to support deterministic TDMA [29]. The gateway supports three main transmission modes: discovery mode, configuration mode and online mode.

The discovery mode takes place during either network setup or joining procedure. The gateway sends superframes with beacons to indicate the discovery mode. When a device wanting to join the network receives such a beacon, it tries to access the transmission medium to send a Discover Response frame to the gateway with its current configuration parameters. The frame will be retransmitted by the device until the gateway receives it or changes its transmission mode.

During network setup or reconfiguration, the gateway is in a configuration mode and indicates this status in the superframe beacon. When the device receives the beacon and gets access to the transmission medium, it sends a Configuration Response frame to the gateway with its current configuration until the gateway receives it or changes its mode. As soon as the gateway receives the Configuration Response frame, it sends a Configuration Request frame with the new device configuration parameters and the device sends an acknowledgement in the next superframe.

In the online mode, devices can send data to the gateway in the timeslots allocated during the configuration mode and the gateway acknowledges the received data in the following superframe.

The superframes are sent to the end devices by the gateway and their structures depend on the current gateway transmission mode. The first slot is designed as the beacon slot (Fig. 11) and is common to all superframe structures. The end devices can detect the start of a new superframe at the reception of this first slot and synchronize themselves with it. Additionally, the beacon specifies the current transmission

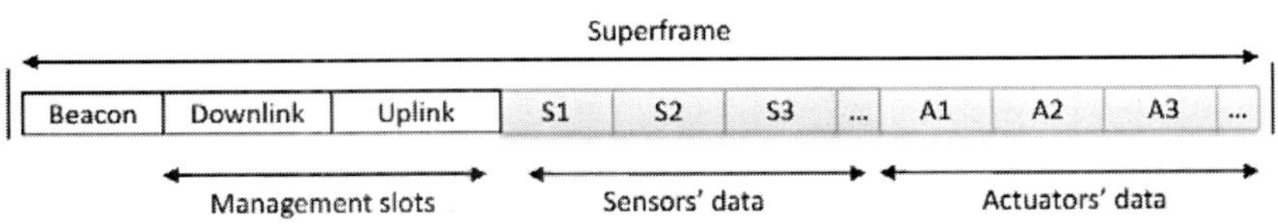

Figure 11. 802.15.4e FA MAC superframe structure [29].

mode and also acknowledgements for data transmitted in the previous superframe. In discovery, configuration and optionally online modes, the beacon is followed by up to two management time slots, which manage the bidirectional transmission between gateway and actuators. During online transmission mode, the next time slots are allocated to sensors. These timeslots can be either dedicated to a particular device or shared by a group of devices using CSMA/CA. In the first case, no addressing information is necessary, whereas the second case requires a simple addressing scheme. Then, actuator time slots are reserved in the superframe. The direction of the communication between actuators and gateway is indicated in the beacon, and each time slot can be either dedicated or shared.

3. QUALITY OF SERVICE

Quality of Service (QoS) refers to "the collective effect of service performance which determine the degree of satisfaction of a user of the service" [30]. More technically, QoS can be defined as the "well-defined and controllable behavior of a system with respect to quantitative parameters" [31]. Interesting parameters to monitor for computer networks can be delay, jitter, throughput, fairness, and packet losses for example. However, the requirements of industrial automation networks are different from usual computer networks. The harsh industrial operation environment may degrade the wireless communication performance due to path loss and shadowing, multi-path propagation, and interference [12]. Moreover, the traffic is mainly composed of short packets containing sensor measurements or actuator commands that need to be delivered timely, instead of large multimedia streams or interactive traffic that prevails in computer networks. We next identify the specific QoS requirements and evaluate the aforementioned selected specifications with respect to their compliance to these requirements. Open issues are identified and discussed based on the results of this evaluation.

3.1. Requirements

Industrial automation applications can be divided into two main categories of systems [32]: close-loop1 and open-loop systems2 as defined in [33]. The applications based on either close or open-loop systems have different QoS requirements, as close-loop systems monitor discrete operations, such as actuator control, and open-loop systems monitor continuous processes, such as cooking raw material.

Table 2. Industrial automation applications classified according to their real-time requirements [35].

	Real-time requirements			
	None	Soft	Hard	Isochronous
Cycle duration		greater than 10 ms	1 ms to 10 ms	250 μs to 1 ms
Jitter				less than 1ms
Application examples	Maintenance	Process automation	Control	Motion control
	Diagnosis	Data acquisition	Machine tools	

In close-loop applications, the sensors generate traffic that needs to be transmitted timely, reliably and accurately [33] to the control system, whereas in open-loop applications, real-time processing is typically not required, but the energy consumption is a crucial point. An additional classification [35] of the industrial applications refines these categories and is based on the degree of the required real-time guarantees. The classification is composed of four classes presenting increasing real-time requirements that are summarized in Table 2. We next focus on the support of real-time traffic and the reliability offered by the selected specifications.

3.2. Evaluation of the Selected Specifications

To evaluate the support of real-time transmission, the medium access control mechanisms including priority management schemes of the selected specifications are compared. Different diversity parameters including frequency and space diversity as well as protocol features such as acknowledgements are additionally considered in order to estimate the reliability provided by the different specifications.

Real-time Support

The selected specifications are mainly based on the IEEE 802.15.4 data link layer, except for the WISA specification relying on the IEEE 802.15.1 standard. As 802.15.4e FA MAC Layer and WISA only address the specifications of the physical and data link layers, the comparisons of the selected technologies are mainly based on the features of their two first layers. To complete this comparison, the 802.15.4 FA MAC needs to be evaluated in the context of a whole system. However, this aspect is outside the scope of this survey.

Medium Access Control Mechanisms

Each standard uses TDMA as medium access control mechanism (Table 3). However, TDMA is not the main medium access scheme in the ZigBee standard, which uses primarily CSMA/CA, but can also provide guaranteed timeslots in its beacon mode. Even if all specifications make use of TDMA access mechanisms, some differences between them can be observed. First of all, the timeslot length can be configured in the ISA100.11a, ZigBee and 802.15.4e FA MAC standards, while it is fixed in the others standards. Adapting the timeslot length to the needs of the applications allows taking into account the specific characteristics and optimizes the timeslot duration. Without such optimization, time can be lost between two successive slots if the slot is longer than the data to send. Configuring the length of the timeslots is thus an important feature to allow for optimizing the real-time support. However, all timeslots inside each superframe have the same length that limits a real adaptation to the requirements of individual applications, as only the overall traffic mix can be optimized. Another feature is the selection of the type of the timeslots, which can be either shared or dedicated. Most of the selected specifications offer both kinds of timeslots. Only the WISA specification uses exclusively dedicated timeslots. The choice between dedicated and shared timeslots is made difficult by the trade-off between real-time support and optimized utilization of the medium. Indeed, the dedicated timeslots are assigned to one particular device only. If this device wants to send commands or measurements, the data are transmitted immediately within the reserved timeslots. Otherwise the timeslot is lost and other devices wanting to transmit information may be constrained to wait longer. In case of shared slots, the devices have to compete to access the medium. The medium utilization is thus optimized because slots can be utilized by the stations in need of bandwidth, but the data transmission cannot always be immediate due to random back-off mechanisms for example.

Superframe Management

Furthermore, the WirelessHART, ISA100.11a and the ZigBee standards allows optimizing the transmission of superframes to fit the real-time communication constraints. The ISA100.11a standard supports insertion, removal and activation of superframes during the operating process, whereas additional superframes can be transmitted in parallel to the mandatory superframe in the WirelessHART standard. In the without beacon mode of the ZigBee standard, no superframe is transmitted and the devices do not have to wait for the next timeslot to transmit their data. However, this solution has also drawbacks. If one device has a great amount of data to send, the channel would be occupied for a long period of time and data requiring real-time transmission could not be send during this period.

Moreover, the time to get access to the channel may be longer than in case of dedicated slots, if all the devices have data to be sent. The management of dedicated and shared timeslots has therefore to be tailored to application characteristics and the traffic patterns.

Priority Management

In addition to medium access control, flow control with assignment of priority to the packets can be introduced to support real-time at higher level. A priority flag indicates to each device on the path if the arriving packet has to be transmitted without delay or can be buffered. The WirelessHART standard defines four main priority levels according to classes of data [36]. Control and configuration information as well as network diagnostic messages have the highest priority, followed by a second category composed of process data measurements and network statistic messages. The lowest priority is assigned to the fourth category, which includes packets reporting information about events and alarms. Additional classes of data are classified into the third category. In comparison, the ISA100.11a standard allows to prioritize the QoS contracts established between the devices and the system manager in addition to

Table 3. Features relevant for real-time operation.

	WISA	WirelessHART	ISA100.11a	ZigBee	802.15.4e MAC
TDMA	x	x	x	x[3]	x
CSMA/CA				x	
FDD	x				
Superframe	x	x	x	x[4]	x
Fixed timeslot duration	x	x			
Duration	64 and 128 μs	10 ms			
Shared timeslots		x	x	x[4]	x
Dedicated timeslots	x	x	x	x[4]	x
Superframe optimization		x	x		
Message-based priority		x	x		
Priority levels		4	2		
Contract-based priority			x		
Priority levels			4		

message priorities. The device wanting to transmit data first sends the required priorities in a contract proposal, as well as other additional parameters including reliability, periodicity and negotiability. The system manager replies to this request in a contract response and indicates the provided QoS for the requested communication. Depending on the traffic conditions, the system manager might not be able to provide the requested QoS. If the device indicates that the contract is negotiable, the system manager can propose another QoS level or postpone the contract to provide the requested QoS level. The contract priority is set during the contract establishment and concerns all messages exchanged during the contract duration. Depending on these priorities, the system manager manages the routing and the load balancing to provide the guarantees defined in the contract. Four levels of contract priority are available at network layer: network control, real time buffer, real time sequential and best effort queued. The first priority level can be used by the system manager to communicate critical information about the network management; whereas the second category is used for periodic data exchanges with buffer overwrite operations in case of fresher messages. The third class, real time sequential, is appropriate for applications requiring sequential and real-time data delivery like video or voice-based applications. The last level is adapted for client-server communications. Within the contract, message priority can also be assigned by setting one bit to either low or high. However, the contract priority takes precedence over the message priority.

Each standard supports real-time communication in a different way. While the WISA specification and the 802.15.4e FA MAC layer target applications with strong real-time requirements, the ZigBee and the WirelessHART standard are more adapted to applications with softer requirements. The recent ISA100.11a standard supports currently soft real-time requirements. More precisely, the evaluation of the selected technologies has shown that:

Dedicated timeslots are supported by all standards. Their utilization is a key feature to fulfill strong real-time requirements and is particularly appropriate, if the complete set of the sensors have a continuous stream of information send. To optimize the transmission, dedicated as well as shared timeslots can be combined to support fluctuations of the traffic load. Such hybrid modes can be envisaged in most of the considered technologies except for the WISA standard, which only support dedicated timeslots.

Tuning the timeslot length is a second key feature that allows to support application-specific traffic as well as optimize the real-time support. Except for the WISA and the WirelessHART standards, all standards offer this option. However, the timeslot length is common to all timeslots within a superframe, which limits the adaptation to variations of traffic as well as real-time requirements. Although this property may be useful for applications with various length of data to send, it would introduce additional overhead and complexity and may therefore unnecessary.

Superframe management is an additional means to maintain real-time communication in case of very high traffic load. However, the superframe optimization proposed by WirelessHART and ISA100.11a remains a minor contribution to the real-time support in comparison to the aforementioned dedicated timeslots and adaptable timeslot lengths.

Priority mechanisms are provided by the WirelessHART and the ISA100.11a standards. Both standards support message-based priority, while the ISA100.11a standard offers contract-based priority additionally. Although the proposed contracts are promising in terms of QoS guarantees, the overhead and the complexity introduced by their management as well as the additional resulting traffic may limit their practical feasibility. Even if message-based priorities may be less efficient than contracts, their utilization allows to reach a balance between overhead and efficiency and provides a complementary method to support real-time traffic.

Reliability Support

Industrial WSNs are located in spaces, where equipment moves, conditions change and interference perturbs the communication. Mechanisms, such as space and frequency diversity as well as acknowledgements (Table 4), are thus required to protect the wireless networks from these disturbances [37].

Space Diversity

Space diversity allows bypassing obstacles and interference by modifying the routing within networks organized in mesh topology. Such ability is however impossible with star topologies, as each device communicates exclusively and directly with a central coordinator. In case of obstacles that block the wireless communication, no alternative path is available and the communication cannot be established. Within the selected standards, the WISA specification and the 802.15.4e FA MAC standard are foreseen to be deployed in star topology only. Their protection against wireless channel obstructions and fluctuations is therefore reduced in comparison with the other standards. However, the maximal distance between the central coordinator and the sensors is shorter than between devices deployed in mesh topology. The probability of an obstacle breaking the communication is therefore low, if careful network planning is performed.

Table 4. Features relevant for robust/reliable features.

	WISA	WirelessHART	ISA100.11a	ZigBee	802.15.4e MAC
Mesh topology		x	x	x	
Channel agility				x[5]	?[6]
Channel hopping	x	x	x		?[6]
Channel blacklisting		x	x		?[6]
DDL acknowledgements	x	x	x	x	x
TL acknowledgements		x	x		
Automatic repeat request	x	x	x	x	?[6]

Frequency Diversity

In addition to space diversity, frequency diversity reduces the effects of the environment on the wireless communication by limiting the interference. Two frequency diversity schemes are possible: channel hopping and channel blacklisting. The channel hopping allows avoiding interference by changing the transmission frequency, while the devices maintain lists of frequencies to avoid due to their significant interference with the channel blacklisting scheme. The WirelessHART and the ISA100.11a standards use both of them and possess therefore an efficient response against interference. The WISA specification uses channel hopping, but does not blacklist faulty channels. In comparison, the ZigBee and 802.15.4e FA MAC standards do not provide any mechanism to avoid potential interference, which may lead to erroneous transmissions or even worse to a total transmission break down. However, the ZigBee PRO standard proposes an enhancement of the ZigBee standard by offering frequency agility. The available channels are scanned during the network setup phase to select a frequency without interference. Then, the frequency is shared by all ZigBee compliant devices and remains unchanged until the next network setup period. Even if the frequency agility may limit the effects of potential interference during the network formation, it can rapidly become inefficient in case of additional sources of interference.

Acknowledgement Management

Interferences or obstacles can also lead to packet loss. To ensure transmission reliability, acknowledgements (ACK) at Data Link Layer (DLL) are supported by all the selected standards, as well as end-to-end acknowledgements at Transport Layer (TL) for the WirelessHART and ISA100.11a standards. Each standard however manages its ACK mechanisms according to its transmission scheme. For example, the WISA specification transmits each ACK in the downlink channel, whereas each ACK is transmitted during the same timeslot as the received data in the WirelessHART standard. Once an ACK is missing, all standards7 support the automatic retransmission of the data.

The comparison of the aforementioned mechanisms summarized in Table 4 shows that:

The technologies targeting process automation applications provide a good protection against potential obstacles and node failures, as mesh topology is supported.

The WirelessHART and ISA100.11a standards offer the most complete set of mechanisms with channel blacklisting and frequency hopping to avoid perturbations caused by interference. While channel blacklisting allows saving time and energy by avoiding scanning channels previously identified as faulty, the obtained benefits are minor in comparison with the frequency hopping capability that is fundamental. Even if the ZigBee PRO version shows some improvements, the frequency diversity proposed by the ZigBee standards is insufficient to fulfill the strong reliability requirements of industrial applications.

All selected technologies use acknowledgements at data link layer and automatic repeat request to ensure reliable transmissions and identify packet losses. At transport layer level, only the WirelessHART and ISA100.11a standards support acknowledgements.

3.3. Open Issues

The previous evaluation has highlighted that the selected specifications differ with respect to the supported QoS requirements for industrial applications, particularly concerning real-time support and reliability.

Open Standard for Factory Automation

The first conclusion that can be drawn is that no standard provides currently an open solution to factory automation with strict real-time requirements. The WISA specification is dedicated to this kind of deployments, but it is based on the IEEE 802.15.1 standard that consumes more energy than the IEEE 802.15.4-based standards. Moreover, the WISA specification is proprietary, thus locking the user into a single vendor as well as the design and maintenance of a proprietary set of interfaces. Even if WISA successfully fulfills the requirements of factory automation, it does not support openness and interoperability. The 802.15.4 Factory

Automation MAC provides a promising perspective, but is still under development. Furthermore, the recently released ISA100.11a standard is expected to be enhanced by future addenda to make it fit to applications with stronger real-time requirements. As a consequence, additional research and development is necessary in order to obtain open and IEEE 802.15.4-based standards, which suit the needs of factory automation.

QoS Support in Heterogeneous Networks

In the process automation domain, the standards WirelessHART and ISA100.11a occupy the first positions and a potential convergence is analyzed by the ISA100.12 Working Group. WirelessHART devices are foreseen to be deployed within ISA100.11a-based WSNs, as illustrated in Fig. 12. In parallel to the ISA100.12 Working Group, the ISA100 Wireless Backhaul Backbone Network Working Group addresses potential interoperability between the ISA100.11a and the ZigBee-based standards. Additionally, the ISA100.11a standard is compatible with existing wired standards including HART, FOUNDATION Fieldbus, Modbus, and Profi bus. The ISA100.11a standard offers therefore a wide panel of compatible standards, which allows manufacturers to reuse existing devices. However, questions regarding the provided QoS capabilities among heterogeneous WSNs have to be raised. Indeed, such compatibility requires translation between the standards at the gateways, which may increase the end-to-end transmission delay of end systems belonging to different kinds of networks. Moreover, options implemented in one standard may not be supported or may have to be disabled to support interoperability. In such case, segments of the network that do not support the same priority scheme or the QoS contracts established in the ISA100.11a standard may ignore the indicated priority leading to a best-effort QoS. Thus, the efforts to support the QoS requirements would be wasted. A central network manager for both or more network segments based on different standards would be a possible solution because options, configuration settings and capabilities could be centrally managed. The network manager would be able to obtain a global view of the network and thus is able to take the appropriate decisions. However, a single central network manager remains a single point of failure that threatens the resilience of the network, if it is not deployed redundantly. In case of malfunctions, the whole network would be affected and could catastrophically fail. Therefore, the support of interoperability between the investigated standards remains an open issue. Although addressing the heterogeneity has a number of interesting research challenges, its practical feasibility may be limited by the technical complexity and financial costs incurred to adapt different technologies.

QoS Support in Multi-hop Networks

The surveyed standards dedicated to process automation can also be deployed in mesh topology, thus raising the question of QoS support over multi-hop routes. The WirelessHART and ISA100.11a standards propose solutions based on priority mechanisms, which are centrally managed by the network manager. While the standards provide the basic mechanisms to support differentiated QoS, the detailed specification to support operation in a multi-hop network is outside the scope of the standard. Although mechanisms that support QoS in multi-hop networks were developed for other applications, e.g. the Internet, their reuse is limited as they do not provide sufficiently strong real-time guarantees and are mainly adapted to multimedia streams, which differ from data streams in sensor networks. Further research is therefore mandatory to provide an exhaustive description of the required mechanisms to support QoS in multi-hop networks.

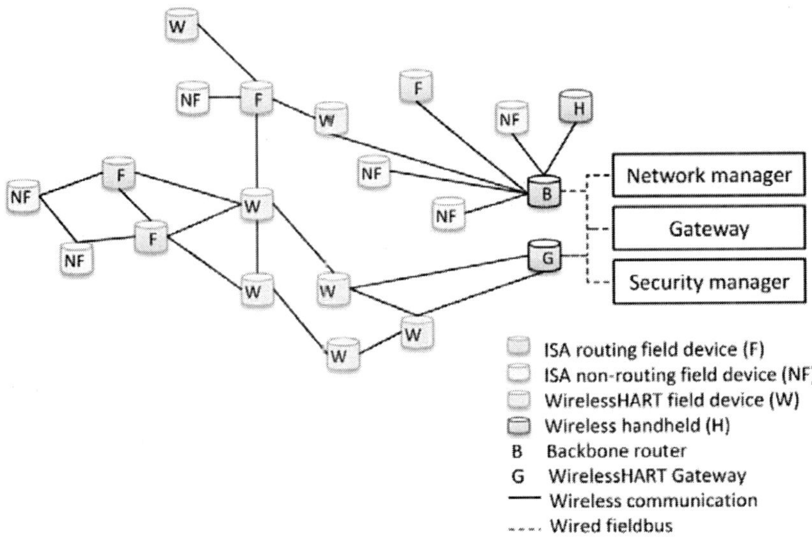

Figure 12. Interoperability of WirelessHART and ISA100.11.a networks [38].

4. SECURITY

In addition to the previously considered quantitative QoS parameters, security guarantees play an important role within industrial WSNs. Indeed, without any protection mechanism, the network could suffer from attacks or malfunctions that degrade the desired QoS by introducing additional delays, or not delivering correctly and timely the needed information. For example, these malfunctions can perturb the production chain, as one of the machines would move at an unexpected time or in the wrong direction. Such perturbations can have important consequences going from delayed and damaged production to broken equipment. Additional costs are not the only consequence; employee's lives can be endangered in the worst case; for example in case of explosions due to false temperature measurements in chemical industries. A threat analysis is conducted in this section followed by an evaluation of the selected standards in order to determine whether the WSNs are protected against the identified threats. Within the scope of this survey, only attackers located within the range of the WSNs and taking advantage of the wireless characteristics of the industrial networks are considered, as many methods like firewalls [12] are efficient to protect the networks against attacks coming from the outside. Moreover, attacks requiring physical capture of sensor platforms are excluded from this analysis.

4.1. Threat Analysis

To protect industrial WSNs efficiently against potential attackers, the following main security criteria have to be considered: confidentiality of information, integrity of information, authentication of communication peers and availability of information [35]. The first criterion ensures that the data access is restricted to authorized parties only, while the second ensures their protection against alteration and modifications by either malicious parties or the harsh surrounding environment. The authentication of communication peers allows guaranteeing that the exchanged data are coming from trusted devices. At last, the information availability ensures that data and services are accessible even in case of attacks. To perturb or even break down industrial WSNs, the attackers can therefore target one or several of the aforementioned criteria and conduct the appropriate attack(s).

Confidentiality of Information

The wireless nature of the communication between the sensors and devices eases these attacks, as there is no strict physical boundary of the transmission medium. An attacker located close to the network can thus easily eavesdrop the communication and threaten the confidentiality of the transmitted information. The content of the packets can be revealed to the attacker, who can benefit from stolen information like network configuration data to conduct further attacks. Eavesdropping can also be coupled with network monitoring to perform traffic analysis. The aim of this attack is to determine the responsibility of each sensor and identify the data sink for example. An analysis of the packet content is not mandatory to success; the amount of exchanged packets can be a sufficient clue [39]. However, an attack directed against the data sink can be very efficient, as the entire data set may be damaged or lost.

Integrity of Information

In addition to the confidentiality of the exchanged information, its integrity can be threatened by attackers adding additional fragments to the packets or manipulating the data. However, malicious behavior is not the only source of packet manipulations; errors due to the harsh industrial environment are also possible. The modifications of the packet content may cause misbehavior of the equipment and thus have inconvenient effects on the production, or even worse.

Authenticity of Communication Peers

Packet manipulation can be one sign that one or several malicious nodes have succeeded in integrating itself with the network. Such intrusions widely open the doors to further attacks like Sybil [40] and node replication [41] attacks. Both attacks profit from weaknesses of the authentication mechanisms to insert malicious nodes. In the former case, these nodes take illegitimately multiple identifiers; while in the latter, they capture and use existing device IDs. The identifier manipulation allows the attacker to modify the content of the traffic exchanged between the devices as well as control messages such as routing messages. These attacks can therefore be the basis of further routing attacks like wormhole [42] or black hole [43] attacks, where the attackers are able to disconnect part of the network or make it totally inoperable.

Availability of Information

Such routing attacks also threaten the last criterion, as the data may not be delivered timely or even at all and the information are therefore not available. Additional attacks can be conducted at different layers to disturb the availability of information. At the physical layer, jamming may cause interference at different frequencies in an intermittent or constant manner that make the communication impossible. Jamming may be caused by malicious attacks or unintentionally by surrounding equipment. To fight against malicious jamming, the physical protection of the industrial sites is one of the first measures to adopt. However, most of the industrial sites still accept external visitors. Even if their visits may be strictly controlled, attackers might benefit from security weaknesses to introduce jammers within the factory. Additionally, uninterrupted transmission of data by the attacker can generate collisions and force retransmissions at data link layer. The energy budget of the node decreases rapidly due to the retransmissions and the sensor is made inoperable. Additional energy consumption can also be caused by flooding the network with many connection requests at transport layer for example.

4.2. Evaluation of the Selected Standards

The selected standards are evaluated to determine how the current industrial WSNs are protected against the aforementioned threats. The set of considered standards is restricted to the WirelessHART, ISA100.11a and the ZigBee standards8.

Confidentiality of Information

The evaluation begins with data confidentiality including protection against eavesdropping and traffic analysis. The most efficient way to protect the industrial WSNs against eavesdropping is to encrypt the exchanged data. The WirelessHART, the ISA100.11a as well as the ZigBee standards use the 128-bit AES encryption [44] coupled with different keys depending on the layer of encryption. For example, WirelessHART uses the session key to encrypt the message at transport layer, while link and network keys are used at data link layer and at network layer respectively in the ZigBee standard. As mentioned by [45], AES remains an efficient mechanism to keep the data secret. Moreover, its efficiency is increased by the utilization of keys with short lifetime and unique for each device such as the session key used in WirelessHART. Eavesdropping is consequently made difficult or even impossible in networks running the three considered standards. However, the confidentiality is not ensured at all layers. For example, even if the packets are encrypted at transport layer, header and payload of packets sent at network layer are transmitted unencrypted in the WirelessHART standard [45]. An eavesdropper can therefore discover the crucial information, such as source and destination addresses that are contained in the network header, and perform traffic analysis easily afterwards. Nonetheless, the traffic analysis attack can also be performed without any message decryption [39].

Integrity of Information

The selected industrial standards benefit from the security mechanisms included in the IEEE 802.15.4 standard that ensure data integrity at data link layer. An additional Message Integrity Code (MIC) is inserted at the queue of the data to protect. The data are signed and the receiver is able to determine whether the data have been tampered with or not. Data integrity protection can be provided in complement with encryption by using the enhanced combined encryption and authentication block cipher mode (CCM*). Depending on the desired security level, the length of the MIC can be set to 32, 64 or 128 bits. The longer the code is, the higher the integrity protection is, but also the greater the overhead is. The length should therefore be selected carefully.

Message and Device Authenticity

In addition to provide hop-to-hop data integrity, the MIC allows to authenticate the packets by using secret symmetric keys known by both sender and receiver. For example, the shared network key and the unique session key are used in WirelessHART to authenticate the messages at data link and network layers respectively [45]. The authenticated packets are thus recognized as originated by authorized members of the network. However, before message authentication can be performed, each device must be first authenticated during the join procedure. Even if the name of the keys may vary between the standards, their functions are similar. Devices willing to participate to the network exchange join requests and join responses with the network manager and use public key and asymmetric private key kept inside the joining devices. These keys are used for computing the data link and network MICs respectively and may be either preloaded in the devices at the factory or distributed by a unique trusted center, which maintains and updates the security keys. In WirelessHART, an additional join key is used to encrypt the join request. Once the device is recognized as authorized member, it can exchange authenticated data with the other members. As each standard is based on a central entity responsible for the network management and keeping tracks of the participating devices, the probability is very low that the attacks such as Sybil and node replication attacks may be performed successfully. For example, in WirelessHART, the network manager links each device with a unique identity [45]. The identification is completed by a list of unique IDs maintained at the gateways. The network manager identifier and the gateway ID are used conjointly with the session key to maintain sessions between the device and the network manager as well as the gateway respectively. Devices claiming the same identity as an existing one or sharing multiple identities would be immediately discovered, as these would already be listed.

Availability of Information

The last threat to be evaluated is the availability of information. First of all, the information availability can be threatened by jamming according to different patterns. In case of continuous jamming with one or several jammed frequencies, channel blacklisting provides an efficient solution, as the jammed channels are eliminated from the set of communication frequencies. In case of intermittent jamming, frequency hopping provides good results and allows keeping sufficient levels of information availability. With both frequency hopping and channel blacklisting features, the WirelessHART and ISA100.11a standards provide therefore a better protection against jamming than the ZigBee PRO standard that only offers frequency agility. At network layer, attacks modifying the routing scheme can be avoided by the authentication mechanisms, as devices would only be able to route packets, if they have been previously identified as reliable and authorized to take part to the WSN. Nonetheless, the selected standards do not provide dedicated mechanisms to avoid the generation of collisions by a malicious source transmitting continuously data, as well as solutions against flooding of connection requests at transport layer.

To summarize, we have shown that:

- Eavesdropping is made difficult or even impossible, but confidentiality is not addressed at all layers.
- Traffic analysis is still possible.
- The information integrity is sufficiently ensured.
- The probability of successful Sybil and node replication attacks are limited.
- The frequency diversity and agility is sufficient to protect the network against intermittent jamming.
- The current mechanisms do not provide protection means against malicious sources transmitting continuously or performing higher layer attacks such as flooding of connection requests at transport layer.

4.3. Open Issues

The selected standards are resistant against most of the considered attacks except for continuous jamming at all frequencies, collision and flooding attacks as well as traffic analysis. However, solutions against such particular kind of jamming and collision attacks are particularly difficult to find because the data transmission is made impossible in both cases. The only solution would necessitate human interventions to eliminate the interference source(s), as soon as a long-term communication breakdown is detected. Flooding of connection request is more delicate to solve, as regular nodes must still be able to send such request in order to join the network and authenticate themselves. Filtering the connection requests by a list of devices susceptible to join the network would not solve the problem, as the network manager would have to receive the requests in order to determine the sender identity. A solution remains therefore to be found. Even if payload encryption provides a partial solution to traffic analysis, it is not sufficient. Indeed, an analysis of the routing paths can be sufficient to determine the traffic scheme. Countermeasures to this analysis could be to insert fake packets and/or randomly distribute the traffic. However, the routing would not be optimized and the network performances including end-to-end delay would be degraded. Moreover, insertion of additional packets would drag the energy budget of the nodes down.

The existing security mechanisms are theoretically sufficient to cover the main attacks. However, the standards only provide specifications and leave design and implementation to the users. For example, the roles of each security key are described, but their management scheme has to be developed by the users, according to their requirements and respecting the standard specifications. To the best of our knowledge, Raza et al. are the first to have proposed a design and an implementation of a security manager adapted to the WirelessHART standard in [47]. The investigation of similar proof of concept for additional standards e.g. the promising ISA100.11a may be an interesting track to follow in order to provide already evaluated implementations that the users can tune to meet their requirements.

5. CONCLUSIONS

Within the scope of this article, we have provided a detailed survey on WSN standards dedicated to industrial automation networks. The standardization efforts are ongoing and targeting different application areas such as factory automation or process automation. We have focused on the IEEE 802.15.1 and IEEE 802.15.4 standard families, which have been adapted to industrial applications in need of short-range communication with high data rate, and energy-aware applications requiring larger coverage, respectively. We have selected the WirelessHART, ISA100.11a, ZigBee and 802.15.4e Factory Automation MAC standards among the IEEE 802.15.4 standard families and the WISA specification among the IEEE 802.15.1-based standards. Except for the WISA and 802.15.4e Factory Automation MAC, all the selected standards target mainly process automation applications. An overview of each standard has been provided with particular focus on the network elements and the features of the protocol stack.

In the next step, we have identified several QoS requirements posed by industrial applications, such as support of real-time communications as well as highly reliable communications. The selected standards have been evaluated to determine how real-time and reliability requirements are supported. The results have revealed that no officially released and open standard is currently able to fulfill the strong real-time requirements of the factory automation domain. Moreover, the questions of QoS provisioning over heterogeneous networks as well as over multi-hop routes in homogeneous networks have been raised.

We have then focused on security issues of the surveyed standards by identifying potential attacks that could threaten the industrial WSNs and affect their operation. The standards have also been evaluated to determine if the proposed security mechanisms are sufficient to protect the WSNs against the derived threats. The evaluation has shown that the standards are resistant against most of the investigated threats, except for continuous jamming at all frequencies, collision attacks and flooding of connection requests. Moreover, we have pointed out that the design and the implementation of the security managers are left to the users or implementers of the standard. Here, the detailed operation of such security and network managers and the corresponding protocol mechanisms are an interesting area for further research.

We conclude that the selected standards fulfill almost completely the identified QoS and security requirements as long as they operate in single-hop mode. However, some aspects that are of high interest for the domain of industrial automation, including multi-hop operation and support of QoS and security over heterogeneous network segments, need further research.

ACKNOWLEDGEMENTS

This work was supported by CASED (www.cased.de) and Siemens AG Corporate Technology. The authors would like to thank the anonymous reviewers for their valuable comments.

REFERENCES

1. Geng, H. Manufacturing Engineering Handbook; McGraw-Hill Professional: New York, NY, USA, 2004.
2. Shell, R.L.; Hall, E.L. Handbook of Industrial Automation; Marcel Dekker: New York, NY, USA, 2000.
3. Low, K.S.; Win, W.N.N.; Er, M.J. Wireless Sensor Networks for Industrial Environments. In Proceedings of the International Conference on Computational Intelligence for Modelling, Control

and Automation and International Conference on Intelligent Agents, Web Technologies and Internet Commerce (CIMCA-IAWTIC), Vienna, Austria, November 2005.
4. Mukhopadhyay, S.C.; Huang, Y.M. Sensors: Advancements in Modeling, Design Issues, Fabrication and Practical Applications; Springer-Verlag: Heidelberg, Germany, 2008.
5. IEEE Computer Society. IEEE Standard for Information Technology, Telecommunications and Information Exchange between Systems, Local and Metropolitan Area Networks, Specific Requirements, Part 11: Wireless LAN Medium Access Control (MAC) and Physical Layer (PHY) Specifications. 2007.
6. Willig, A.; Matheus, K.; Wolisz, A. Wireless Technology in Industrial Networks. Proceedings of the IEEE 2005, 93, 1130–1151.
7. IEEE Computer Society. IEEE Standard for Information Technology, Telecommunications and Information Exchange between Systems, Local and Metropolitan Area Networks, Specific Requirements, Part 15.1: Wireless Medium Access Control (MAC) and Physical Layer (PHY) Specifications for Wireless Personal Area Networks (WPANS). 2002.
8. IEEE Computer Society. IEEE Standard for Information Technology, Telecommunications and Information Exchange between Systems, Local and Metropolitan Area Networks, Specific Requirements, Part 15.4: Wireless Medium Access Control (MAC) and Physical Layer (PHY) Specifications for Low Rate Wireless Personal Area Networks (LR-WPANs). 2007.
9. Scheible, G.; Dzung, D.; Endresen, J.; Frey, J.E. Unplugged but Connected - Design and Implementation of a Truly Wireless Real-time Sensor/Actuator Interface. IEEE Industrial Electronics Magazine 2007, 1, 25–34.
10. ABB. Reliable Factory Automation: Wireless Automation with WISA. http://www.abb.com (accessed on November 2009).
11. Steigmann, R.; Endresen, J. Introduction to WISA and WPS. http://www.abb.com (accessed on November 2009).
12. Willig, A. Recent and Emerging Topics in Wireless Industrial Communications: A Selection. IEEE Transactions on Industrial Informatics 2008, 4, 102–124.
13. ODVA. The DeviceNet Specification, Common Industrial Protocol (CIP) Specification. http://www.odva.org (accessed on November 2009).
14. Modbus Organization. Modbus Application Protocol Specification. http://www.modbus.org (accessed on November 2009).
15. Dzung, D.; Apneseth, C.; Endresen, J.; Frey, J.E. Design and Implementation of a Real-time Wireless Sensor/Actuator

Communication System. In Proceedings of the IEEE Conference on Emerging Technologies and Factory Automation (ETFA), Catania, Italy, September 2005.

16. IEEE 802.15.1 Standard. http://www.ieee802.org/15/pub/TG1.html (accessed on December 2009).
17. De Biasi, M.; Snickars, C.; Landernäs, K.; Isaksson, A.J. Simulation of Process Control with WirelessHART Networks Subject to Packet Losses. In Proceedings of the Conference on Automation Science and Engineering (CASE), Washington, DC, USA, August 2008.
18. Lennvall, T.; Svensson, S.; Hekland, F. A Comparison of WirelessHART and ZigBee for Industrial Applications. In Proceedings of the IEEE International Workshop on Factory Communication Systems (WFCS), Dresden, Germany, May 2008.
19. Griessmann, J.L. WirelessHART, an Overview. http://www.hartcomm.org (accessed on December 2009).
20. HART Communication Foundation. WirelessHART™ Technical Data Sheet. http://www.hartcomm.org (accessed on December 2009).
21. HART Communication Foundation. TDMA Data Link Layer Specification. http://www.hartcomm.org (accessed on December 2009).
22. Song, J.; Han, S.; Mok, A.K.; Chen, D.; Lucas, M.; Nixon, M. WirelessHART: Applying Wireless Technology in Real-Time Industrial Process Control. In Proceedings of the IEEE Real-Time and Embedded Technology and Applications Symposium (RTAS), St. Louis, MO, USA, April 2008.
23. Tanenbaum, A.S. Computer Networks; Prentice-Hall: Upper Saddle River, NJ, USA, 2002.
24. International Society of Automation. ISA-100.11a-2009,Wireless Systems for Industrial Automation: Process Control and Related Applications. http://www.isa.org (accessed on November 2009).
25. International Society of Automation. ISA100.11a Status. http://www.isa.org (accessed on December 2009).
26. Montenegro, G.; Kushalnagar, N.; Hui, J.; Culler, D. RFC 4944: Transmission of IPv6 Packets over IEEE 802.15.4 Networks. http://www.ietf.org/rfc/rfc4944.txt (accessed on November 2009).
27. Baronti, P.; Pillai, P.; Chook, V.W.; Chessa, S.; Gotta, A.; Hu, Y.F. Wireless Sensor Networks: A Survey on the State of the Art and the 802.15.4 and ZigBee Standards. Computer Communications 2007, 30, 1655–1695.
28. Galeev, M. Home Networking with ZigBee. http://www.media.mit.edu (accessed on December 2009).

29. Winkel, L.; Bahr, M.; Vicari, N. 15-08-0572-00-004e Proposal for Factory Automation. http://www.ieee802.org/15/pub/TG4e.html (accessed on December 2009).
30. International Telecommunication Union. E.800 - Terms and Definition Related to Quality of Service and Network Performance Including Dependability. http://www.itu.int (accessed on November 2009).
31. Schmitt, J. Heterogeneous Network Quality of Service Systems; Kluwer Academic Publishers: Norwell, MA, USA, 2001.
32. Mathiesen, M.; Thonet, G.; Aakwaag, N. Wireless Ad-hoc Networks for Industrial Automation: Current Trends and Future Prospects. In Proceedings of the IFAC World Congress, Prague, Czech Republic, July 2005.
33. DiStefano, J.J.; Stubberud, A.R.; Williams, I.J. Schaum's Outline of Feedback and Control Systems; McGraw-Hill Professional: New York, NY, USA, 1994.
34. Liu, X.; Goldsmith, A. Wireless Communication Tradeoffs in Distributed Control. In Proceedings of the IEEE Conference on Decision and Control (CDC), Maui, HI, USA, December 2003.
35. Neumann, P. Communication in Industrial Automation: What is going on? Control Engineering Practice 2007, 15, 1332–1347.
36. Kim, A.N.; Hekland, F.; Petersen, S.; Doyle, P. When HART goes Wireless: Understanding and Implementing the WirelessHART Standard. In Proceedings of the IEEE International Conference on Emerging Technologies and Factory Automation (ETFA), Hamburg, Germany, September 2008.
37. HART Communication Foundation. Why WirelessHART? The Right Standard at the Right Time. http://www.hartcomm.org (accessed on December 2009).
38. Sereiko, P. The ISA100 Standard: Characteristics and Benefits of the Standard, Latest Developments and Progress. http://www.isa.org (accessed on December 2009).
39. Walters, J.; Liang, Z.; Shi, W.; Chaudhary, V. Wireless Sensor Network Security: A Survey. Security in Distributed, Grid, Mobile, and Pervasive Computing 2007, 367–405.
40. Newsome, J.; Shi, E.; Song, D.; Perrig, A. The Sybil Attack in Sensor Networks: Analysis & Defenses. In Proceedings of the International Symposium on Information Processing in Sensor Networks (IPSN), Berkeley, CA, USA, April 2004.
41. Parno, B.; Perrig, A.; Gligor, V. Distributed Detection of Node Replication Attacks in Sensor Networks. In Proceedings of the IEEE Symposium on Security and Privacy (S&P), Oakland, CA, USA, May 2005.

42. Hu, Y.; Perrig, A.; Johnson, D. Packet Leashes: a Defense against Wormhole Attacks in Wireless Networks. In Proceedings of the IEEE Conference on Computer Communications (INFOCOM), San Francisco, CA, April 2003.
43. Karlof, C.; Wagner, D. Secure Routing in Wireless Sensor Networks: Attacks and Countermeasures. Ad Hoc Networks 2003, 1, 293–315.
44. Schneier, B. Applied Cryptography, 2nd ed.; John Wiley & Sons: Hoboken, NJ, USA, 1996.
45. Raza, S.; Slabbert, A.; Voigt, T.; Landernäs, K. Security Considerations for the WirelessHART Protocol. In Proceedings of the IEEE International Conference on Emerging Technologies and Factory Automation (ETFA), Mallorca, Spain, September, 2009.
46. Phan, R. Impossible Differential Cryptanalysis of 7-round Advanced Encryption Standard (AES). Information Processing Letters 2004, 91, 33–38.
47. Raza, S.; Voigt, T.; Slabbert, A.; Landernäs, K. Design and Implementation of a Security Manager for WirelessHART Networks. In Proceedings of the IEEE International Conference on Mobile Adhoc and Sensor Systems (MASS), Macau, China, October 2009.

8
CHAPTER
RECONFIGURABLE MICROSTRIP DOUBLE-DIPOLE ANTENNAS FOR PERSONAL WIRELESS COMMUNICATIONS

Abdelnasser Eldek, Abubaker Abdallah, Mahmoud Manzoul

ABSTRACT

In this paper, a number of reconfigurable antennas are designed to support multiband and wideband wireless applications in different frequency bands in the range between 1 GHz and 6 GHz. The proposed designs consist of microstrip-fed two dipoles through simplified balun. By controlling the lengths of the dipoles by connecting or disconnecting slots, we are able to change the antenna operating frequency as well as the operation mode. Two modes are supported by all of the proposed antennas: Multiband and Broadband modes.

Keywords: Reconfigurable Antennas, Dipole, Wideband, Multiband

1. INTRODUCTION

As communication devices become smaller due to greater integration of electronics, the antenna becomes a significantly larger part of the overall package volume. This results in a demand for similar reductions in antenna size. However, reducing antenna size without significantly impacting gain and efficiency is a challenging task. As integration increases, a single antenna is often required to support two or more of the many wireless services across a broad frequency range. Multiband, wideband and reconfigurable antennas are being developed to meet this need [1]. Multi-band and wideband antennas require complicated filters with stringent requirements to improve their out-of-band noise-rejection performance. These filters, moreover, are generally bulky and expensive. In contrast, frequency-reconfigurable antennas have great potential for reducing production cost and offer better out-of-band noise rejection [2]. In addition, compared to multiband and broadband antennas, reconfigurable antennas offer the advantages of compact size, similar radiation pattern for all designed frequency bands, efficient use of electromagnetic spectrum and frequency selectivity useful for reducing the adverse effects of co-site interference and jamming [3]. Dual frequency reconfigurable microstrip antennas can offer additional advantages of frequency reuse for doubling the system capability and polarization diversity for good performance of reception and transmission or to integrate the receiving and transmitting functions into one antenna for reducing the antenna size [4,5].

Microstrip antennas are attractive candidates for reconfigurable antenna technology because of their inherent low profile, light weight, low cost, and ease of fabrication and installation [2,6-13]. Number of attractive reconfigurable antenna designs were presented in [2] with the objective of reduction of the number of MEMs utilizing antenna symmetry to decrease the cost. The concept of a frequency-reconfigurable rectangular ring slot antennas fed by slotlines or CPW is presented in [7]. A reconfigurable patch antenna is obtained in [8] by inserting slits at the non-radiating edges of the patch. A reconfigurable Yagi antenna is presented in [9] to operate at 2.4 and 5.78 GHz for wireless communications. A reconfigurable slot dipole antenna was presented in [10] in the X-band. Patch antenna with polarization diversity using switchable L-shaped slots is presented in [11]. A reconfigurable dipole antenna with harmonic trap was presented in [12] where the operating frequency is controlled by a reconfigurable dipole length, while the higher-order modes are eliminated using a reconfigurable l/4 open circuit stub. A compact multiband antenna has been redesigned to be reconfigurable with embedded switches to allow selective addressing of these bands [13].

In this paper, three microstrip-fed double-dipole antenna designs [14-16] to support multi-band, wideband, and single-band operation modes are presented for personal wireless communications. The reconfiguration for the microstrip antennas is carried out by connecting or disconnecting (switching on or off) appropriate rectangular strips to model the MEM switches. The shape and number of these strips along with the switching state and position is used to control the operating frequency of the antenna without changing the feeding mechanism. The antennas' geometries and their results will be explained in the next two sections. The full-wave electromagnetic simulations and analysis for the presented antenna are performed using the commercial computer software package Ansoft High Frequency Structure Simulator (HFSS) [17], which is based on the finite element method. Measurements of the return loss and radiation patterns are also conducted to verify the simulation results and demonstrate the feasibility of the proposed configurations.

2. ANTENNAS' GEOMETRIES AND PARAMETRIC STUDY

Three double-dipole antennas were designed for different frequency bands and depicted in Figure 1. The double-dipole antenna consists of two dipoles with similar or different lengths depending on the required operating frequencies. A simplified balun is used to feed the dipoles. In each dipole, the left half is printed on the top substrate layer and fed by a microstrip line, while the right half is printed on the bottom layer and connected to the truncated ground plane.

The length of the dipole is approximately half the wavelength at the operating frequency, while its distance from the truncated ground plane is quarter wavelength for the best performance. Thus, if the antenna is required to operate at a lower frequency (f_L) and a higher frequency (f_H), then the shorter dipole will be closer to the ground plane. In this case, the antenna will operate at multiple bands. A broadband design can be achieved by making the longer and the shorter dipoles to resonate at closer frequencies. This can be achieved by reversing the positions of the two dipoles. With the contribution of the coplanar strip (CPS) line length to the antenna length, the difference between their lengths become small, and the two operating bands will be close and can merge in one relatively wide frequency band.

The next step is to merge these two modes in one so that the length of the closer dipole is controlled by the lower operating frequency in the wideband mode, and the length of the second dipole is controlled by the higher operating frequency in the multiband mode. Next, a proper matching circuit between the two dipoles is designed to eliminate the mutual effect between the two modes. Therefore, the shape of the CPS connecting the two dipoles is modified as needed by the design to obtain the required modes. Finally, when adding the ground plane, it will have an effect on the performance; therefore, the last step is to shape the ground plane to reduce this effect as shown in Figure 1. A smooth transition from the microstrip line mode to the two CPS line mode is used for this purpose.

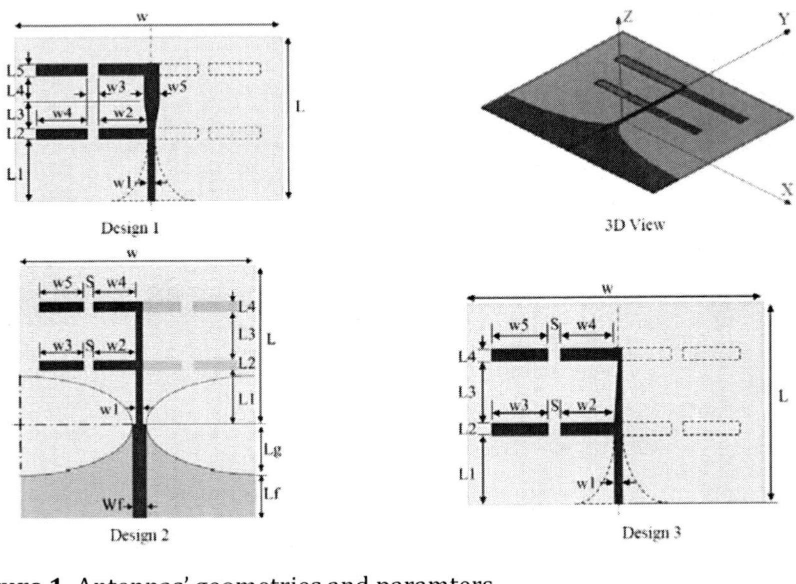

Figure 1. Antennas' geometries and paramters.

Parametric study is performed on each antenna and depicted in Figures 2 to 11 for Design 1. This study is done in the multiband mode to know the effect of each parameter, and select the best dimensions, and tune the antenna to obtain the best bandwidth in the broadband mode without affecting the multiband mode. The dimensions of the three designs resulted from the parametric study are shown in Table 1.

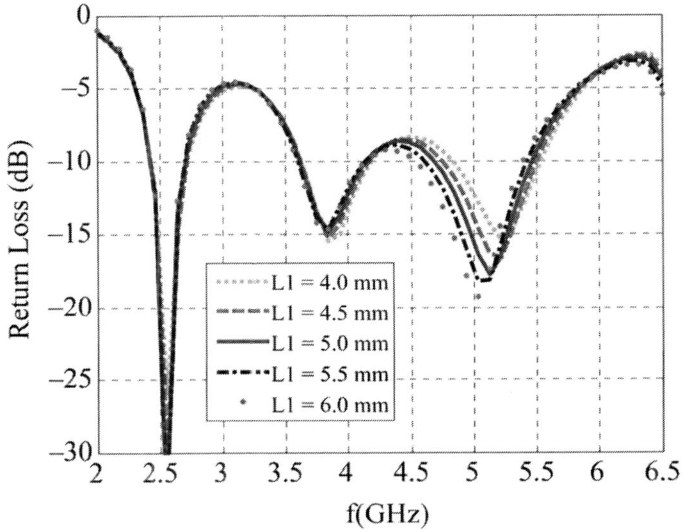

Figure 2. Effect of L1.

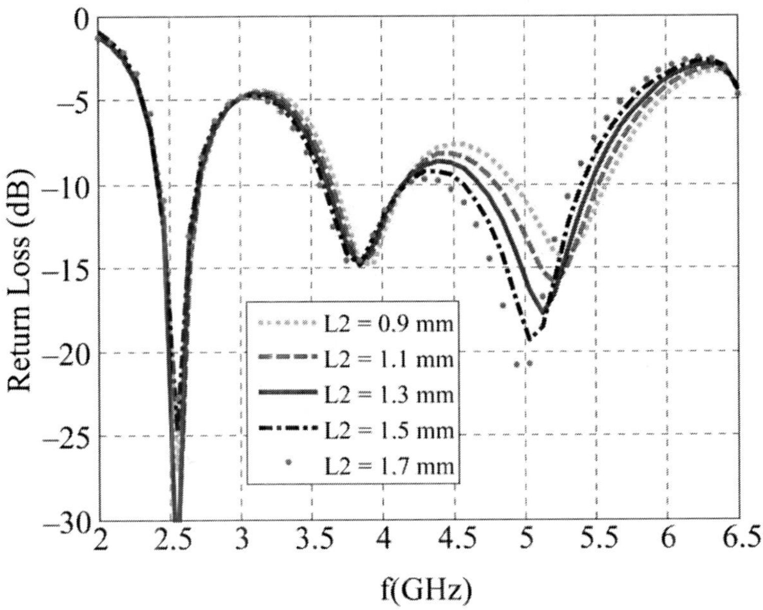

Figure 3. Effect of L2.

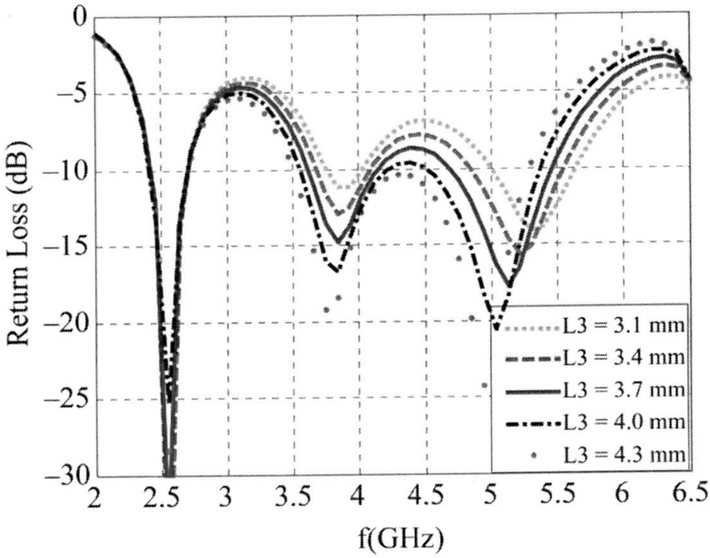

Figure 4. Effect of L3.

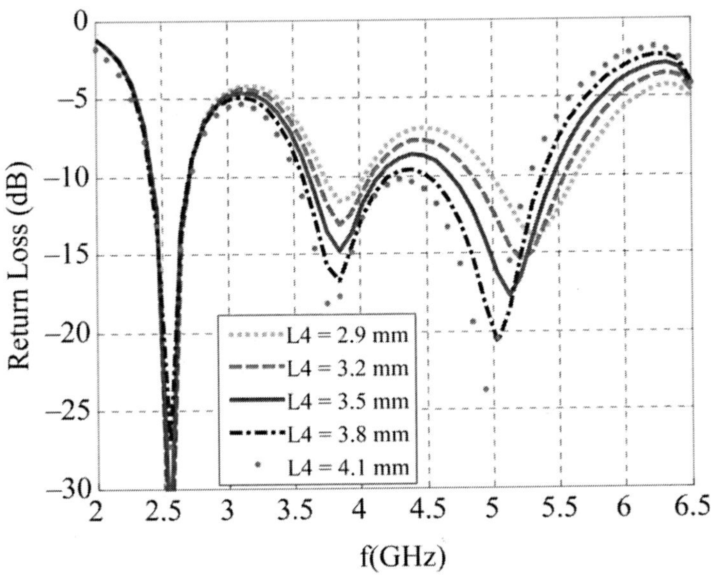

Figure 5. Effect of L4.

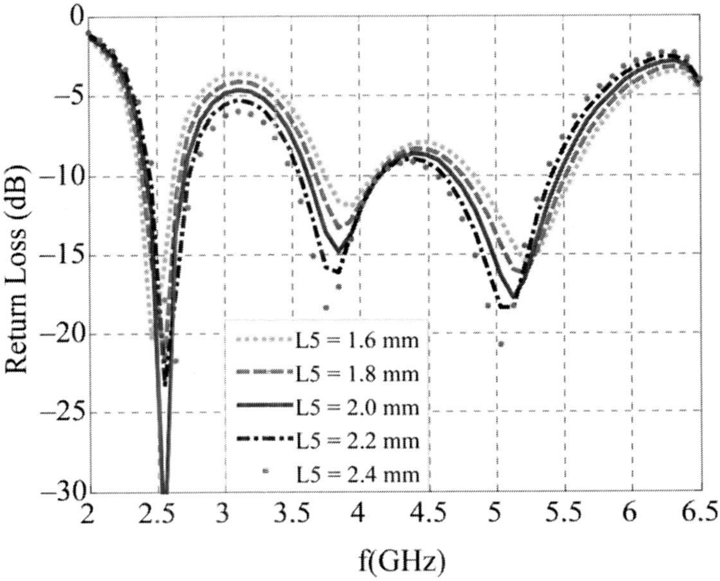

Figure 6. Effect of L5.

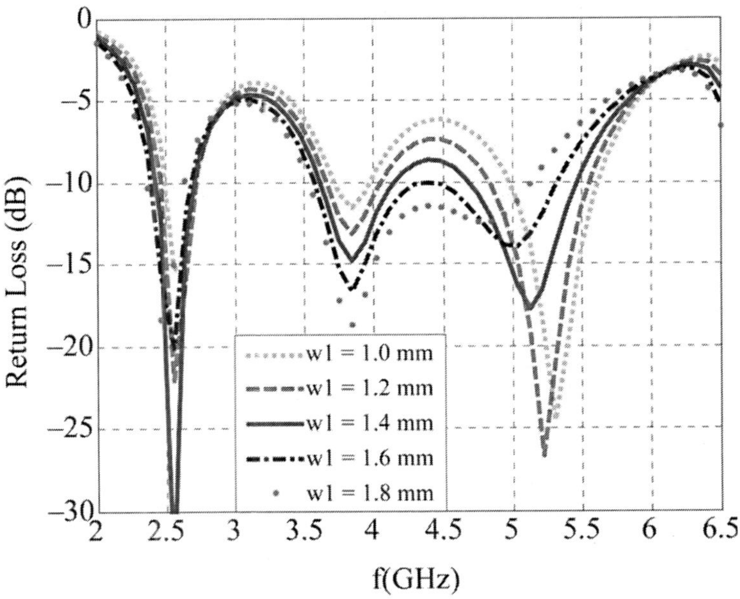

Figure 7. Effect of w1.

3. RESULTS

The return loss and radiation patterns are computed using Ansoft HFSS for Designs 1, 2 and 3. The reconfiguration is carried out by connecting or disconnecting appropriate rectangular strips to model the MEM switches. The rectangle width is "S" and its height is the same like the strip on which Pin diode or MEMs can be installed. Since we have four switches, we can create 16 cases starting from Case 0000 when all switches are open, to Case 1111 when all switches are closed. However, only 10 different combinations are considered after excluding the similar cases such as 0010 and 0001.

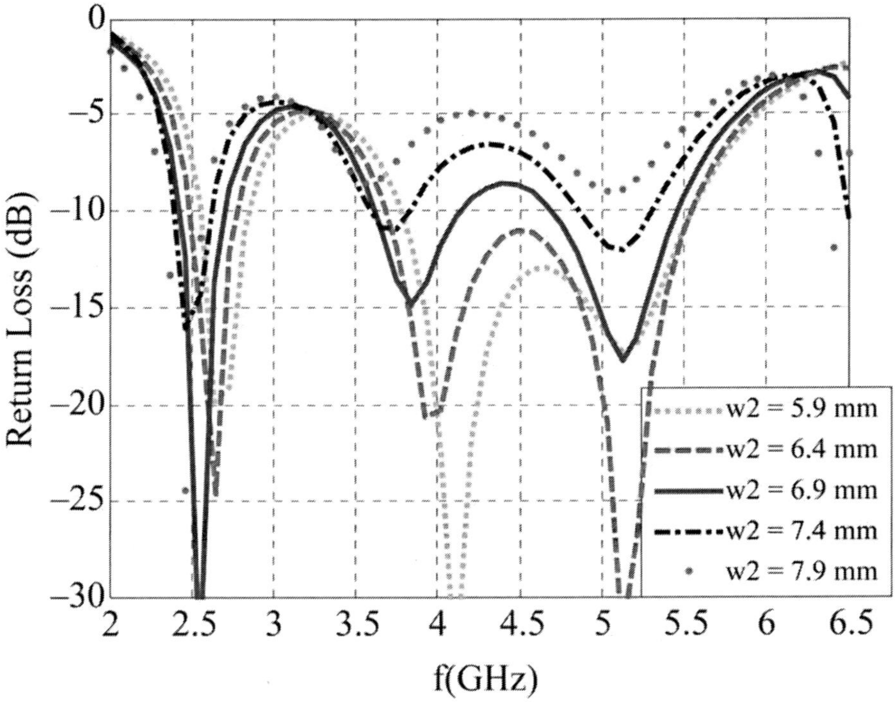

Figure 8. Effect of w2.

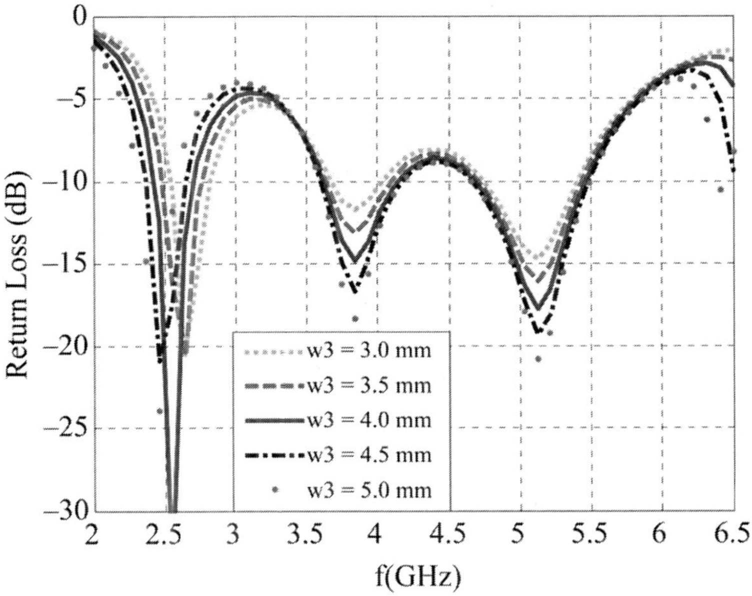

Figure 9. Effect of w3.

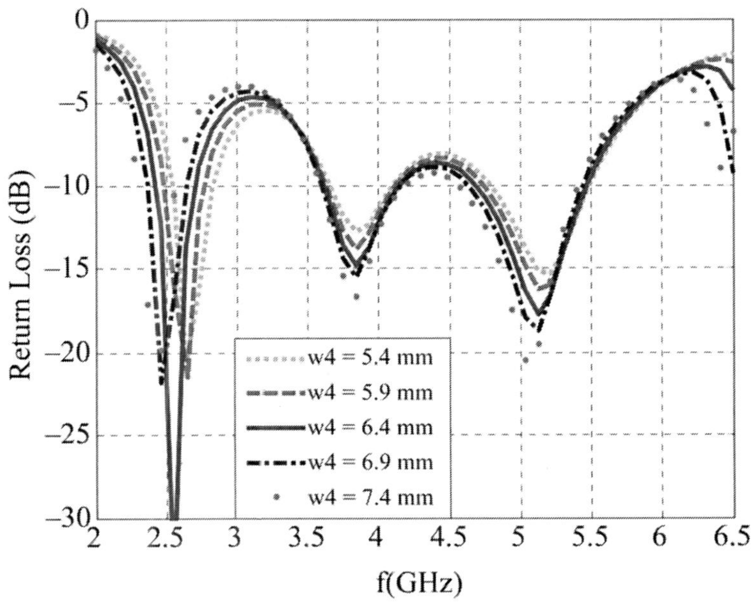

Figure 10. Effect of w4.

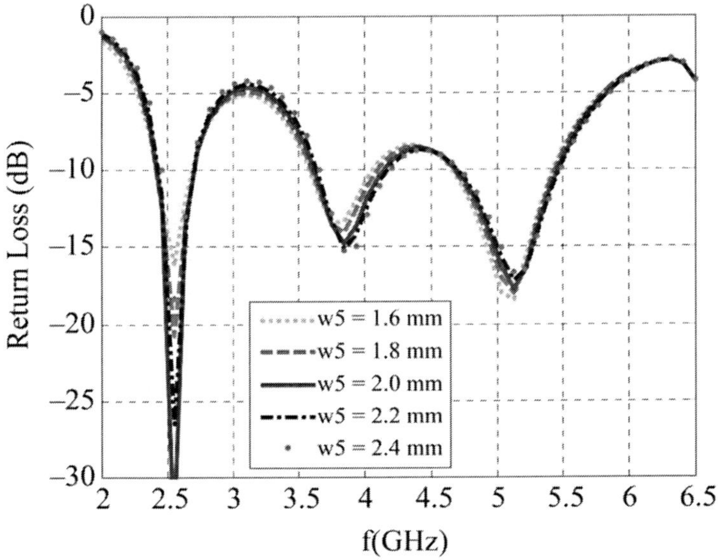

Figure 11. Effect of w5.

Table 1. Dimensions (mm) of the proposed designs.

	Design 1	**Design 2**	**Design 3**
w	46.2	93.5	144.5
L	18.9	46.3	89
w1	0.6	1.3	3.5
w2	7	9	22
w3	1	16.8	31
w4	9.8	19.8	32
w5	1.2	12.8	30
L1	5	12	42
L2	1.3	4.3	3
L3	3.1	13	36
L4	2.9	5	3
L5	1.6	N/A	N/A
S	N/A	1	1

Figure 12 shows the multiband (0011) and broadband (1100) cases in addition to the cases when all switches are ON/OFF (0000 and 1111). These cases are the symmetrical cases where S1 and S3 are the same as S2 and S4, respectively. The operating bands of the three designs in this figure are summarized in Tables 2, 3, and 4 for Designs 1, 2, and 3, respectively. With only four cases out of ten, Design 1 covers 90% of the band between 2.22 and 6.28 (95% bandwidth), Design 2 can operate in 90% of the band between 1.64 and 3.2 (64% bandwidth), and Design 3 also covers almost all the band between 0.83 and 2.34 (95% bandwidth). In the broadband Case, the three Designs provide from 22% to 56% bandwidth. Figure 13 shows some unsymmetrical cases which show that we still can make these antennas operate at other frequencies.

These results show many advantages for these designs. First, we can switch between different frequency bands without changing the antenna. Second, we can also change between multi-band and broadband modes. In addition, the overall antenna operating bandwidth is at least 1.7 times the maximum bandwidth of all cases.

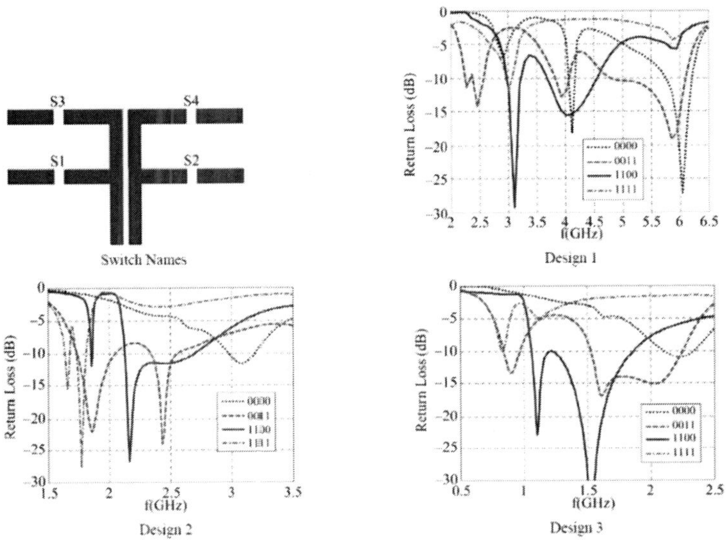

Figure 12. Return Loss (dB) for the symitrical cases.

Table 2. Operating frequency bands and bandwidth for Design 1 in the symitrical cases.

	Case	Band 1	Band 2	Band 3
Design No. 1	0000	4.02 - 4.15 (3%)	5.77 - 6.28 (8.5%)	N/A
	0011	2.22 - 2.55 (14%)	3.82 - 4.05 (6%)	4.75 - 6.10 (25%)
	1100	2.96 - 3.25 (9%)	3.65 - 4.55 (22%)	N/A
	1111	2.95 - 3.10 (5%)	N/A	N/A

Table 3. Operating frequency bands and bandwidth for Design 2 in the symitrical cases.

	Case	Band 1	Band 2
Design No.2	0000	2.95 - 3.20 (8%)	N/A
	0011	1.73 - 2.05 (17%)	2.36 - 2.53 (7%)
	1100	1.85 - 1.86 (0.5%)	2.13 - 2.73 (24%)
	1111	1.64 - 1.68 (2.4%)	1.73 - 1.80 (4%)

Table 4. Operating frequency bands and bandwidth for Design 3 in the symitrical cases.

	Case	Band 1	Band 2
Design No.3	0000	2.1 - 2.34 (11%)	N/A
	0011	0.83 - 0.98 (17%)	1.52 - 2.25 (39%)
	1100	1.05 - 1.86 (56%)	N/A
	1111	0.83 - 0.84 (1%)	N/A

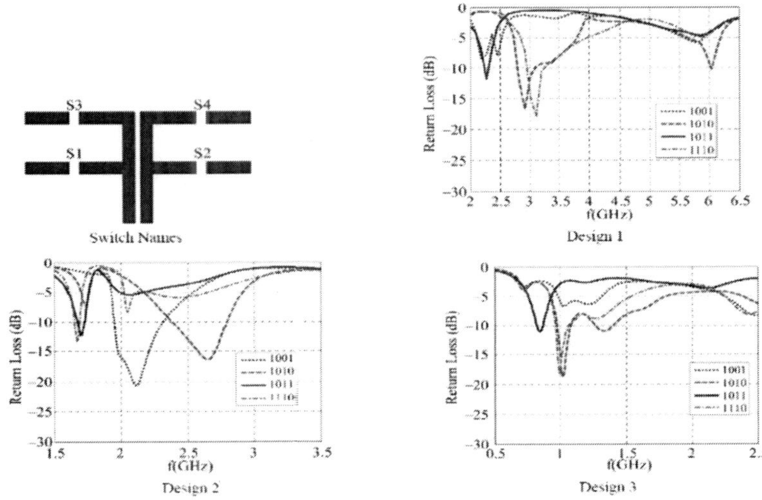

Figure 13. Return Loss (dB) for the unsymitrical cases.

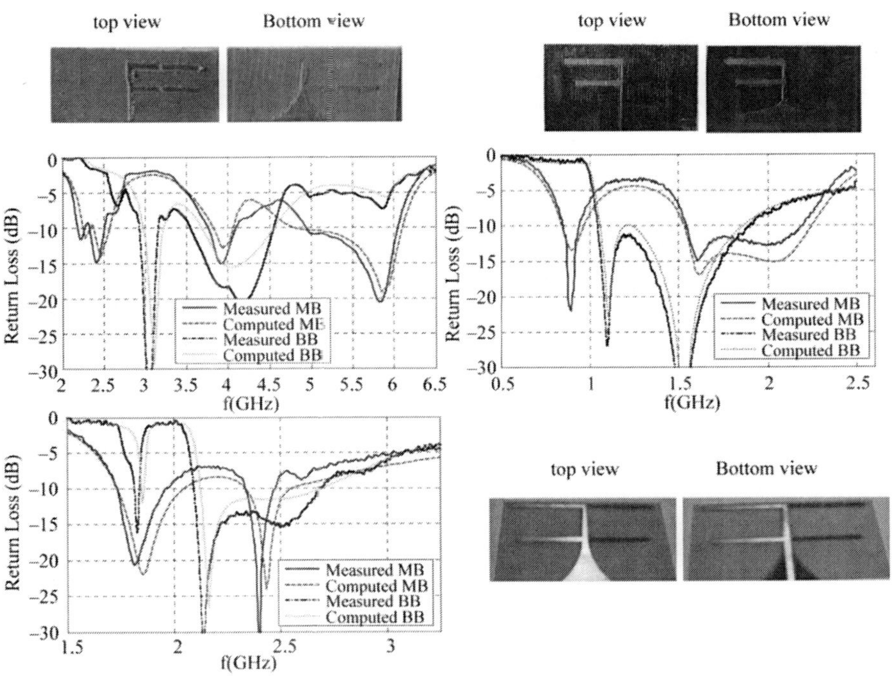

Figure 14. Measured and computed return loss and antenna prototypes.

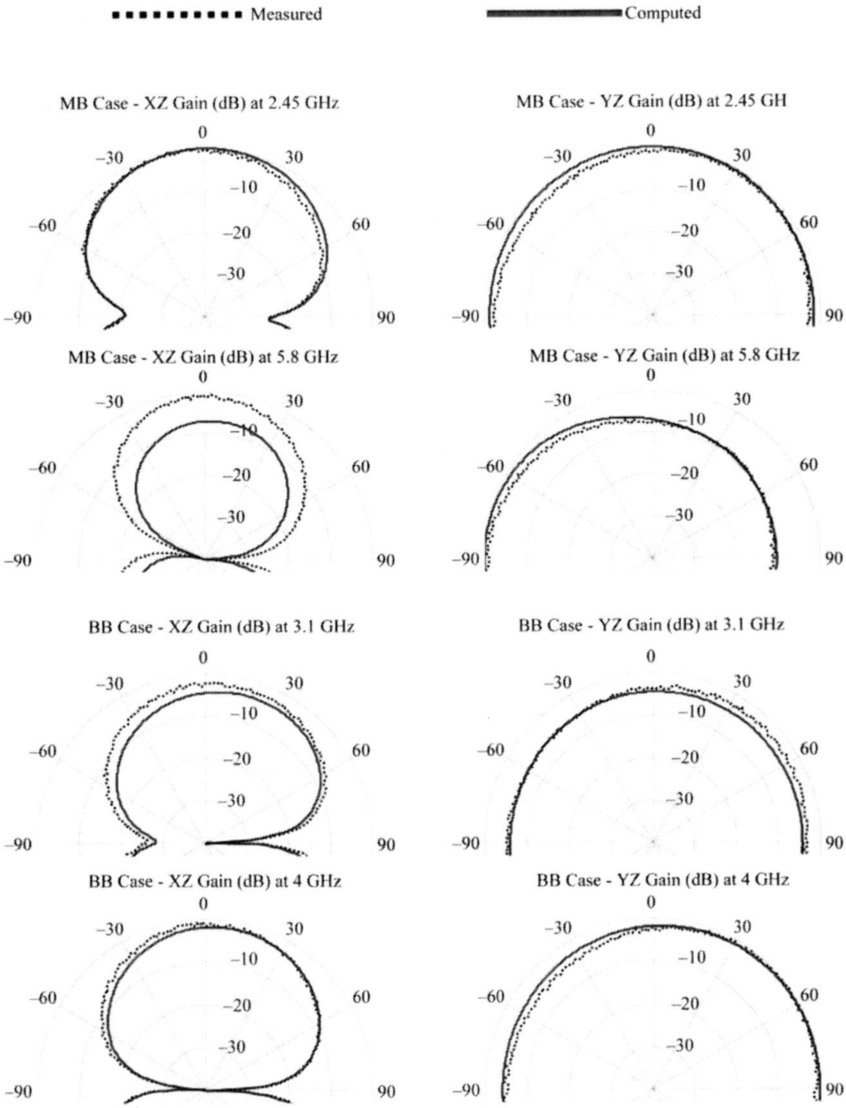

Figure 15. Radiation Patterns for Design 1 in multiband and broadband modes.

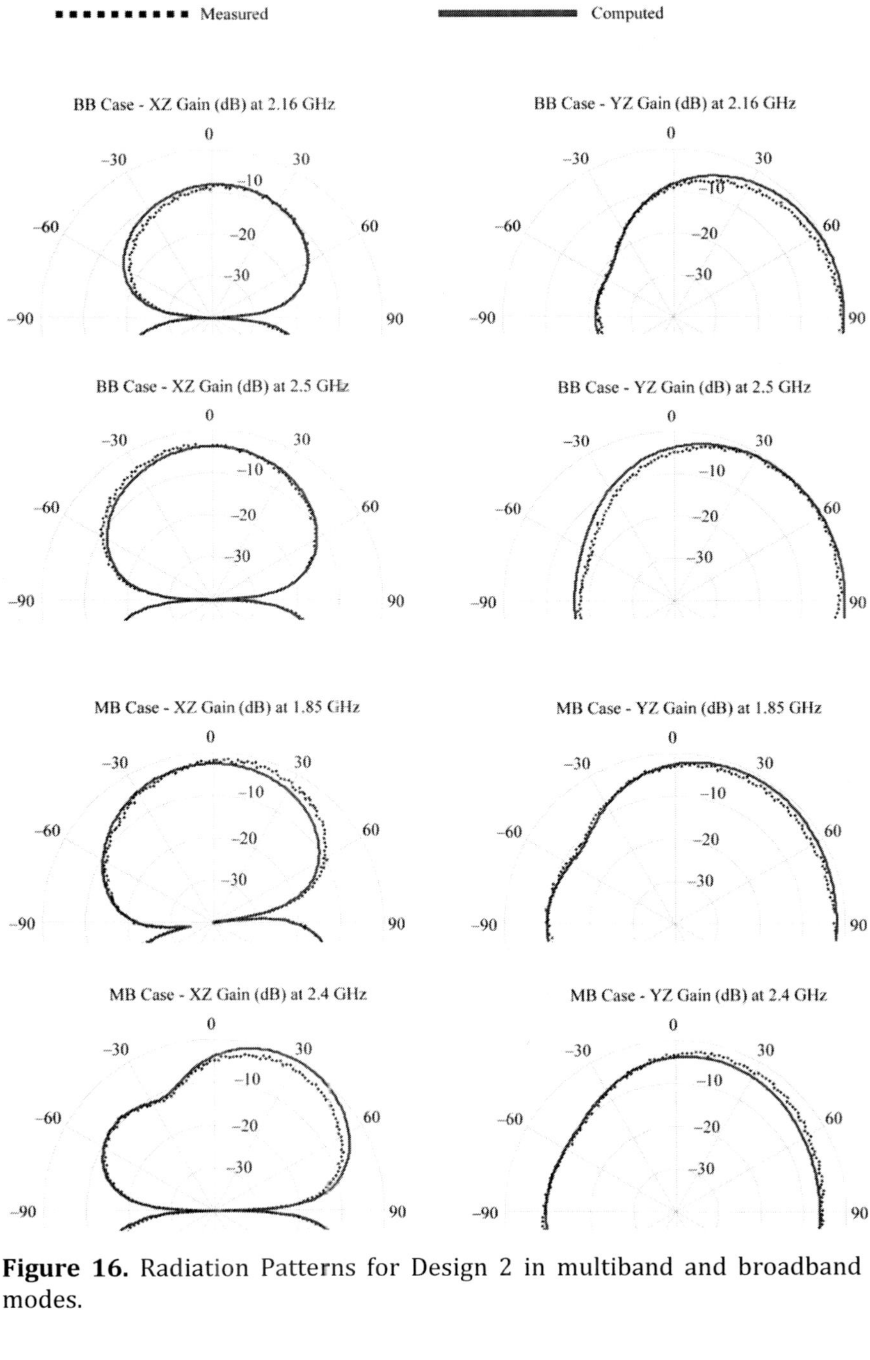

Figure 16. Radiation Patterns for Design 2 in multiband and broadband modes.

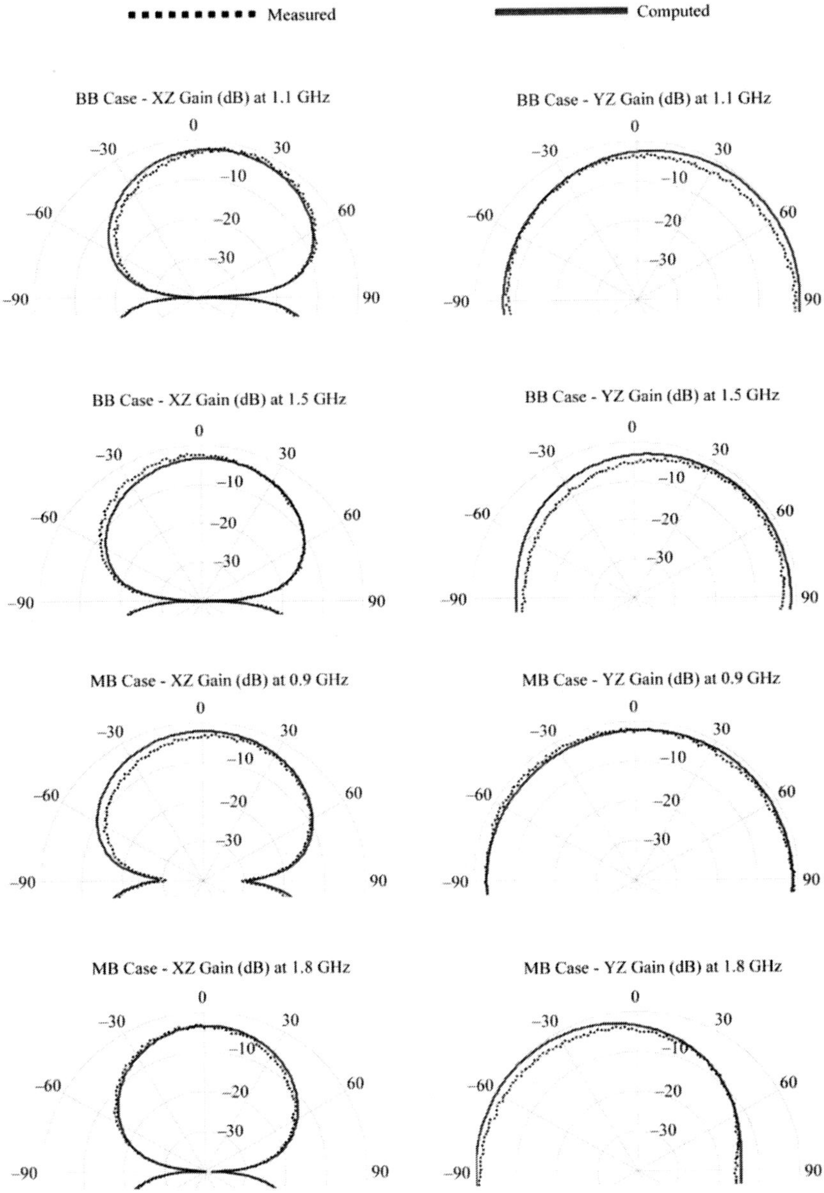

Figure 17. Radiation Patterns for Design 3 in multiband and broadband modes.

Prototypes from the three designs are fabricated using LPKF milling machine. The return loss is measured using vector network analyzer for the different reconfigurable cases. Figure 14 shows a good agreement between the measured and computed results in the multi-band (MB) and broadband (BB) modes. Finally, the radiation patterns are computed and measured using anechoic chamber. Figures 15 to 17 show the co-polarized fields at selected resonance frequencies for Designs 1, 2 and 3, respectively, for the MB and BB cases. The lower half is cropped because of the symmetry. An acceptable match is obtained between the measured and computed results. Currently, we are in the process of sending antenna prototypes to specialized for integrating reconfigurable enabling technology such as MEMs in these designs.

4. CONCLUSIONS

Three novel reconfigurable microstrip antenna designs are presented for wireless communications applications. The proposed antennas are combination of two dipole antennas of similar and different lengths. By changing the dipoles' lengths, we were able to change the antenna

operating frequency as well as the operation mode. The antenna is merging two operating modes: Multi-band and Broadband in a single design. The presented designs cover different frequency bands: 0.83 - 2.34 GHz, 1.64 - 3.2 GHz, and 2.22 - 6.28 GHz. A wide overall bandwidth of 95% was obtained and a wide single case operating bandwidth of 56% was also provided.

5. ACKNOWLEDGEMENTS

The authors would like thank the Air Force Research Laboratory (AFRL)—Minority Leaders Program (MLP), and Clarkson Aerospace Corp for funding this research through contract number FA8650-05-D-1912. Many thanks are also to Mr. William Munn and Mr. Christopher Munn from Jackson State University for their help in building the proposed antennas.

REFERENCES

1. C. B. Dietrich, Jr., R. M. Barts, W. L. Stutzman and W. A. Davis, "Trends in Antennas for Wireless Communications," Microwave Journal, vol. 46, no. 1, January 2003, p. 22.

2. S. N. Yang, C. N. Zhang, H. K. Pan, A. E. Fathy and V. K. Nair, "Frequency Reconfigurable Antennas for Multiradio Wireless Platforms," IEEE Microwave Magazine, vol. 10, no. 1, February 2009, pp. 66-83.

3. D. Peroulis, K. Sarabandi and P. B. K. Katehi, "Design of Reconfigurable Slot Antennas," IEEE Transactions on Antennas and Propagation, Vol. 53, No. 2, 2005, pp. 645- 654.

4. S. V. Shynu, G. Augustin, C. K. Aanandan, P. Mohanan, and K. Vasudevan, "A Reconfigurable Dual Frequency Slot Loaded Microstrip Antenna Controlled by Pin-Diodes," Microwave and Optical Technology Letters, vol. 44, No. 4, 2005, pp. 374-376.

5. S. V. Shynu, G. Augustin, C. K. Aanandan, P. Mohanan, and K. Vasudevan, "Design of Compact Reconfigurable Dual, Frequency Microstrip Antennas Using Varactor Diodes," Progress in Electromagnetics Research, Vol. 60, 2006, pp. 197-205.

6. W. H. Weedon, W. J. Payne and G. M. Rebeiz, "MEMsSwitched Reconfigurable Antennas," IEEE Antennas and Propagation Society International Symposium, Boston, vol. 3, 8-13 July 2001, pp. 654-657.

7. K. C. Gupta, J. Li, R. Ramadoss and C. J. Wang, "Design of Frequency-Reconfigurable Rectangular Slot Ring Antenna," IEEE Antennas and Propagation Society International Symposium, Salt Lake City, vol. 1, 16-21 July 2000, p. 326.

8. S. Xiao, B. Z. Wang and X. S. Yang, "A Novel Frequency-Reconfigurable Patch Antenna," Microwave and Optical Technology Letters, vol. 36, no. 4, February 2003, pp. 295-297.

9. P. F. Wahid, M. A. Ali and B. C. Deloach, "A Reconfigurable Yagi Antenna for Wireless Communications," Microwave and Optical Technology Letters, vol. 140, no. 2, July 2003, pp. 140-141.

10. A. A. Eldek, A. Z. Elsherbeni, C. E. Smith and K. F. Lee, "Wideband Slot Antennas for Radar Applications," Proceedings of IEEE Radar Conference, Huntsville, 5-8 May 2003, pp. 79-84.

11. S. Raghavan, T. Shanmuganantham and M. S. K. Kumar, "Reconfigurable Patch Antenna with Switchable L-Shaped Slots for Circular Polarization Diversity," Microwave and Optical Technology Letters, vol. 50, no. 9, September 2008, pp. 2348-2350.

12. A. Mirkamali, P. S. Hall and M. Soleimani, "Reconfigurable Printed-Dipole Antenna with Harmonic Trap for Wideband Applications," Microwave and Optical Technology Letters, vol. 48, no. 5, May 2006, pp. 927-929.

13. C. N. Zhang, S. N. Yang, H. K. Pan, A. E. Fathy, S. El-Ghazaly and V. K. Nair, "A Reconfigurable Multiband Patch Antenna for Wireless Applications Using MEMS Switches," Microwave and Optical Technology Letters, vol. 51, no. 8, August 2009, pp. 1892-1896.

14. F. Tefiku and C. A. Grimes, "Design of Broad-Band and Dual-Band Antennas Comprised of Series-Fed PrintedStrip Dipole Pairs," IEEE Transactions on Antennas and Propagation, vol. 48, no. 6, 2000, pp. 895-900.

15. A. A. Eldek, "Pattern Stability Optimization for Wideband Microstrip Antennas for Phased Arrays and Power Combiners," Microwave and Optical Technology Letters, vol. 48, no. 8, August 2006, pp. 1492-1494

16. A. A. Eldek, "Design of Double Dipole Antenna with Enhanced Usable Bandwidth for Wideband Phased Array Applications," Progress in Electromagnetics Research, Vol. 59, 2006, pp. 1-15.

17. Ansoft Corporation, "HFSS: High Frequency Structure Simulator Based on the Finite Element Method," Version 12, 2008.

9

CHAPTER

CRITICAL INFRASTRUCTURE SURVEILLANCE USING SECURE WIRELESS SENSOR NETWORKS

Michael Niedermeier [1,*], Xiaobing He [1], Hermann de Meer [1], Carsten Buschmann [2], Klaus Hartmann [3], Benjamin Langmann [3], Michael Koch [4], Stefan Fischer [5] and Dennis Pfisterer [5]

ABSTRACT

In this work, a secure wireless sensor network (WSN) for the surveillance, monitoring and protection of critical infrastructures was developed. To guarantee the security of the system, the main focus was the implementation of a unique security concept, which includes both security on the communication level, as well as mechanisms that ensure the functional safety during its operation. While there are many theoretical approaches in various subdomains of WSNs—like network structures, communication protocols and security concepts—the construction, implementation and real-life application of these devices is still rare. This work deals with these aforementioned aspects, including all phases from concept-generation to operation of a secure wireless sensor network. While the key focus of this paper lies on the security and safety features of the WSN, the detection, localization and classification capabilities resulting from the interaction of the nodes' different sensor types are also described.

Keywords: wireless sensor networks; security; functional safety; networking

1. INTRODUCTION

Wireless sensor networks are networking structures comprised of many small and low-cost sensor nodes, which have limited computational power and energy supply. The main goal of these networks is sensing, actuating and sending of environmental information to a data sink, which then processes it. During the last years, the application areas have evolved from military surveillance to environmental and animal monitoring. While these networks have a high potential—both from a research and economic perspective—they are currently rarely deployed, especially in high-security applications. The reason behind this is that the biggest advantage of WSNs—their autonomous and unattended operation—in turn also opens up many attack possibilities, like, e.g., tampering, physical manipulation and node compromise, due to the unavoidable disappearance of a security perimeter.

To prove that WSNs are applicable, even in high-security areas, this paper describes the comprehensive development of a secure WSN, including hardware design, software implementation as well as practical tests in laboratory and field environments. To cope with the aforementioned security issues, the system not only implements features from the field of communication security, but also includes functional safety and self-protection functions. All of these are fused in the system, creating a comprehensive security concept, which guarantees confidentiality, integrity, and availability of the WSN at any time. In addition, the reporting and logging mechanisms allow for users to constantly monitor the activities inside the sensor network, both in online and offline. Moreover, the usability of the system is not impaired by the inherent complexity of the WSN, as it can be configured and operated from a central command center. The combination of these features creates a uniquely secure, flexible and usable system.

The remainder of this paper is structured in a way that describes all stages of the system realization: Section 2 explains the requirements of our system. Section 3 covers all aspects of the system development and Section 4 demonstrates the abilities of the utilized sensors. Section 5 shows both the laboratory assessments as well as the field trials of the system. Section 6 presents results of the aforementioned tests. The final two sections, Section

7 and Section 8 present related work to protect critical infrastructures and a conclusion to the paper.

2. REQUIREMENT ANALYSIS

As stated in Section 1, the application areas of WSNs have broadened from solely scientific tasks to non-critical, yet productive applications and are currently on the verge to be considered as solution to achieve critical objective, such as the protection of critical infrastructures, too. In this paper, a WSN is presented that fulfills the requirements required for the latter purpose. The WSN is capable to protect both the borders of an arbitrary object (against intruders) as well as itself (against manipulations). While the WSN can be used to surveil any kind of bordered area, this combination of features ideally fits its usage in critical infrastructures protection.

While this work's general focus are the security and safety features, the main functional goals of the system are defined by three tasks of the WSN: (i) detection; (ii) localization; and (iii) classification of objects in a predefined area. The system must be able to detect and locate persons or cars with certain accuracy. In our case, all values for deviations or timings are chosen in a way that the goal on the user's side—which is to be able to organize a well-directed response to a penetration of the surveillance area—is possible. For the given case, a deviation of $\leq 5m$ is therefore considered acceptable.

In addition, it is required to calculate the trajectory of the trespassing object to indicate its direction. In order to provide not only accurate but also timely data, the system is required to fulfill certain real-time criteria. The delay between object-entry and the signaling of the event to the user was fixed at a maximum delay of $3s$. Due to the special hardware used in WSNs, several additional system requirements have to be fulfilled. Among them is primarily a reliable, scalable and efficient communication architecture ensuring that all events occurring inside the monitored area are not only registered, but also that the messages of these events are transferred reliably to the base station [1].

The security of the system is of major importance, especially when the unattended operation of the system is taken into account. The system's overall security concept, which is described in Section 3.3.2, therefore comprises not only IT security measures but also functional safety aspects. This is required to protect the system from coincidental or targeted physical damage and hardware failures. Additionally, because of the hardware of the sensor nodes, which only relies on limited battery supply, energy efficiency has to be carefully considered in both the hardware and software design [2].

The following Section 3 shows how the aforementioned requirements are fulfilled and implemented in a prototype system.

3. NETWORK DESIGN AND ARCHITECTURE

In this section, the development of the wireless sensor network is described in detail. This includes an overview of the overall concept of the sensor network, a fine-grained presentation of the hard- and software concepts and the proposed secure detection, localization and classification algorithms.

3.1. Network Overview

This section presents the general structure of the system to give an overview of it's functionalities. As depicted in Figure 1, the system is based on a hierarchical structure, that consists of a single command center c, several clusterheads ch i ch_i on the intermediate level and for each clusterhead ch_i multiple sensor nodes n_{ij} on the bottom level. In the prototype system, whose realization is described later on, a total number of 100 nodes is used (10 clusterheads ch i ,1≤i≤10 with each 10 sensor nodes n i j , 1≤i≤10 , 1≤j≤10). Due to the system's scalability, the system can however also be employed using a much larger number of nodes.

The functions of the system—detection, localization and classification of trespassers—are done centrally at the command center c, using data provided by the sensor nodes n i j . The clusterheads ch i fulfill managing and information-aggregating functions in the data delivery process from n i j to c. The details concerning the hard- and software realization is covered in the following Section 3.2 and Section 3.3.

The sensor nodes are equipped with different sensor combinations as shown in Table 1. A more detailed description of the sensors is presented in Section 4.

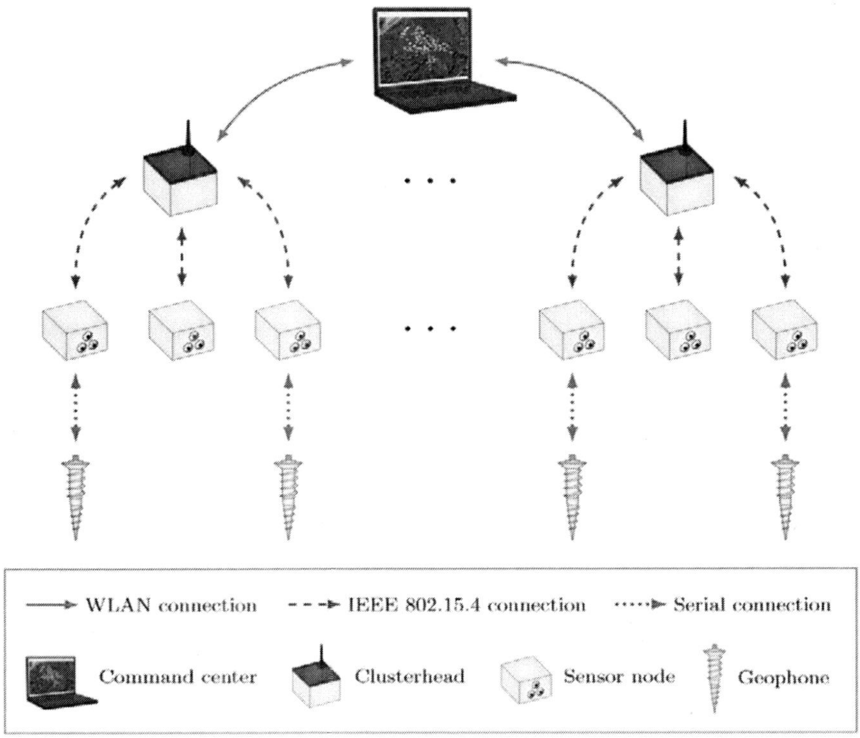

Figure 1. Overall system structure.

Table 1. Used sensor node types.

Sensor Type	Number of Nodes Equipped	Node Type Equipped
AMR	90	Sensor nodes
Accelerometer	90	Sensor nodes
Single-PIR	45	Sensor nodes
Multi-PIR	45	Sensor nodes
Longrange-PIR	10	Sensor nodes
Geophone	50	Sensor nodes
GPS	10	Clusterheads

3.2. Hardware

In this section, the hardware concept is described in the order of the previously shown three levels: command center, clusterheads, and sensornodes.

3.2.1. Command Center

The command center c aggregates data from the clusterheads ch_i and displays the information to the user (cf. Figure 2). This information is presented using the detection algorithms described in Section 3.4 and delivered to a network visualization software called Spyglass [3]. The existing open-source Spyglass was extended to have a control window that allows it to send commands to the network, like, e.g., restarting nodes, sending messages to a number of nodes, or re-programming nodes.

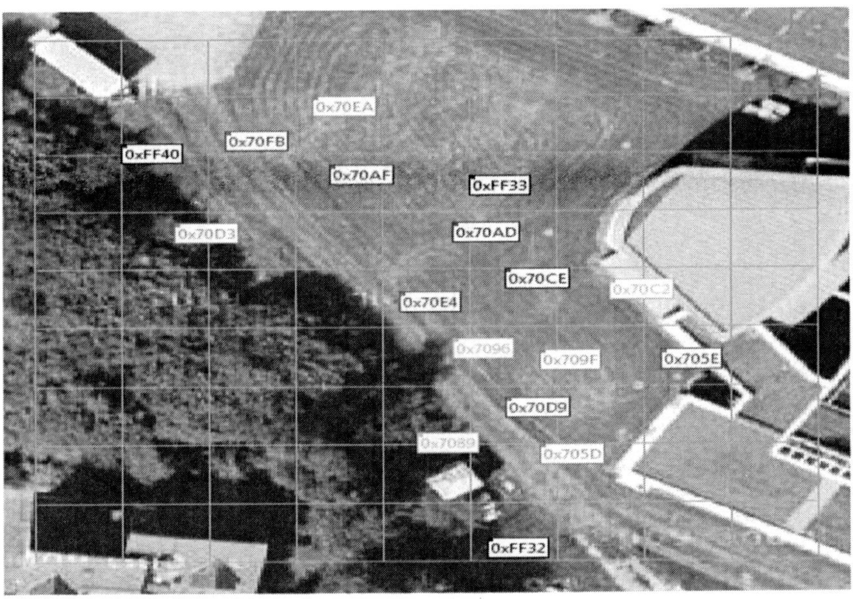

Figure 2. The visualization program Spyglass.

3.2.2. Clusterheads

Each clusterhead ch_i consists of two components: an embedded-PC board (ARM-based CPU with 800MHz , 512MB RAM, 512MB flash memory, power requirement of about 1.5W) and an iSense sensor node. The embedded PC maintains the WiFi connection to the command center c, preprocesses sensor data, and performs some network management tasks. To communicate with the sensor nodes, an iSense sensor network device is attached to the CPU board. It consists of an iSense Core Module (CM30I, [4]), an iSense Gateway Module (GM20-P, [5]) for USB connection to the embedded PC, and an iSense GPS Module (GPSM10S, [6]) for time synchronization purposes. All aforementioned components are depicted in Figure 3.

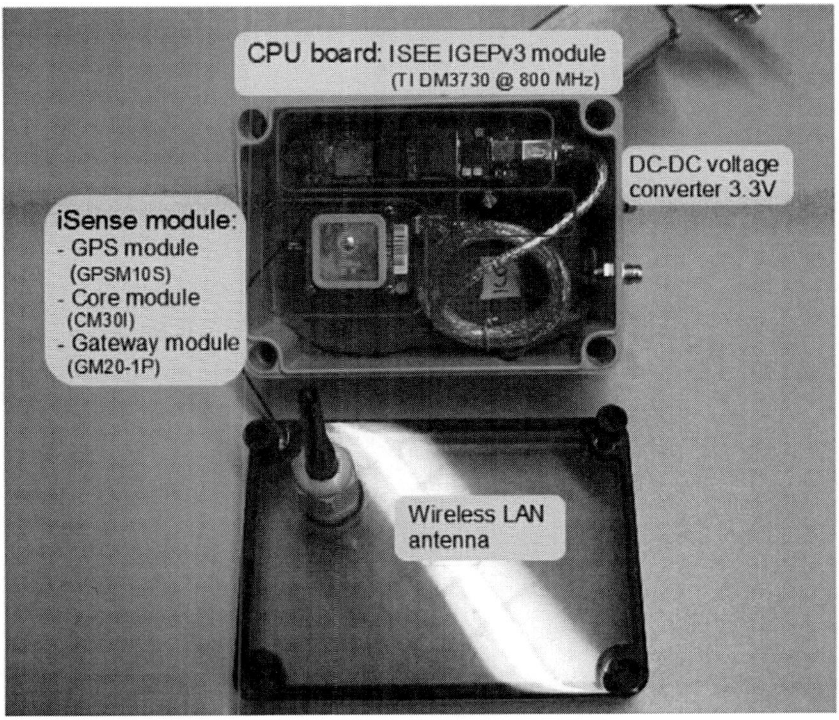

Figure 3. Hardware components of a clusterhead.

3.2.3. Sensor Nodes

Several different configurations of sensor nodes have been designed, which differ in the type of attached sensors and housing (Figure 4). The main type of housing (Fibox PC121210) is used for the majority of sensor nodes. These consist of 3 or 4 different hardware modules of the iSense modular sensor network hardware platform [7].

The Core Module includes a Jennic JN5148 32-bit RISC controller [8] operating at a frequency of 16MHz and provides 192kB of ROM, 512kB of flash memory, 128kB of RAM for instructions and data, and an IEEE 802.15.4 [9] compliant radio interface. The radio chip operates at a frequency of 2.4GHz , offers 16 different radio channels, provides a data transfer rate of 250kB/s and includes a hardware AES [10] engine. Besides those, the Core Module contains an ultra-stable real-time clock (RTC) (typical 6ppm), a voltage regulator and a 34-pin connector on each side to connect other modules. The second type of housing contains long-range Siemens IS392 [11] PIR sensors that are attached to the iSense Sensor and Core Module. In addition, an accelerometer is included in the housing.

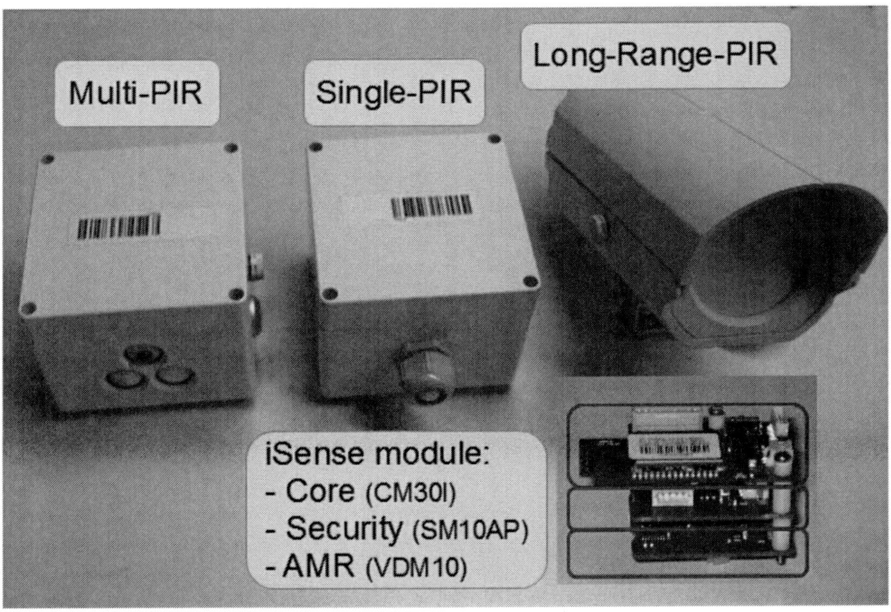

Figure 4. Overview of the different sensor nodes and iSense module components.

3.2.4. Sensors

In the sensor network, the previously described four types of sensors are utilized for various purposes. The general detection of activity in the observed area is performed by PIR sensors, which can be differentiated into three types: Single-PIRs (Panasonic AMN34111, [12]) with a range up to 10m , Multi-PIRs (AMN31111, [12] + 2 × AMN33111, [12]) with a range of approximately 5m and Long-Range-PIRs (Siemens IS392) with a range up to 50m . With the Multi-PIRs, the directions of moving objects can be estimated and Long-Range-PIRs play a significant role in the detection of new objects entering the surveillance area. Figure 11 shows detection results for these different PIR sensors.

The second type of sensors used for detection purposes are geophones. They are comprised of three seismic capsules (SM-24, Sensor Nederland, [13]) in an orthogonal arrangement, an analog signal conditioning part and a microcontroller-based (ARM-Cortex M3 MCU) signal processing unit. The complete geophone is housed in a protection enclosure and is connected to the sensor node via a communication cable that also provides the required power. For a sufficient coupling to the ground, a soil drilling tool in combination with a screw-shaped housing is used.

The Vehicle Detection Modules [14] are used for object classification, since only moving metallic objects influence their measurements. They are based on a 2-axis Phillips KMZ52 anisotropic magneto-resistive sensor bridge that is combined with two amplifier stages as well as circuitry for de-gaussing and earth-magnetic field compensation. It exploits the fact that large ferro-magnetic objects distort the earth-magnetic field to detect such objects by observing field changes. To achieve a detection range of more than 5m , it amplifies the bridge output by a factor of approximately 40,000 and features a sensitivity of 786.2mV/(kA/m) at a bandwidth of 1kHz . Because the module typically has a current consumption of 20–25mA during operation, it is activated only on PIR sensor events within less than 180ms . Figure 14 and Figure 15 show the typical output of the module when a car passes by in a distance of 4m . The vehicle detection itself is done in software on the sensor node's controller by using two ADC inputs to sample the module's output on both channels.

The last sensor is an accelerometer, which is part of every sensor node. Its sole purpose is the detection of physical tampering with the sensor nodes, since even small movements of the device trigger the accelerometer.

3.3. Software

In this section, we describe the software architecture of the sensor network, which consists of three parts:

The hierarchical communication structure.

The security concept for providing communication security and functional safety (guaranteeing the detection of malfunctions in the sensor or node hardware).

The detection algorithm. This uses data gathered by the sensor nodes and relates between the sensor clusters to derive a trespasser's position and current direction of movement.

3.3.1. Hierarchical Communication Structure

As discussed earlier, the hierarchical communication structure ensures network scalability and is comprised of the command center, clusterheads (embedded PC and sensor node), and the sensor nodes which are associated with the clusterheads. The communication between command center and clusterheads is realized via standard TCP/IP-based connections over WiFi while the communication in the sensor network is based on the energy-efficient IEEE 802.15.4 standard. This structure is also depicted in Figure 1: sensor readings from the sensor nodes are forwarded to the sensor node attached to the clusterhead, are pre-processed, and finally forwarded to the command center, where further data aggregation and preprocessing occurs before data is visualized and presented to the operator. Vice versa, there is a data flow from the command center downstream to send configuration instructions, to reprogram the sensor node parts that are attached to the clusterhead PCs, and to wirelessly reprogram the sensor nodes.

In order not to build a proprietary system, we re-use existing open-source software and implement our additional functionality as plug-ins. The software Testbed Runtime (https://github.com/itm/testbed-runtime) is a result of the EU-project WISEBED [15] and allows to operate wireless sensor network testbeds. Its architecture is depicted in Figure 5.

It consists of a number of sensor nodes attached to gateways (e.g., via USB). The gateways communicate over TCP/IP with a Testbed Server that exposes the functionality of the network (e.g., sending messages to attached sensor nodes or reprogramming them, receiving messages from nodes emitted via USB, etc.) via SOAP-based web services. In addition, it supports so-called filters through which all data is passed that is exchanged between (i) the sensor nodes and gateway and (ii) the Testbed Server and the web service clients. A filter may alter, drop, or fabricate messages (e.g., for data aggregation).

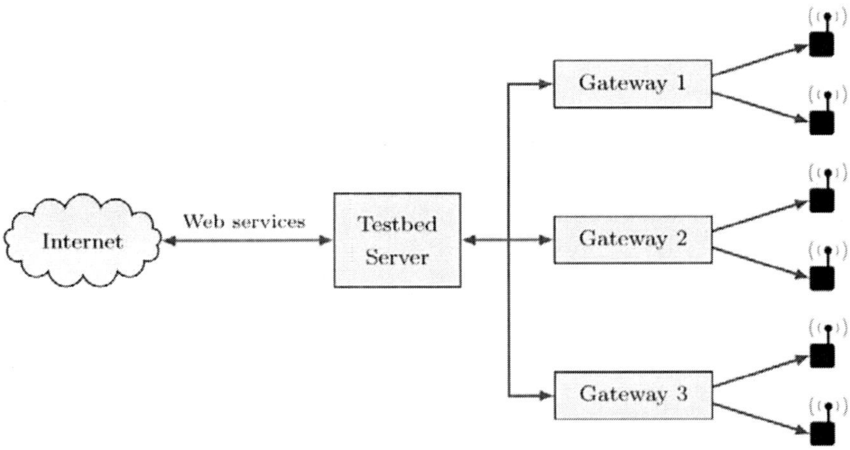

Figure 5. Architecture of the Testbed Runtime software.

This structure blends well with the overall communication infrastructure (cf. Figure 1). To implement our hierarchical communication structure, we use the Testbed Runtime software as the basis and implement our data aggregation and trespasser detection schemes as filters. To control the network, a client for the SOAP-based web services is required and we add this feature to the visualization framework Spyglass, which acts as the control center. In order to send messages to the sensor network, data is sent to the gateway node, which receives this information from Testbed Runtime via USB. For the wireless sensor network, we design a protocol on the USB connection so that data can be forwarded securely to/from the wireless sensor nodes. To initialize the communication between the clusterhead and sensor nodes, firstly, the clusterhead broadcasts a heartbeat solicitation (HBS) message to all sensor nodes. After checking the authentication and validity of the clusterhead, the sensor node replies with a heartbeat advertisement (HBA) message (with 1 status Byte and 1 signal quality Byte to indicate the status of the sensor node and the signal quality of the received HBS, respectively), as well as a clusterhead solicitation (CHS) message. The clusterhead will also check the authentication and validity of a sensor node and then sends a cluster advertisement (CHA) and a timestamp message to the sensor node. After that, the communication for the wireless sensor network is established and of course, all the messages exchanged for communication initialization are secured with the strategies described in Section 3.3.2.

3.3.2. Security

To fulfill high-security requirements (e.g., protecting critical infrastructures) IT technologies require special communication security features to guarantee the necessary protection level during the system's operation. However, securing the communication alone is not sufficient if the devices should also be transportable, usable in infrastructure-less areas, and shall be suited for unattended operation. As previously described, the combination of these requirements requires novel security concepts, which are explained in the following.

Communication Security

While the IEEE 802.15.4 standard already offers several security features, these were not used to secure the data exchange for two reasons. First, the security functions should be kept separate from the underlying communication protocol to offer flexibility. Second, the security features offered by 802.15.4 are not sufficient for the critical application scenario of the system. The system implements additional security functions to detect integrity breaches, e.g., replay attacks as well as arbitrary and targeted jamming. The communication security itself is structured in two security layers, which bundle certain security functions. The communication stack of the system is made up of four layers, from top to bottom: application-layer, security-layer, AFH-layer (Adaptive Frequency Hopping) and IEEE 802.15.4 -layer. This structure guarantees a mutual independence of the sensor data gathering and evaluation, the AFH scheme and the other security features. the AFH-layer implements the adaptive frequency hopping, time synchronization, management of the currently allowed communication channels, and the procedure for the initial joining of nodes to clusters. The security layer is described next.

Security-Layer: The security-layer implements all communication security features except for the AFH mechanism, which can be easily found in [16], and we are not going to repeat it here but how AFH works in this system is tested in Section 6. The goal of the security-layer is to achieve confidentiality, integrity, and availability. To do so, a combination of several security measures is used.

To prevent messages from being forged or manipulated, the message payload is encrypted using the AES-CBC-128 cipher, which is supported by a crypto co-processor available in all sensor nodes, which makes it very fast and energy-efficient. In addition, a 4Byte timestamp, a 16Byte CBC-MAC AES, and a 1Byte sequence number is added to the payload. These are used to achieve resistance against, e.g., replay attacks, delayed message sending, or manipulation of the message order.

All security measures realized by the security layer are depicted in Figure 6. The cryptographic secret key k_e is also a part of the security layer and it is used with the CCM* cipher algorithm [17]. To prevent the system from being compromised by brute-force attacks, k_e is altered before each communication round by re-encrypting the currently used key: $k_{e,t_{n+1}} = AES(k_{e,t_n})$.

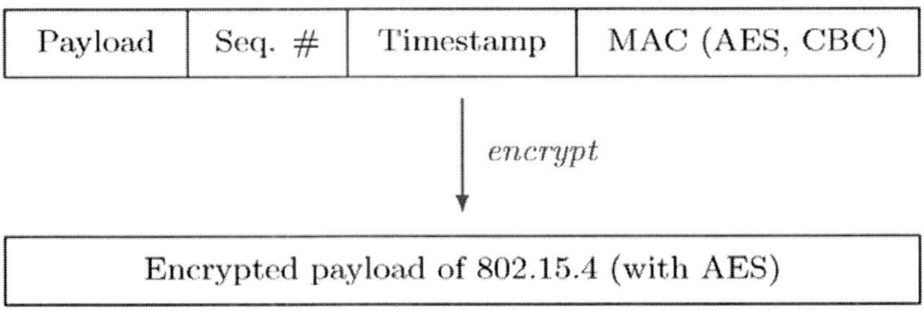

Figure 6. Communication security measures.

The communication between the clusterheads and the command center c is not realized with IEEE 802.15.4 but with standard IEEE 802.11 g WiFi offering WPA2 security. As WPA2 offers all necessary features for a high protection level of the communication [18], no further security measures are applied here.

Functional Safety

The integrity features of the nodes and their respective sensors are key components of the functional safety of the wireless sensor network. The constant monitoring of the components' integrity allows to detect both random, transient as well as static errors and malfunctions in the components. The error detection then allows to control the error and thereby achieve a safe state in the sensor network. The reliability of the system is highly important, as undetected sensor or system malfunctions could easily lead to detection holes in the surveillance area. Because of that, numerous functional safety features are implemented, to uncover any system failures and induce accordant countermeasures to guarantee the proper operation of the system.

RAM: To exclude hardware-related failures due to RAM malfunctions, an integrity check for the hardware is implemented that checks the correct functionality of the RAM each time a node is started or restarted. This test is implemented using the "Checkerboard" scheme, that is outlined in Code 1.

The test writes hexadecimal 5 s (0×55) in all even and hexadecimal As (0×AA) in all odd memory locations in a checkerboard pattern. After the specified parts of the RAM have been written and an additional wait time of ≥ 10ms (which are needed to make sure that the remaining capacitance is gone before the reading starts) the patterns are read and compared. That comparison is done by XOR ing the ith word ($0 \leq i < \lceil n2 \rceil, n=$ number of words in memory) in the memory with the $(n-i)$ th word. The expected result of this operation is identical for all words in the memory: A word consisting of only 0×FF . If the result differs from the expected value, an error is assumed. After this, the test is repeated using complementary patterns (0×55 in odd locations, 0×AA in even locations).

If the RAM test fails, the node is restarted to avoid the spreading of possibly falsified information in the network. After the restart, the hardware testing starts again, until the RAM is working correctly again. If the error is permanent, the node tries to restart until its energy supply is drained.

Program data: To protect the integrity of the application running on the nodes, a checksum mechanism was implemented that checks the code inside the RAM and compares it to the original image's code stored inside the flash memory. The code check uses Fletcher's checksum that is both effective and computationally efficient, which is important due to the energy restrictions of the sensor nodes. The quality of this checksum algorithm (diagnostic coverage value) is similar to that of CRC (Cyclic Redundancy Check) [19]. The pseudo code in Code 2 shows the basic scheme of Flechter's checksum. This check ensures that the code is not altered on its way from the flash memory to the RAM.

CPU: As the key component of every sensor node, the CPU is included in the functional safety tests. For a complete verification of the CPU, a very complex, and thereby energy- and time-consuming, procedure would be required that exhaustively tests every CPU instruction. Because of that, an extensive assessment is not implemented. Instead, a correctly working CPU is assumed, if the self-tests of both the RAM and the flash memory are finished without errors and the self-test algorithm is completed and exits as expected. To check that, the error flag is set before the test routine begins and is only reset if the program finishes at the expected exit point. If this is not the case, a CPU error is indicated, leading to a node reset, as previously explained in the RAM test procedure.

Code 1 Pseudocode of Checkerboard RAM test

```
//Checkerboard RAM test
integer i, j, k, n

while i is odd and j is even do
  write 0×55 in cell[i]
  write 0×AA in cell[j]
end while

pause(10 ms)

//Check if sum of even and odd locations is 0×FF
for k = 0 to ⌈n/2⌉ - 1 do
  sum[k] = cell[k] + cell[n - k]
  if not_equal(sum[k], 0×FF) then
    FAILURE
  else
    //Complement cells if their sum is 0×FF
    complement(cell[k])
    complement(cell[n - k])
  end if
end for

pause(10 ms)

//Check complemented cells
for k = 0 to ⌈n/2⌉ - 1 do
  sum[k] = cell[k] + cell[n - k]
  if not_equal(sum[k], 0×FF) then
    FAILURE
  end if
end for

PASS
```

Sensors: The most important part of the integrity checking process is the testing of the sensors to prevent two main issues that can disturb the proper function of the system: partial or complete sensor failure. While the complete failure is easier to detect (no reaction from a certain sensor), the partial defect is even more dangerous for two reasons. On the one hand, it can cause the sending of false data to the clusterheads, causing false positives or negatives. On the other, the development of effective detection of partial sensor failures is more complex than a simple test for a sensor's reactivity. The following paragraphs describe the testing routines of the sensors in detail.

Code 2 Pseudocode of Fletcher's checksum

```
//Computation of Flechter's checksum
integer i, sum1, sum2
byte flash[image_length]
sum1 = 0
sum2 = 0

for i from 0 to (image_length - 1) do
  sum1 = (sum1 + flash[i]) modulo 0xFF
  sum2 = (sum2 + sum1) modulo 0xFF
end for

//Binary complement of sum1 and sum2 as reference
check1 = 0xFF - ((sum1 + sum2) modulo 0xFF)
flash[image_length] = check1

//Binary complement of sum1 and check1 as test checksum
check2 = 0xFF - ((sum1 + check1) modulo 0xFF)
flash[image_length + 1] = check2
```

GPS module: The GPS module is checked each time it is activated for time synchronization purposes. After its activation, the GPS module receives a timing signal after a defined amount of time (one tick each second). If the timing signal is not received in this timeframe, a time-out event occurred. This leads to an error of the GPS module, that is reported by an error flag to the command center c. If the next timing signal is received correctly in time, the error flag is cleared again.

AMR sensor: Before each use of the AMR sensor inside of the vehicle detection module, it has to fulfill a calibration self-test mechanism. This test works by actively compensating the earth's magnetic field by sending current through coils. The test routine uses the calibration to find errors in the sensor, which can be found by comparing the current to a reference value. If the difference exceeds a predefined value, an error is reported by setting an error flag. The error flag can be removed again if a following calibration returns an acceptable value.

Acceleration sensor/accelerometer: The acceleration sensor located on the security module of the node is a key element in the tamper resistance concept of the sensor nodes. The test for this sensor is periodically triggered and uses an internal self-test algorithm of the security module. The test uses the apparent gravity to measure if the acceleration sensor is working correctly. If the self-test fails, which is the case if the difference of the current test result and the reference value stored in the test-algorithm exceeds a certain threshold, the error is reported by an error flag that is sent to the clusterhead. Again, if the test is completed successfully afterwards, the error flag is revoked.

Geophones: The test routine of the geophones could not be built upon an internal test routine. Therefore, the testing mechanism can only verify if a geophone, which is connected to a sensor node, is able to receive and send messages. This is done by sending a wake-up message to the geophone. If the device is answering that message with an verification message in a certain amount of time, it thereby confirms to be working in unimpaired condition. Again, if either the verification message is not received by the sensor node in time or not at all, an error flag is set. It is revoked the next time a verification message is received within the expected time.

PIR sensors: Due to the nature of the PIR sensor, there is no self-testing procedure available, due to the lack of an reference value. The only malfunction of a PIR sensor which can be detected is the case when it is always on, because during normal operation, the sensor automatically deactivates directly after being triggered.

Reporting of and Reaction to Errors

To ensure a controllable system behavior, it is not only necessary to detect hardware malfunctions and tampering attempts, but also to report and react to these issues in an appropriate way. Because of that, a multi-level reporting and logging system is realized. The real-time reporting fulfills two important tasks: On the one hand, any detected error is reported to the command center c, where the user can see and if necessary, react to any integrity failures in the network. These are reported using an status byte consisting of 8 signaling flags (cf. Figure 7) which is sent from every sensor node to its clusterhead ch i with each heartbeat message.

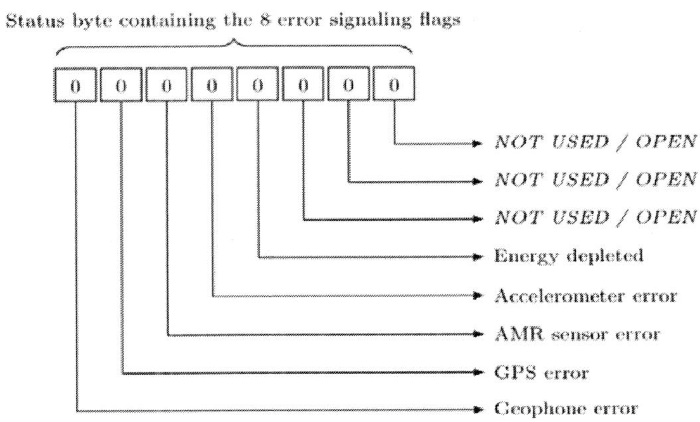

Figure 7. Status byte used for error flag signaling.

An error is indicated by setting the signaling flag from 0 to 1 at the appropriate position, as described in Section 3.3.2. If an error is sent from a sensor node n_{ij}, it is passed through the clusterhead ch_i to the command center c, where a message indicating the error is displayed to the user. On the other hand, certain errors have to cause an immediate reaction in the network, as an delayed reaction by the user would cause an overhead that is unacceptable. For example, a CPU or RAM error could endanger the correct function of the overall system by sending false data, leading to either false positive or false negative detection results. Another possibility would be that an impaired node sends too many messages leading to a reduced bandwidth availability for the other nodes. Therefore, nodes that do not pass the CPU or memory tests are prevented from joining the network without prior user interaction.

Moreover, the system includes a logging function, which is—in contrast to the real-time reporting—used to keep track of the events in the WSN over a longer period of time. This enables an user to assess actions in the surveyed area and detection results of the system in a time-delayed manner. The mechanism included uses a database to document all events coming to and all messages sent by the command center c.

A similar error indication signaling is used in the AFH mechanism. Here, a jammed channel is represented by setting the corresponding bit in the channel mask (cf. Figure 8) from 1 to 0. The jamming detection is done by the clusterheads, which distribute the information of the currently usable channels to their sensor nodes with each heartbeat message.

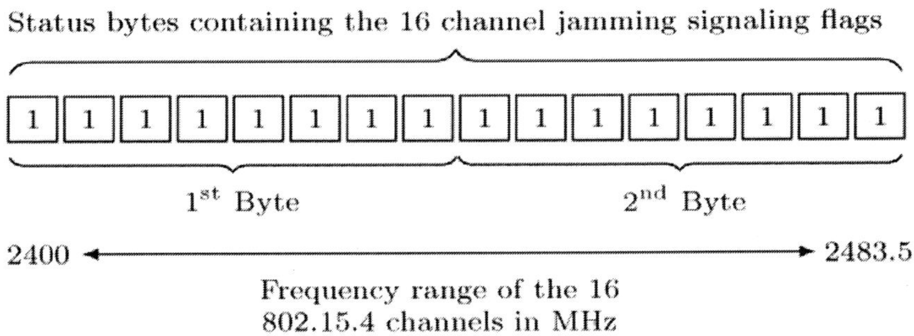

Figure 8. Channel mask used to signal jammed frequencies.

3.3.3. Energy Supply

As stated previously, the energy consumption needs to be highly restricted in sensor nodes, as the energy supply is very limited. This also applies to our system. In addition to the overall energy shortage, this issue is intensified by the usage scenario of the system: The security and reliability is one of the key features that must not be compromised. Because of that, a sudden failure of nodes due to energy drain has to be prevented. To do so, an early warning system is integrated into the sensor nodes that sends an alarm if the energy supply of a node drops lower than 10%, leaving the user enough time to change the node's power supply or replace it with another spare node. To measure the remaining energy, a energy consumption model is implemented that calculates the overall drained and remaining energy supply. This is done by measuring the time the system is in "awake" (sensors and node components activated) and "sleep" (sensors and node components deactivated) state. By doing that, it is possible to approximate the remaining system lifetime by subtracting the consumed energy (sum of energy used in both "awake" and "sleep" state) from the overall battery capacity. The remaining energy value is stored in the flash memory in addition to the RAM. This prevents miscalculations in the case of system restarts, as the remaining energy value is preserved in the flash memory. Tables Table 2 and Table 3 show the approximate energy demands of the different components and node types used in the WSN.

Table 2. Approximate energy demand of different components used.

Component	Awake Demand	Sleep Demand
Accelerometer	650 μA	1 μA
Single-PIR sensor	300 μA	0 μA
Multi-PIR sensor	900 μA	0 μA
GPS receiver	50 mA	0 mA
AMR sensor	40 mA	0 μA
Core module	6 mA	40 μA
FM transceiver	16 mA	0 mA
Geophone (GP)	10 mA	5 mA
IGEP module	600 mA	480 mA

Table 3. Approximate energy demand of different node types used.

Node Type		Awake Demand	Sleep Demand
Clusterhead		672 mA	480 mA
Node with	Multi-PIR, no GP	64 mA	41 μA
	Multi-PIR, with GP	74 mA	5 mA
	Single-PIR, no GP	63 mA	41 μA
	Single-PIR, with GP	73 mA	5 mA

3.4. Secure Detection, Localization and Classification Algorithms

In order to achieve a robustness and reliability suitable for long term surveillance, the detection, tracking and classification methods are implemented in a simple, yet effective, way in the hierarchical network. Figure 9 shows information flow of proposed detection, localization and classification algorithms.

Sensor node: Located a the lowest level, sensor nodes are responsible for the acquisition of sensor data, with the equipped PIR sensors, magnetic sensors, Geophones and accelerometers. For PIR sensors this is a cone of a specific length whereas for all other sensors, this is a radial symmetric area around the sensor node with a certain radius. After the WSN is started, the detection algorithm first remains in a waiting state, until a significant amount of sensor events are registered, which indicates an object entering the area. If the amount of events remains low, it is assumed that these are caused by random noise or other sources, like wind. Additionally, it is required that there is a local accumulation of events. The amount of random noise depends on the sensor type as well as on other factors like the position of the sensor, the time of the day and the weather conditions. If the sensor data satisfies these conditions, it is inferred that an object is inside the surveillance area and an initial position is determined. Each sensor event suggests possible locations of the object.

Clusterheads: The event data from neighboring sensor nodes is further processed at clusterheads. As each sensor node suggests a possible position for a object, at the clusterhead, The current position of the object is then determined by averaging the suggested positions of the sensor events while factoring in the previous position of the object. Afterwards, only sensor events in a vicinity of the current position of the detected object are considered, until the object leaves the surveillance area.

Command center: As the control center of WSNs, command center classifies different objects (person, person carrying metal object, car or unknown) by an analysis of the AMR and geophone events. Magnetic sensor events in the vicinity of the object suggest a car or a motorbike, and geophone events indicate footsteps, but there need to be an accumulation of these events, which exceeds the noise level. Table 4 shows the classification matrix of events detected by different sensor types. For clarification, an example is given here: Assuming a Single-PIR event is triggered, at first, an entity is detected to be present in the WSN's vicinity, however a classification is not yet possible, therefore it is marked as "unknown". However, if the entity also triggers an AMR event, yet no geophone event, it is classified as a vehicle. Experiments show that the AMR sensors are surprisingly affected by wind (possibly due to movement of cables or the sensor node itself). An object in the surveillance area is considered to have left this area if the last known position is near the border of the area and further sensor events are below the detection level for some time.

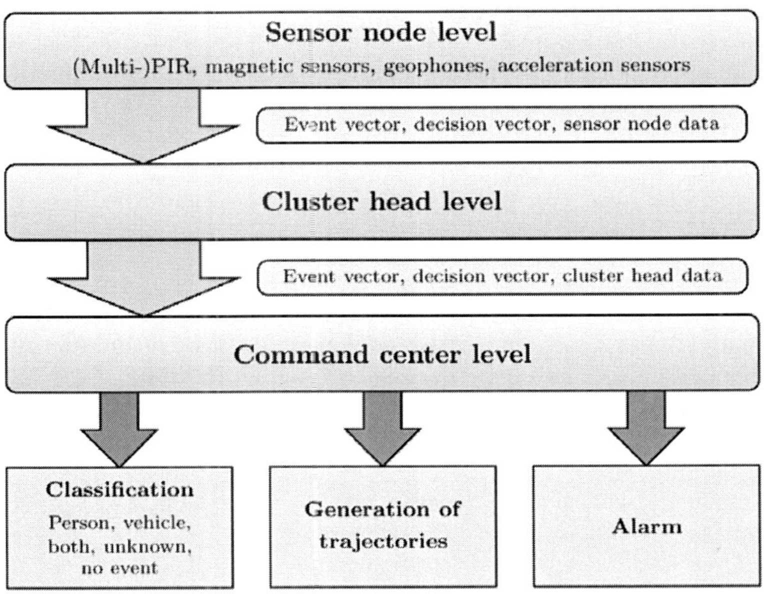

Figure 9. Information flow of detection, localization and classification algorithms.

Table 4. Classification matrix of events by different sensor types.

Component	Event Detected	
	Yes	No
Single-PIR	Person/vehicle	—
Multi-PIR	Person/vehicle	—
Long Range-PIR	Person/vehicle	—
AMR sensor	Vehicle	—
Geophone (GP)	Person	—

4. SENSOR EVALUATION

To achieve the desired quality of the surveillance, the utilized sensors need to be analyzed, since datasheets often do not provide information in the way or detail required. Additionally, the preprocessing algorithms of the geophones and PIR sensors need to be evaluated and characterized. In the following an excerpt of the sensor related experiments is given.

4.1. Geophones

In order to be able to detect intruders and therefore to provide a basis for tracking a robust algorithm is required which filters the incoming signal at the geophone to isolate those representing human footsteps. To overcome the problem of varying environmental conditions (mostly soil quality and humidity as well as ground coverage with plants) and of disturbing seismic noise, a procedure with an adaptive threshold is developed and implemented. It is based on the so-called "sta/lta-picking" which is used in earthquake location scenarios [20]. The decision whether there is an event is made by comparing the ratio of a short and a long time average of the seismic signal with an empiric constant. By introducing an adaptive constant an accommodation to varying conditions is achieved. An example is shown in Figure 10.

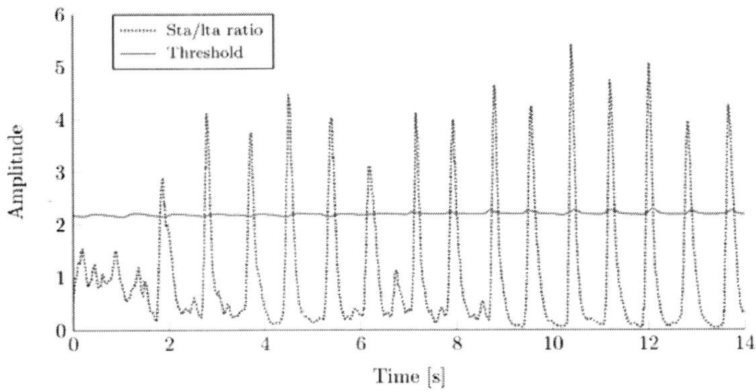

Figure 10. Results of the geophone event picking algorithm.

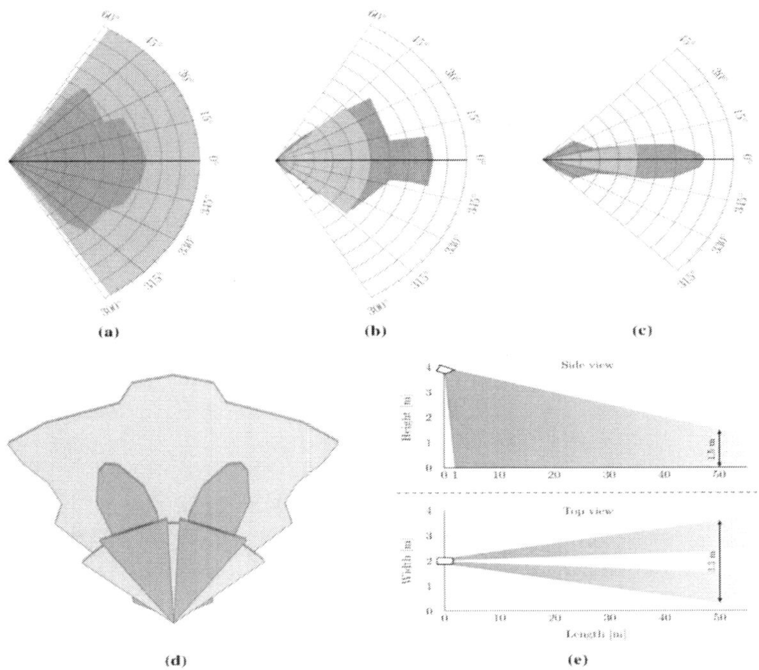

Figure 11. Experimental evaluation of the detection ranges of the different PIR sensor types. Green marks the detection area specified in the datasheet and red the actual detection area for walks with normal speed. (a) AMN34111 (Single). (b) AMN31111 (Multi). (c) AMN33111 (Multi). (d) Multi-PIR. (e) Long-Range-PIR: IS392.

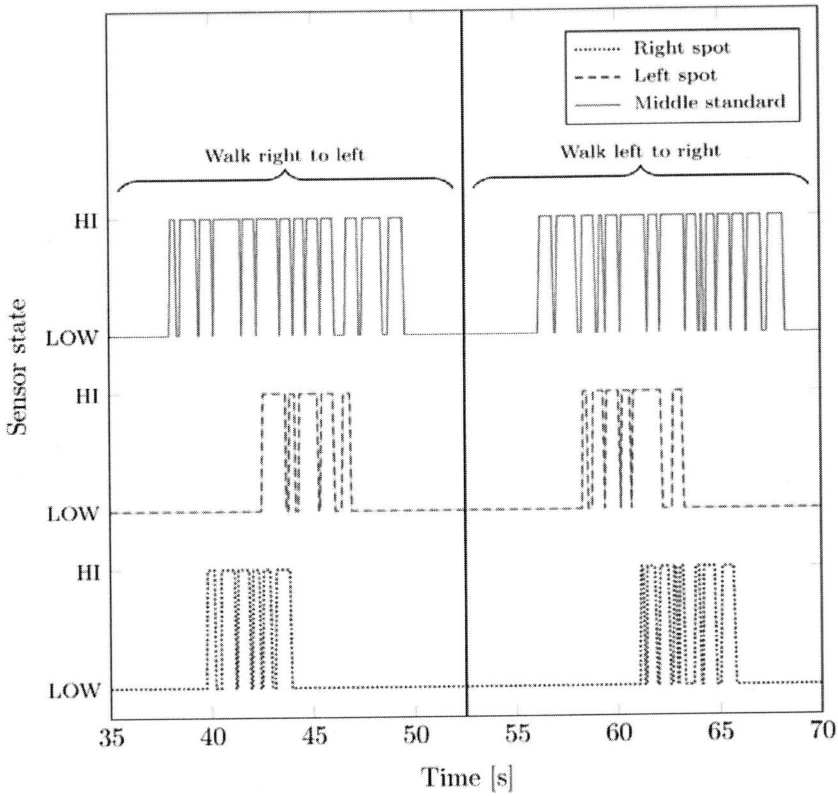

Figure 12. Typical PIR sensor activity of a Multi-PIR sensor node when an object passes the node from right to left and vice versa.

The output event of a geophone reliably signals a person within the detection range. By calculating the geometric center of gravity of one or more geophones reporting events with the same timestamp, it is possible to locate (and track) an intruder within the sensor network.

4.2. PIR Sensors

For the PIR sensors, the region in which objects produce sensor events must be known in order to perform a localization of intruding objects. It is common to conduct simple walking and driving tests for this purpose, for which a path is defined and repeatedly driven, or walked, respectively. In Figure 11 the results for normal walking are given. Green marks the detection area specified in the datasheet, while red marks the region in which sensor events actually occur in practice. In Figure ??the joint detection region of a Multi-PIR sensor node is visualized. Here green marks the region where the two spot type PIR sensors (AMN33111) are sensitive and blue the region of the standard PIR type (AMN31111).

In Figure 12, the exemplary behavior of a Multi-PIR sensor node is shown for a walk-by test in which a person crosses the detection area from right to left with normal walking speed and vice versa. It has been found that the direction of the object can be determined reliably simply using the 3-PIR sensors included in a Multi-PIR sensor node. To do so, the time at which each PIR sensor activates is measured and compared. This is not possible using the Single-PIR sensors, but promises more accurate localization and tracking of objects.

4.3. AMR Sensors

For the vehicle detection based upon the AMR sensors, the activation procedure, the sensor range and detection algorithms were tested prior to the implementation of the final system. The according test setup is shown in Figure 13.

Upon a sensor event from its PIR sensor, the sensor node sends a wireless message to the two AMR sensor nodes, simulating a delay as it would occur in the final installation, where the communication would work via the clusterhead, instead of directly between the sensor nodes. Upon reception of that message, the AMR nodes activate their vehicle detection modules, de-gauss and calibrate the sensors to compensate for the static earth magnetic field, start sampling the two sensor module channels, and forward the live data to a forth sensor node that is connected to a PC to record timestamps and sensor data for all three nodes.

Two sensor nodes with a vehicle detection module are positioned at opposite sides of a road in a distance of 8m from each other. A sensor node with a PIR sensor is placed next to the road 15m before the AMR nodes. A vehicle then passes the installation at speeds of 50km/h or 5km/h , respectively.

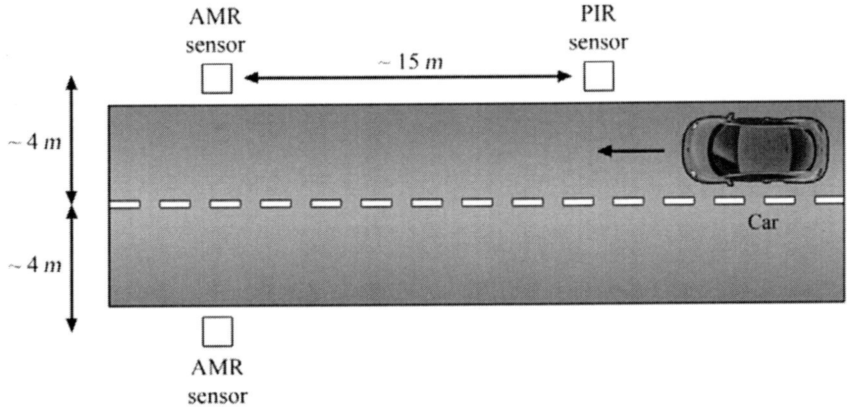

Figure 13. Field test of the AMR sensor.

Figure 14 shows the sensor readings of one of the AMR nodes for a vehicle speed of 50km/h . The time axis of the diagram is set to start at 0 at the time when the PIR sensor event occurred. The y-axis shows the sensor signal as ADC digits, where 1 digit represents 0.59mV or a field change of 786.2mV/(kA/m) . It is visible that the sensor signal starts around t= 0.2s , as the calibration process commences before that.

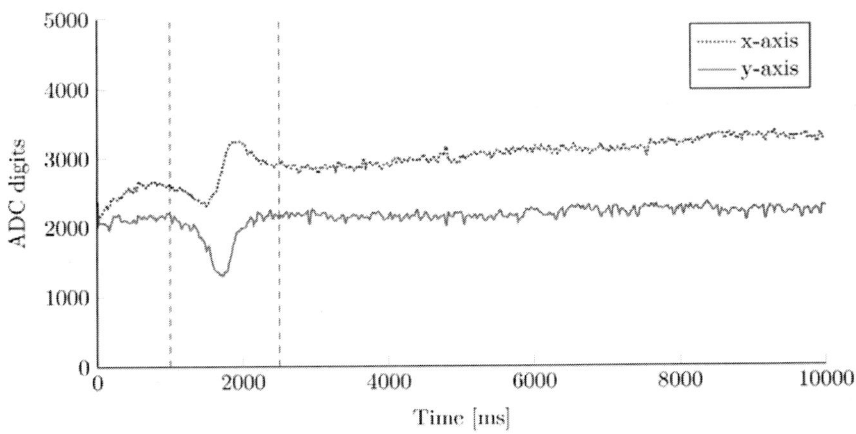

Figure 14. Sensor data with fast car movement (50 km/h).

The AMR sensor starts up and calibrates fast enough to be in normal operation by the time the vehicle passes by the sensor (characteristic signal shape between t= 1s and t= 2.3s). Consequently, the system would still work properly if the PIR sensor is placed at a distance of only 3m to the AMR sensors.

Figure 15 shows the sensor readings of one of the AMR sensors for a vehicle speed of 5km/h . As expected, the signal occurs much later, between t= 9s and t= 16s . Here, the challenge is rather to develop a detection algorithm that detects vehicles from both the signal patterns that result from 5km/h and 50km/h , while still being robust against the signal drift that occurs even though no vehicles passes.

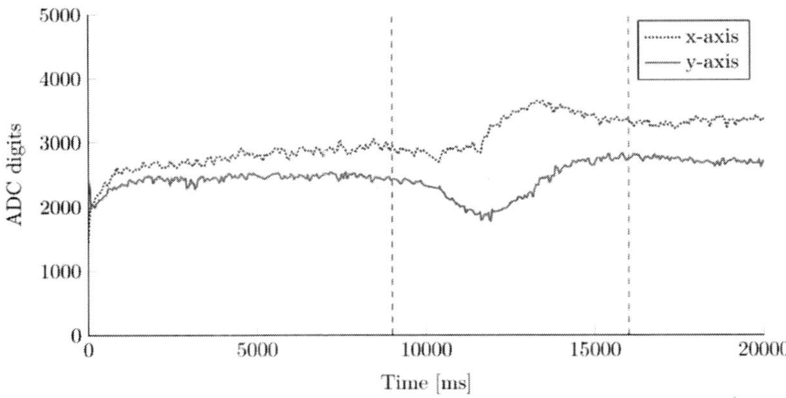

Figure 15. Sensor data with slow car movement (5 km/h).

5. NETWORK IMPLEMENTATION AND ASSESSMENT

In this section, the tests performed to evaluate the overall system performance in the areas of communication, object detection as well as security and safety are described. This is done by two different testing methods: tests in a laboratory environment on the one hand and live field trials under working conditions on the other. The testing methodology to verify the different parts of the sensor network is described in the following.

5.1. Laboratory Tests

5.1.1. Functional Safety

To ensure the proper function of all critical system components, the functional safety tests are conducted in a laboratory test to ensure a controlled environment. The assessment includes the manipulation of all protected sensors and a test for energy drainage, that is run to verify the expected life time of the sensor nodes. Tests for memory errors in RAM or flash, or CPU failures are not done, as causing such errors can only be achieved by either willingly damaging the hardware or by environmental factors (like radiation) that can only be simulated using highly specialized equipment that is not available in a normal laboratory. A detailed listing of the tests is given in Table 5.

Table 5. Functional safety tests details.

#	Function	Action Performed	Time
1	AMR	Removal of module	Before start
2	AMR	Removal of module	During operation
3	Acceleration	Removal of module	Before start
4	Acceleration	Removal of module	During operation
5	Geophone	Disconnection of cable	Before start
6	Energy supply	Battery draining	During operation

On each sensor module, a connection plug is located in the upper right area of the module. In the tests, the module-removal is done by pulling the module off the rest of the sensor node stack. The geophone is disconnected by pulling the connection-cable and thereby interrupting the connection with the sensor node.

5.1.2. Adaptive Frequency Hopping and Communication Robustness

AFH mechanism is also verified in a separate laboratory test. The test arrangement for the different assessments is shown in Figure 16.

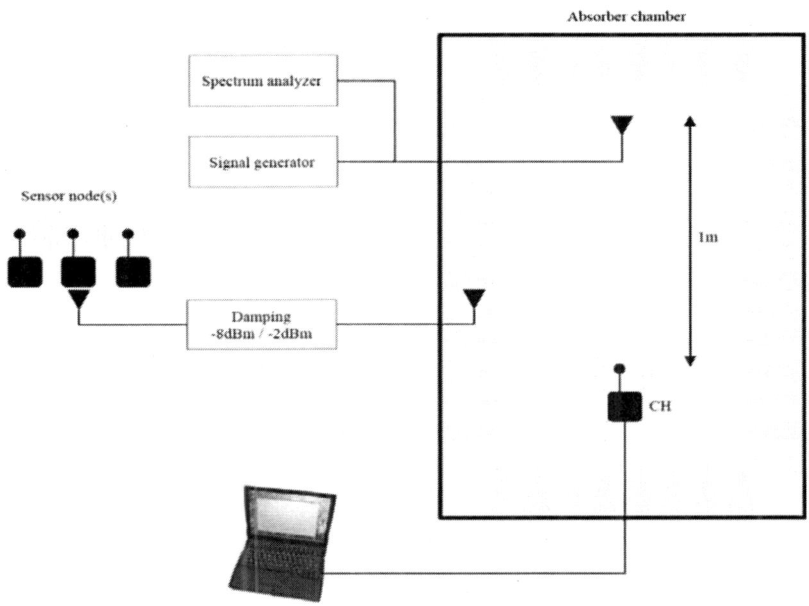

Figure 16. Adaptive frequency hopping and communication robustness test setup.

Figure 17. Absorber hall with jamming device and clusterhead.

As depicted in Figure 17, the sensor node part of a clusterhead is placed inside of an absorber chamber and linked to a laptop via a serial connection to enable the logging of the test results. Additionally, a FM jamming signal emitter is placed at a fixed distance of 1m inside the chamber. The jamming signal's frequency can be remotely configured from the outside. To measure the signals inside the absorber chamber, a spectrum analyzer is used.

The signals of several sensor nodes—which are located outside of the absorber chamber—are routed through an antenna to the inside, simulating a distance of approximately 10m between sensor nodes a clusterhead by damping the signal. In all tests, the sensor node(s) send a 10Byte long packet every 50ms (with an additional sending jitter of 10ms to avoid synchronization between the nodes), leading to an overall data rate of 200Byte/s . As the AFH switches between different channels every 250ms , 5 packets are sent on each channel. A jammed channel is assumed if the data receive rate drops below 75% in the tests.

During the tests, different channels are jammed by an FM jammer. It operates at a bandwidth of 1MHz around the center frequency of the according channel at an output power of −50dBm with a NF-frequency of 400Hz .

5.2. Field Trial

In the course of the system development, two field tests of the whole WSN were conducted. Figure 18 shows an illustration of the first testing terrain painted by the Spyglass application. The first assessment consisted of 100 sensor nodes and 10 clusterheads organized in 10 clusters, where one cluster was reserved for energy consumption monitoring. The surveillance area was a field with a path and had a maximum width of 125m and a length of 50m at the largest extend. The nodes were distributed arbitrarily in distances of 5m to 10m from each other. After this initial setup, the sensor clusters were defined by joining 10 sensor nodes n_{ij} , $1 \leq j \leq 10$ nearest to ch_i together in one cluster cl_i . After the WSN was fully configured, several aspects of the whole system, and their interactions, had to be verified: First, the general system functions, like network connectivity, message sending, relaying and receiving as well as displaying of system events at the command center. Second, the detection, localization and classification had to be assessed and third a security evaluation to test the WSN's security and safety features was performed.

In the second field test of the WSN a total number of 17 sensor nodes partitioned in 2 clusters were utilized. The surveillance area was a small field surrounded by bushes and buildings with a path going through it. Again, a set of walking and driving detection test were performed.

A complete list of the test scenarios performed during the first and second field trial, including both detection and security tests, is given in Table 7.

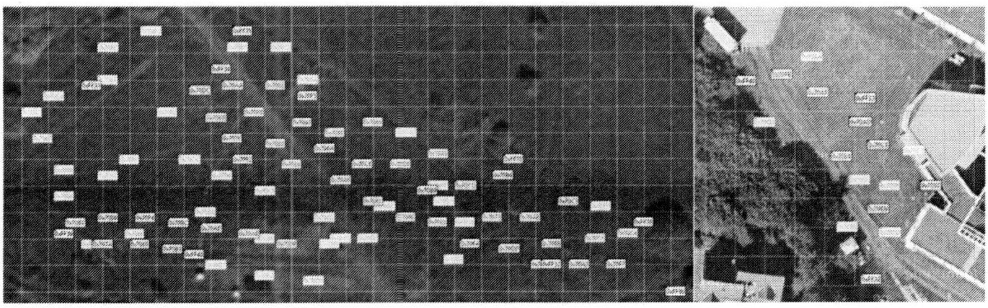

(a) (b)

Figure 18. Spyglass renderings of the maps for the field tests performed within the scope of the system development (grid size is 5m ×5m) (a) Large field test; (b) Small field test.

6. RESULTS

6.1. Laboratory Tests

6.1.1. Adaptive Frequency Hopping and Communication Robustness

Figure 19 below shows the results of the AFH communication tests. It depicts the sequence of channel jamming, jamming detection and the closing of channels by the clusterhead, as well as the amount of data from the three sensor nodes employed during testing arriving at the clusterhead.

At t= 34s , the jammer blocks channel 3. At t= 37s , t= 41s and t= 45s , the clusterhead tries to send data using the blocked channel. As a result, the data sent on that channel is lost, the arriving data rate drops, and the clusterhead detects the jamming. Consequently, it closes channel 3 at t= 45s (cf. "number of closed channels" goes to 1 in Figure 19). After that, the data rate stabilizes again.

The same procedure happens from t= 68s to t= 76s with channel 8 and from t= 89s to t= 97s with channel 12.

The fact that the data rate stabilizes shortly after the jamming of a channel starts shows that the AFH scheme implemented in the system effectively allows to detect channel jamming, and that the channel closing algorithm significantly helps to preserve communication despite of jamming attacks.

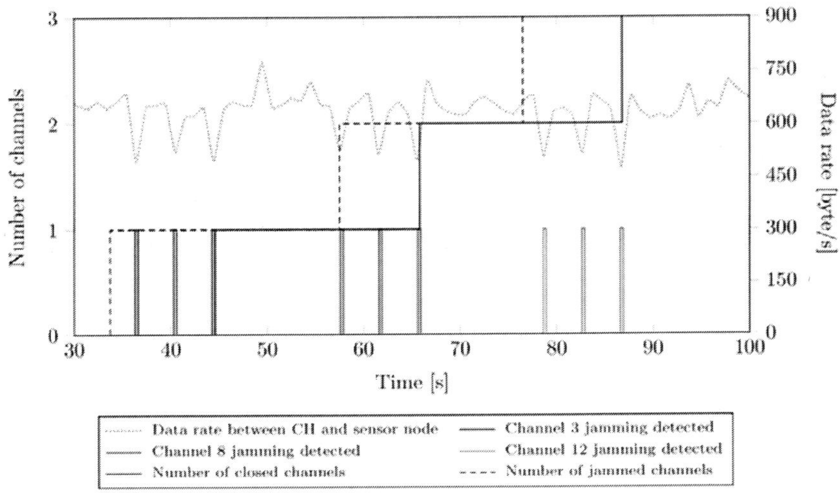

Figure 19. Events occurred during the AFH and jamming tests.

6.1.2. Functional Safety

The functional safety tests described in Table 5 all show the expected results. As stated in Section 5.1, the tests are done by logging the debug messages sent by the sensor nodes. This is visible in Code 3, where the accelerometer is removed during the operation of an arbitrarily chosen sensor node. As line 4 of the log shows, a hardware failure is detected.

Code 3 Log of the test #4

```
[2011-09-13 16:25:55.669] pir2 down at 27:14.754
[2011-09-13 16:25:55.669] pir2 up at 27:14.757
[2011-09-13 16:25:55.825] pir0 up at 27:14.899
[2011-09-13 16:25:55.825] HW-FAILURE DETECTED. ID: 3
[2011-09-13 16:25:55.872] pir0 down at 27:14.952
[2011-09-13 16:25:55.981] HW-OK DETECTED. ID: 2
```

The ID of the error is the position of the error flag in the status byte described in Figure 7. The results of the other sensors behave in a similar way to the assessment of the accelerometer. As an exhaustive listing of all results would exceed the length of this paper, as a second example, the evaluation of the AMR sensor can be seen in Code 4.

Code 4 Log of the test #2

```
[2011-09-21 17:20:40.387] pir0 down at 27:01.567
[2011-09-21 17:20:40.402] calibrate AMR
[2011-09-21 17:20:40.418] JI2C: No ack rx
[2011-09-21 17:20:40.433] AMR flip
[2011-09-21 17:20:40.449] JI2C: No ack rx
[2011-09-21 17:20:40.449] AMR flip
[2011-09-21 17:20:40.465] JI2C: No ack rx
[2011-09-21 17:20:40.480] AMR flip
[2011-09-21 17:20:40.480] JI2C: No ack rx
[2011-09-21 17:20:40.496] AMR flip
[2011-09-21 17:20:40.511] HW-FAILURE DETECTED. ID: 2
```

Another part of the evaluation is the test of the tamper resistance feature based on the accelerometer. To trigger this manipulation alarm, the sensor node is lifted off the ground and then shook several times. The result is a visible indicator ("Accelerometer event" in Table 6) at the command center, signaling a possible manipulation.

Table 6. Event symbols and classification colors in Spyglass.

Symbol	Meaning	Symbol	Meaning
	Geophon event		AMR event
	Longrange-PIR event		Multi-PIR event
	Accelerometer event		Single-PIR event
	Node communication issue		Node failure
	Energy drain alert		
Color	**Classification**	**Color**	**Classification**
	Human		Vehicle
	Human with metal object		Unknown

6.2. Field Trial

To verify the functions assessed in the field trials, multiple verification techniques were used. For the detection, localization and classification, a Mobotix IP-camera was used in combination with the Spyglass user interface. By doing so, it is possible to check if objects were on the one hand detected and localized, on the other if the classification works correctly and the real-time requirement of the system is fulfilled.

As an example, Figure 20 shows one of the walking tests done during the second field trial. As the figure depicts, the person walking on the grass was detected and classified correctly, which is indicated by the green field in the middle of the grid. Similar results were also achieved for other walking styles, like crawling or running, as well as tests with other trespassing objects, like e.g., cars (the classification changed in those tests respectively).

Figure 20. Example of a walking test during the second field trial.

The field trials were also used to optimize the system parameters, especially the calibration of the sensor sensitivity posed a challenge during the first assessments due to the complex processing and classification applied to the sensor events. The geophones are particularly sensitive to parameter changes, as they support a very broad range of sensitivity values. These values have to be adjusted in a way that they are on the one hand reactive enough to detect even a crawling human trespasser, while on the other they are not overreactive, which would lead to a too high number of geophones responding to seismic events at once. This, in turn, would disturb the localization capabilities of the overall system as the geophone reaction would not be localized enough.

Table 7. List of performed field trial tests.

Detection Test Scenarios					
Test #	Distance to Sensor Nodes	Terrain	Type of Trespasser	Employed # of Clusters	Penetration
1	2 m	Short grass	Person	1	Yes, shortest path
2	2 m	High grass	Person	1	Yes, shortest path
3	2 m	Stony ground	Person	1	Yes, shortest path
4	4 m	Short grass	Person	1	Yes, shortest path
5	4 m	High grass	Person	1	Yes, shortest path
6	4 m	Stony ground	Person	1	Yes, shortest path
7	10 m	Short grass	Person	1	Yes, shortest path
8	10 m	High grass	Person	1	Yes, shortest path
9	10 m	Stony ground	Person	1	Yes, shortest path
10	Variable	Short grass	Person carrying metal object	Arbitrary	No
11	Variable	Short grass	Person carrying metal object	Arbitrary	Yes
12	Variable	Stony ground	Car	Arbitrary	No
13	Variable	Stony ground	Car	Arbitrary	Yes, 10 km/h
14	Variable	Stony ground	Car	Arbitrary	Yes, 40 km/h
15	Variable	Stony ground	Person	Arbitrary	Yes, running
16	Variable	Short grass	Person	Arbitrary	Yes, duck gait
17	Variable	Short grass	Person	Arbitrary	Yes, creeping
18	Variable	Short grass	Person	1	Yes, zigzag pattern
19	Variable	Short grass	Person	2	Yes, zigzag pattern
20	Variable	Short grass	Person	1	No
21	Variable	Short grass	Person	2	No
22	Variable	Stony ground	Bicycle	Arbitrary	Yes
23	Variable	Stony ground	Motorcycle	Arbitrary	Yes, slow
24	Variable	Stony ground	Motorcycle	Arbitrary	Yes, fast
25	Variable	Stony road	Person	1	Yes, shortest path
26	Variable	Stony road	Person	2	Yes, running
27	Variable	Stony road	Car	Arbitrary	Yes, slow
28	Variable	Short grass	Parson	Arbitrary	Yes, shortest path
29	Variable	Short grass	Person	Arbitrary	No, U pattern

Security Test Scenarios			
Test #	Terrain	Action Performed	Type of Trespasser
30	Short grass	Shaking of a sensor node	Person
31	Short grass	Removal of a sensor node out of the test field	Person
32	Short grass	Reactivation of a sensor node in the test field	Person
33	Short grass	Measurement of the energy consumption of a cluster	—

In the energy endurance test (Test #33 in Table 7) performed during the first field trial, it became clear that a glitch in the clusterhead software hindered the GPS module to enter standby mode. This caused a significant energy drain, resulting in a severely shortened lifetime of only six days. This problem was fixed by implementing an adaptive standby algorithm, which compares the internal clock $T_{internal}$ of the clusterhead with the GPS time signal T_{gps} in increasingly large time intervals T_{check}. As long as the time difference $\Delta T=|T_{internal}-T_{gps}|<1ms$, T_{check} is doubled every time the time difference test is run, starting at 20s and ending at a maximum of 3600s. As a result, the lifetime of the clusterheads was extended to 17 days on average. The energy consumption of the sensor nodes was also assessed. In the field trials, which lasted for more than two weeks, no energy-related unavailability could be discovered. After an extensive analysis with energy consumption models (Section 3.3.2), an average node lifetime is expected to be 28 days (depending on the sensors used and the amount of registered events).

Additionally, the detection tests showed that it is possible to detect, locate and classify persons and vehicles (cars & motorbikes) with high confidence (Tests #1–3 and #12–29 in Table 7). Furthermore, it was evaluated if a classification of a person carrying metal objects is possible, which would be especially useful to detect weapons like firearms. While the detection works with large ferromagnetic objects, like e.g., fire extinguishers or crowbars, as long as the distance between the object and the sensor node is ≤ 2m , it is not possible for firearms. This is due to the low mass of ferromagnetic parts in modern firearms, resulting in a very low amplitude of the AMR sensor's readings, which does not trigger an event. All classification grid colors and symbols used in Spyglass are shown in Table 6.

To assess the limitations of the system, its behavior in borderline cases was analyzed by running several tests that evaluated the detection and localization limits of the sensors. As the localization is mainly done by the PIR sensors and geophones, this scenarios simulated the failure of several nodes in the same area, leaving gaps of 4m (Tests #4–6 in Table 7), or 10m (Tests #7–9 in Table 7) respectively, between the sensor nodes. While a detection and localization of a trespasser was still possible at a sensor node distance of 4m without a major impairment, this changed when the distance was increased to 10m . Due to the limited range of the PIR sensors (~6.5m , cf. Figure 11), in the test cases #7–9, where a 10m distance was used, both the localization and classification capabilities suffered a severe impact. The network was however still able to identify that an object was present in the surveillance area.

To test the system's ability to detect unavailable sensor nodes (e.g., due to physical removal or damaged wireless interfaces), one test scenario included the removal and shutdown of a single node out of the sensor network (Test #31 in Table 7). The removal first resulted in a tampering alarm, because of the necessary physical manipulation of the node by removing it. As a wireless communication is prone to data loss, the indication of node unavailability was realized using two stages, which indicate a temporally limited or permanent loss of node availability. At first, the status of the node is set to "Node communication issue" (cf. Table 6), which is triggered if a node's heartbeat message is not received three times in a row. If a node resumes communication within 7s after its status was set to "Node communication issue", it is reset to unimpaired status. If no heartbeat is received within 7s , the node is marked as "Node failure" (cf. Table 6), meaning it is permanently unavailable. In the test, the described behavior could be observed.

Moreover, the varying weather conditions during the field trials, ranging from sunshine to thunderstorms, proved the system's adaptability and its resistance against hazardous environments.

7. RELATED WORK

The presented work is manifold combining multiple research fields, e.g., communication security, functional safety but also energy efficient communication and the fusion of WSN data. Therefore, this section first covers the state-of-the-art in conventional (non-networked) solutions for secure infrastructures. Second, related work on systems that rely on wireless communication and work in critical environments is described. Finally, recent projects that partly cover the goals of our work are introduced and compared.

Conventional solutions to secure critical infrastructures range from very simple setups involving mechanical barriers, like doors, fences and trenches or security personnel [21] to complex—however non-networked—equipment, including e.g., passive infrared (PIR), sound and seismic sensors, or video surveillance. While these devices allow detecting certain influences, they lack the possibility to cooperatively analyze a situation [22], as no data transfer between them is possible.

However, over the last years, advances in the area of networked sensor technologies have opened up new application areas for these sensor devices [23]. With new technological paradigms, like the Internet of Things (IoT) or Automated Living, the use of WSNs is rapidly growing, as networked sensors are vital in these scenarios. However, as future technologies more and more depend on sensor networks, the requirements for these—now critical infrastructures—rise in a similar way, especially in the fields of security and safety. An already realized example in this area is the SmartSantander project [24], where more than 10,000 IoT devices are deployed to build a so-called "Smart City". In such scenarios, strict security requirements apply, as personal and critical information is transferred. While there are research efforts in the areas of access protection, secure reprogramming over wireless connections and secure communication, a complete security solution for a wireless network does not yet exist. This example already characterizes several important aspects of WSN security: Due to the usage in unsupervised or even hostile environments while performing critical tasks, it is important to provide measures ensuring integrity and confidentiality of data (communication security) and protection of the devices against malfunctions that either occur randomly or due to tampering [25,26].

The POmSe ("Personen- und Objektdetektion mit mobilen Sensoren") project [27], evaluates the general applicability of WSNs for the protection of critical infrastructures and environments with a focus on the applicability for trespasser detection due to their inherent advantages compared to conventional security techniques (like e.g., security doors, fences, etc.). In contrast to our work, POmSe performs a general analysis of WSNs including the required security features in critical applications. However, the development, application and evaluation of a WSN is not included in the project, neither is a classification of trespassing objects performed.

A complementary project, named Personen- und Objekterkennung basierend auf Trittschall (POT) [28], evaluates the suitability of seismic sensors for the detection, classification and localization of humans and animals. It turns out that a reliable detection algorithm can be provided whereas the exact spatial localization of a person strongly depends on the environmental conditions as, e.g., the type and humidity of the surrounding soil.

There are also some other works, such as FleGSens [29], Line in the Sand [30], Border Sense [31] etc, leverage WSNs for border intrusion detection and surveillance [32]. FleGSens is able to to detect movement in critical areas and considers authenticity and data integrity of alarm signals. Other security issues, like jamming and functional safety, are not addressed. FleGSens uses a flat network hierarchy, which limits system scalability due to message collision. The used detection algorithm only considers the position of trespassers and—in contrast to our work—does not include a classification algorithm. Other works can either detect, classify and (or) track metal and nonmetal objects or detect passing ships and their speeds. However, security issues, like security for network topologies and security for data communication, are not tackled. As argued in [32], few WSN-based mission-critical applications are really experimented and deployed. Compared with the previous works, the main difference is the implementation of a comprehensive security solution covering all necessary aspects, while still maintaining a high usability despite the high system complexity. To ensure its applicability for critical infrastructure monitoring, in this paper, the involved sensors are comprehensively evaluated and tested in both laboratory environment and live field trials. This paper improves our previous work [33] by implementing functional safety / self-protection functions and adding energy supply awareness. Further, this paper also includes laboratory tests for adaptive frequency hopping, communication robustness and functional safety.

8. CONCLUSIONS AND FUTURE WORK

As this paper demonstrated, the developed system is a WSN solution that is capable of monitoring and securing critical infrastructures. To do so, it offers security by using different sensors to detect, locate, track and classify various types of trespassers while operating in a secure and safe way. This is done by employing several functional safety and security features which guarantee high levels of confidentiality, integrity and availability. In addition to the features described before, the system operates in real-time using its efficient detection algorithms and network hierarchy. It enables its user to keep track of the events inside the sensor network by logging all events in a database for later analysis and, at the same time, reporting it in real-time to the command center. Moreover, the system uses a scalable, hierarchic network structure making it easily extendable for a wider area, if necessary. To adapt the system to changing usage conditions, it is possible to dynamically reconfigure the whole system in the field by using an over-the-air-programming (OTAP) function. This unique combination of features makes the system ideal for unattended operations in hazardous environments.

While the described features already enable a secure detection of trespassing objects, future work could go in three directions. One possibility would be to include additional hardware in the concept, e.g., IP-cameras, to document trespassing. These devices would be integrated as a type of additional sensor into the existing sensor nodes and be activated if an object is detected by the other sensors. Of course, these new nodes would need to be carefully analyzed, especially their energy consumption and network traffic generation would need to be carefully balanced with the achievable visual quality. Another possible method would be to treat the sensor nodes with visual capabilities separately (and not integrate them into the existing network hierarchy). The visual data would e.g., be sent directly to the command center using WLAN. This would on the one hand save energy and network bandwidth, by bypassing the clusterheads and IEEE 802.15.4 network altogether, however, the hierarchical network structure would be broken using that concept. Therefore, a critical analysis would need to be performed before deciding which strategy to choose.

Another option would be to extend the software used for object detection. Currently, the software is able to monitor and detect the trespassing of a single target in the surveillance area ("Single target"). In the future, this software could be extended to enable the monitoring of multiple targets at once ("Multi target"). In addition, the current trespassing decision method offers a binary approach: either the target is trespassing or not. This method could be enhanced by adding a risk estimate that is based on the trespasser's speed and direction. Another promising direction is to extend the current system by integrating mobile sensors. There are various forms of these devices, such as sensor-equipped cell phones, moveable sensor nodes or other mobile, networked devices using appropriate sensors. This would have the benefit that mobile sensors can provide context awareness based on the location and trajectory of a trespassing entity.

ACKNOWLEDGMENTS

The research leading to the results presented in this paper was supported by the "Security, Educatuion and Competence for Bavarian IT" (SECBIT) project, which is co-funded by the European Union's European Regional Development Funds - Regional Competitiveness and Employment, the European Commission's Project No. 608090, HyRiM (Hybrid Risk Management for Utility Networks) under the 7th Framework Programme (FP7-SEC-2013-1), as well as the project MOVEDETECT supported by the German Federal Office for Information Security (BSI—Bundesamt für Sicherheit in der Informationstechnik).

AUTHOR CONTRIBUTIONS

Michael Niedermeier and Hermann de Meer coordinated the project leading to the results in a technical and organizational manner. They also contributed the security and safety solution utilized by the WSN. Michael Koch performed all technical and practical work related to geophones; Stefan Fischer and Dennis Pfisterer were responsible for the Wisebed and Spyglass applications; Klaus Hartmann and Benjamin Langmann investigated and chose the different sensor types employed in the sensor nodes. They also provided major contributions in the sensor data aggregation and interpretation. Carsten Buschmann provided insight knowledge on sensor nodes and their communication abilities. The writing of the paper was mainly performed by Michael Niedermeier and Xiaobing He, with editorial support by Hermann de Meer.

REFERENCES

1. Estrin, D.; Govindan, R.; Heidemann, J.; Kumar, S. Next Century Challenges: Scalable Coordination in Sensor Networks. In Proceedings of the 5th Annual ACM/IEEE International Conference on Mobile Computing and Networking (MobiCom 99), Seattle, WA, USA, 15–19 August 1999.
2. Asada, G.; Dong, M.; Lin, T.S.; Newberg, F.; Pottie, G.; Kaiser, W.J.; Marcy, H.O. Wireless integrated network sensors: Low power systems on a chip. In Proceedings of the 24th IEEE European Solid State Circuits Conference (ESSCIRC 98), The Hague, The Netherlands, 22–24 September 1998.
3. SpyGlass, a Modular and Extensible Visualization Framework for Wirelesssensor Networks. Institute of Telematics, University of Lübeck: Lübeck, Germany, 2006; Available online: https://github.com/itm/spyglass (accessed on 23 October 2015).
4. coalesenses GmbH. iSense Core Module 3 Data Sheet. 2010. Available online: http://www.coalesenses.com/ download/data_ sheets/DS_ CM30X_1v3.pdf (accessed on 23 October 2015).
5. coalesenses GmbH. Gateway Module Data Sheet. 2010. Available online: http://www.coalesenses.com/download/gateway_module_data_sheet_1v2.pdf (accessed on 23 October 2015).
6. coalesenses GmbH. GPS Module Product Brief. 2010. Available online: http://www.coalesenses.com/download/product_ briefs/ProductBriefGpsModule.pdf (accessed on 23 October 2015).
7. Buschmann, C.; Pfisterer, D. iSense: A Modular Hardware and Software Platform for Wireless Sensor Networks. In Technical Report, 6. Fachgespräch Drahtlose Sensornetze der GI/ITG-Fachgruppe Kommunikation und Verteilte Systeme; RWTH Aachen: Aachen, Germany, 2007.
8. NXP Laboratories UK. Data Sheet: JN5148-001 IEEE802.15.4 Wireless Microcontroller. 2010. Available online: http:// www.jennic.com/ files/product_briefs/JN-DS-JN5148-1v6.pdf (accessed on 23 October 2015).
9. IEEE-SA Standards Board. IEEE Standard for Information Technology—Telecommunications and Information Exchange between Systems—Local and Metropolitan area Networks Specific Requirements-Part 15.4: Wireless Medium Access Control (MAC) and Physical Layer (PHY) Specifications for Low-Rate Wireless Personal Area Networks (LR-WPANs); Technical Report; IEEE Computer Society: New York, NY, USA; 8; September; 2006.

10. The Federal Information Processing and Announcing. Announcing the Advanced Encryption Standard (AES); FIPS 197. Federal Information Processing and Announcing: Gaithersburg, MD, USA, 26 November 2001.
11. SIEMENS. IS392, IS392H Outdoor Passive Infrared Detector. 2005. Available online: http://www.ntcbpl.ru/pdfinfo/Data-sheetIS392 en.pdf (accessed on 23 October 2015).
12. Panasonic. MP Motion Sensor (AMN1,2,4). 2004. Available online: http://datasheet.octopart.com/AMN34111-Panasonic-datasheet-110118.pdf (accessed on 23 October 2015).
13. Input/Output, Inc. SM-24 Geophone Element. 2006. Available online: http://cdn.sparkfun.com/datasheets/Sensors/Accelerometers/SM-24%20Brochure.pdf (accessed on 23 October 2015).
14. coalesenses GmbH. Vehicle Detection Module Data Sheet. 2010. Available online: http://www.coalesenses.com/download/product_briefs/ProductBriefVehicleDetectionModule.pdf (accessed on 23 October 2015).
15. Coulson, G.; Porter, B.; Chatzigiannakis, I.; Koninis, C.; Fischer, S.; Pfisterer, D.; Bimschas, D.; Braun, T.; Hurni, P.; Anwander, M.; et al. Flexible experimentation in wireless sensor networks. Commun. ACM 2012, 55, 82–90.
16. Mpitziopoulos, A.; Gavalas, D.; Konstantopoulos, C.; Pantziou, G. A survey on jamming attacks and countermenasures in WSNs. IEEE Commun. Surv. Tutor. 2009, 11, 42–56.
17. Whiting, D.; Housley, R.; Ferguson, N. Counter with CBC-MAC (CCM). 2003. Available online: http://tools.ietf.org/html/rfc3610 (accessed on 23 October 2015).
18. Lashkari, A.H.; Danesh, M.M.S.; Samadi, B. A survey on wireless security protocols (WEP, WPA and WPA2/802.11i). In Proceedings of the 2nd IEEE International Conference on Computer Science and Information Technology (ICCSIT 2009), Beijing, China, 8–11 August 2009; pp. 48–52.
19. Fletcher, J.G. An arithmetic checksum for serial transmissions. IEEE Trans. Commun. 1982, 30, 247–252.
20. Havskov, J. Instrumentation in Earthquake Seismology (Modern Approaches in Geophysics); Springer Science + Business Media B.V.: Dordrecht, The Netherlands, 2004.
21. Practices for Securing Critical Information Assets; Critical Infrastructure Assurance Office: Gaithersburg, MD, USA, 2000.
22. Intanagonwiwat, C.; Govindan, R.; Estrin, D. Directed diffusion: A scalable and robust communication paradigm for sensor networks. In Proceedings of the 6th Annual ACM/IEEE International

Conference on Mobile Computing and Networking (MobiCom 00), Boston, MA, USA, 6–11 August 2000; pp. 56–67.
23. Akyildiz, I.F.; Su, W.; Sankarasubramaniam, Y.; Cayirci, E. Wireless sensor networks: A survey. Comput. Netw. 2002, 38, 393–422.
24. The SmartSantander Consortium. SmartSantander. 2010. Available online: http://www.smartsantander.eu/ (accessed on 23.10.2015). [Google Scholar]
25. Roman, R.; Zhou, J.; Lopez, J. On the security of wireless sensor networks. In Proceedings of the International Conference on Computational Science and Its Applications—ICCSA 2005, Suntec City, Singapore, 9–12 May 2005; pp. 681–690.
26. Perrig, A.; Stankovic, J.; Wagner, D. Security in Wireless Sensor Networks. Commun. ACM 2004, 47, 53–57.
27. Bonitz, F.; Ghobadi, S E.; Hartmann, K.; Hauff, H.; Herrmann, R.; Kargel, C.; Löpprich, O.E.; Heckmann, D.; Maisch, M.M.; Seidl, A.; et al. Personen- und Objektdetektion mit mobilen Sensoren—Abschlussbericht zum Archbeitspacket 3 (Teil B)—Bericht zum Meilenstein 5; Internal End Report; University of Passau: Passau, Germany, 2010.
28. Koch, M.; Hubert, C. Personen- und Objekterkennung Basierend auf Trittschall (POT); Internal Technical Report; University of Passau: Passau, Germany, 2011.
29. Rothenpieler, P.; Krüger, D.; Pfisterer, D.; Fischer, S.; Dudek, D.; Haas, C.; Zitterbart, M. Flegsens—Secure area monitoring using wireless sensor networks. In Proceedings of the 4th Safety and Security Systems in Europe, Micro Materials Center Berlin at Fraunhofer Institute IZM and Fraunhofer ENAS, Potsdam, Germany, 4–5 June 2009.
30. Arora, A.; Dutta, P.; Bapat, S.; Kulathumani, V.; Zhang, H.; Naik, V.; Mirral, V.; Cao, H.; Demirbas, M.; Gouda, M. A line in the sand: A wireless sensor network for target detection, claffisication, and tracking. Comput. Netw. 2004, 46, 605–634.
31. Sun, Z.; Wang, P.; Vuran, M.C.; Al-Rodhaan, M.A.; Al-Dhelaan, A.M.; Akyildiz, I.F. BorderSense: Border patrol through advanced wireless sensor networks. Ad Hoc Netw. 2011, 9, 468–477. [Google Scholar] [CrossRef]
32. Felemban, E. Advanced Border Intrusion Detection and Surveillance Using Wireless Sensor Network Technology. Int. J. Commun. Netw. Syst. Sci. 2013, 6, 251–259.
33. Langmann, B.; Niedermeier, M.; de Meer, H.; Buschmann, C.; Koch, M.; Pfisterer, D.; Fischer, S.; Hartmann, K. MOVEDETECT—Secure detection, localization and classification in wireless sensor networks. In Internet of Things, Smart Spaces, and Next Generation

Networking; Balandin, S., Andreev, S., Koucheryavy, Y., Eds.; Springer Berlin Heidelberg: St. Petersburg, Russia, 2013; Volume 8121, pp. 284–297.

INDEX

5G wireless, 63, 64, 65, 66, 67
5GBeyond 2020, 189

A

Actuators, 224, 226, 227
Automation, 223, 224, 226, 259, 260
Automation MAC Layer, 224

C

Cognitive radios, 7

D

Dipole, 261, 279

F

Fingerprinting, 100, 117
functional safety, 281, 282, 312, 317, 318, 319

H

HetNets mmWave Massive, 64

L

Localization, 97, 98, 111, 128, 137, 138, 300

M

MIMO C-RAN, 64
Multiband, 261, 262, 279
Multilateration, 101, 102

Q

Quality of Service, 223, 224, 240, 259

R

Range-Based, 97, 112

Reconfigurable Antennas, 261, 278

S

Security, 25, 50, 52, 219, 223, 320, 323
Software defined, 47, 56, 64, 78

T

Trilateration, 99, 100

V

Virtualization, 14, 16, 32, 35, 61, 64, 204, 221

W

waveform Dense networks, 64
Wideband, 136, 137, 261, 278, 279
Wireless Sensor Networks, 97, 323
WirelessHART, 224, 226, 229, 255, 258, 259, 260

Z

ZigBee, 28, 137, 224, 226, 229, 255, 258